CW01464014

The Open University

ANIMALS

EDITED BY CAROLINE M. POND

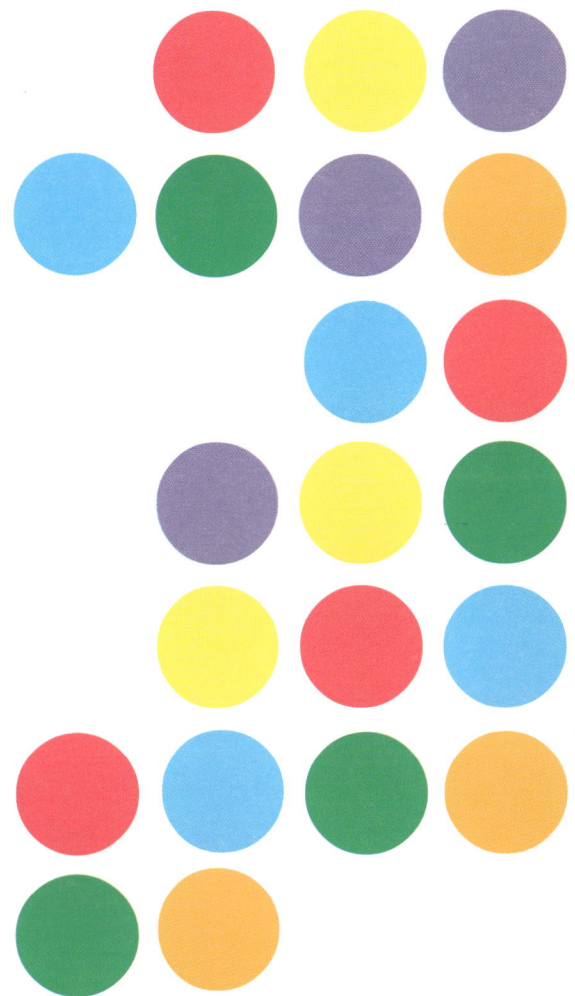

PHOTOS ON COVER ① ② ③ ④ ⑤

① Ant

False-colour scanning electron micrograph of the head and anterior thorax of a velvety tree ant, *Liometopum occidentale* (Order Hymenoptera, family Formicidae) showing parts of the two antennae (top), the left compound eye, the stout serrated mandibles (mouthparts) and parts of the maxillae and legs. The velvety tree ant is common in California and the western states of the US. It lives in cracks in tree bark and feeds on insects and honeydew but also can be a pest, building its elaborate nest in human habitations. Courtesy of David Scharf/Science Photo Library.

② Frog spawn

Light micrograph of freshly-laid eggs (frog spawn) of the European edible frog, *Rana temporaria*. The dark circle at the centre of each egg is the developing embryo, protected from harmful sunlight by granules of melanin pigment. *R. temporaria* lays its eggs in freshwater ponds during February and March in lowland areas and as late as June in the mountains. *R. temporaria* eggs are amongst the first kind of frog spawn to hatch and the tadpoles often feed on the spawn of other species of amphibians. They metamorphose into frogs in two to three months, and as adults, feed on insects that they catch mainly at night in damp woods and meadows. Courtesy of Dr Jeremy Burgess/ Science Photo Library.

③ Sea horse

X-ray (in false colours) of a sea horse, *Hippocampus* sp. A sea horse is a type of teleost fish. Sea horses swim slowly using fins (not seen here) and can use their tails to attach to seaweed or coral. The male sea horse incubates the fertilized eggs in a brood pouch on his abdomen. Courtesy of D. Roberts/Science Photo Library.

④ Sunflowers

A group of blooming sunflowers, *Helianthus* sp., against the sky. The plant produces composite flowers: the small, true flowers are clustered into a compact, round head which resembles a single flower. The plant is widely cultivated for its seeds which are used to make cooking oil and as food for livestock. Courtesy of David Nunuk/Science Photo Library.

⑤ Viruses

Illustration based on a transmission electron micrograph of human immunodeficiency viruses (HIV) (shown in yellow) budding out of a human T cell. HIV causes AIDS (Acquired Immune Deficiency Syndrome) by damaging T cells — the white blood cells that play a crucial role in the immune system. Courtesy of Chris Bjornberg/Science Photo Library.

The Open University, Walton Hall, Milton Keynes, MK7 6AA

First published 2001, reprinted with corrections 2006

Edited, designed and typeset by The Open University.

Printed in the United Kingdom by CPI, Glasgow.

ISBN 978 0 7492 1870 6

This publication forms part of an Open University course S204 *Biology: Uniformity and Diversity*. The complete list of texts which make up this course can be found at the back. Details of this and other Open University courses can be obtained from the Student Registration and Enquiry Service, The Open University, PO Box 197, Milton Keynes, MK7 6BJ, United Kingdom: tel. +44 (0)870 333 4340, email general-enquiries@open.ac.uk

Alternatively, you may visit the Open University website at http://www.open.ac.uk where you can learn more about the wide range of courses and packs offered at all levels by The Open University.

To purchase a selection of Open University course materials visit http://www.ouw.co.uk, or contact Open University Worldwide, Michael Young Building, Walton Hall, Milton Keynes MK7 6AA, United Kingdom for a brochure. tel. +44 (0)1908 858785; fax +44 (0)1908 858787; email ouwenq@open.ac.uk

2.1

The S204 Course Team

SKILLS

Patricia Ash

Hilary Denny

CD-ROM AND VIDEO PRODUCTION

Mary Bell

Gail Block

Jacki Brown

Phil Butcher

Hilary Denny

Mike Dodd

Susan Dugher

Anna Furth

Phil Gauron

Michael Gillman

Tim Halliday

Nicky Heath

Caryl Hooper

G. D. Jayalakshmi

Juliet Kauffmann

Martin Kemp

William Kirk

Hilary MacQueen

Derek Martin

Mark Murphy

Phil Parker

Caroline Pond

Brian Richardson

Irene Ridge

Jerry Roberts

David Robinson

Jill Saffrey

Liz Sugden

Andrew Sutton

Margaret Swithenby

Gary Tucknott

Verina Waights

Darren Wycherley

CONTENTS

INTRODUCTION TO ANIMALS

Animals are by far the most structurally diverse kingdom: the number of species so far described, well over a million, is believed to be only a fraction of the total. One reason why our knowledge of animal diversity is so incomplete compared to that of plants is that many animal habitats are inaccessible to biologists except with special equipment. Animals are common where lack of light excludes plants and photosynthetic protoctists and microbes, including the deep sea, caves and soil, and as internal and external parasites.

Animal hard parts (e.g. teeth and skeletons) fossilize more readily than do soft tissues, so animals have a much more extensive fossil record than does any other kingdom. Information from fossils is combined with studies of comparative and functional anatomy, embryology and molecular biology to elucidate the origins of new tissues and organs, the interrelationships between groups, and the physiology of extinct and living but inaccessible species. Palaeontology shows clearly that animals first evolved in the sea, and the majority of phyla are still entirely marine. While at least a few marine species, living or extinct, are known from each animal phylum, only a few lineages have any entirely terrestrial members, though some, notably arthropods, molluscs and chordates, are very diverse on land and in freshwater.

Biochemists and molecular biologists emphasize the uniformity of cellular mechanisms in all animals, regardless of their size, habit or phylogenetic affiliation. This book tries to answer the obvious question: if animals' basic molecular tool kits are so similar, how can their structures and habits become so diverse?

○ What is the main basis upon which animal phyla are distinguished? How do the criteria differ from those for phyla of other eukaryote kingdoms?

● Each phylum of animals has a different body plan. Life cycle, mode of reproduction and cellular organization are more important features in plants, fungi and protoctists.

Zoologists now recognize about 35 living phyla but this figure, together with the arrangements of classes and orders, is constantly being revised in the light of new discoveries and interpretations. As recently as 1983, minute animals that bear no resemblance to any other group were found in marine gravel collected off the coast of Brittany (north-western France) and a new phylum (Loricifera) was established to accommodate them. Since then, a further two dozen species of loriciferans have been described from similar habitats in both shallow and deep seas throughout the world.

The kingdom Animalia comprises a large number of phyla. In addition, a few classes, notably Insecta and Teleostei, include a great many species. Are some body plans much more efficient and adaptable than others? If so, why do the phyla with fewer members persist in the face of competition from the many different species of the more numerous phyla? The answer lies in understanding the functional, as well as phylogenetic and developmental implications of the various body plans, which is the main topic of Chapter 1 and features prominently in Chapters 2–5.

In this book, we only have space to discuss a few of the most familiar and abundant animal phyla. Chapter 5 is about the genetic, molecular and cellular mechanisms that generate new proteins, cell types and body plans, and allow certain species and families to evolve adaptive assemblages of characters. Zoologists have discussed the origins and implications of body plans for more than two centuries, taking advantage of new techniques as they arose to develop new kinds of evidence. Many different kinds of investigation may provide evidence for ancestry and interrelationships between groups. In many cases, molecular biology suggests relationships quite different from those established from comparative anatomy or palaeontology, so the taxonomy of many kinds of animal is hotly disputed, with a wide range of schemes proposed. Beyond mentioning some of the evidence, this book does not favour one scheme over another, but aims to illustrate the range of animal structures and habits.

○ What methods of investigation would provide information about body plans?

● The gross arrangement of organs can be established by examining whole and dissected animals, or, in the case of small specimens, by studying histological sections of whole bodies. Particular types of cell such as secretory cells and muscle cells, for example, can also be identified by histology.

Histology and more elaborate microscopical techniques enable biologists to follow the course of embryonic development that leads to the adult body plan. Since the 1970s molecular biology has allowed direct study of the genes and their regulators that direct the synthesis of molecules and their assembly into cells and tissues. The genetic and cellular mechanisms that build bodies to a clear plan are discussed in Chapter 5; in Chapter 1 we deal only with the use of body plans as a means of defining phyla, and how body plans determine the way animals move and feed.

It is the discoveries from laboratory animals, especially those with obvious medical relevance, that grab the headlines; information about the majority of less thoroughly studied animals is usually relegated to text books. The comparative study of invertebrates is often perceived as stagnant, boring and irrelevant to progress in medicine and 'basic' science. In fact, exploration of the deep sea, arctic regions and remote mountains, combined with the application of modern techniques of cell biology and biochemistry, has recently stimulated much new research on invertebrates, some of which has caused fierce controversy by appearing to undermine long-established theories. This book offers you the opportunity to read several recent reports of original research that has led to startling discoveries, and to assess for yourself the validity of the methods used and the conclusions claimed.

SUMMARY OF INTRODUCTION

Animals are very diverse and have a long and abundant fossil history. Many animals live in obscure and inaccessible habitats, so species new to science are frequently discovered. About 35 phyla are currently recognized, defined mainly by body plans, but recent discoveries in molecular biology point to revision of many established taxonomies.

THE DIVERSITY OF INVERTEBRATES

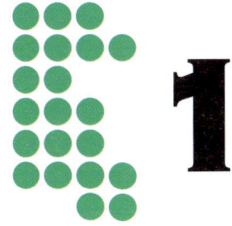

1.1   INTRODUCTION

The great majority (over 95%) of all animals are invertebrates, which are very diverse in structure, habits and habitats. We do not have space to describe more than a few of the most familiar groups, ranging from the simplest (Section 1.2) to those that rival vertebrates in anatomical complexity and physiological capacities (Section 1.4).

1.2   STRUCTURALLY SIMPLE INVERTEBRATES

The Porifera (sponges), Cnidaria (jellyfish, corals and sea anemones) and Ctenophora (sea-gooseberries) are the major phyla with the simplest body plans. They are also among the most ancient and most persistently successful animals. The fossil record indicates that they appeared either before, or very early in, the Cambrian (545–495 Ma); at periods in the remote past these phyla represented a larger proportion of the fauna than they do now, although they were not necessarily more diverse. Many species of these phyla are still common in marine habitats, including some of the most demanding of all, shallow seas and the intertidal zones of beaches.

1.2.1   SPONGES

All sponges live in water, the great majority in the sea, and many are sessile except during a brief larval stage: they live encrusted on, or protruding from a fixed object, or, quite commonly, on the outer surface of a mobile animal such as a crab or snail. Typical sponges are sessile filter feeders (Figure 1.1a and b, overleaf), extracting tiny particles from the water that flows past their cells. Any of the cells (except the gametes) are capable of taking in and digesting food particles by phagocytosis. Some species probably also feed by osmotrophy, as did many of their protoctistan ancestors (Ridge, 2001).

Sponges consist of three main kinds of cell (Figure 1.1c). The most distinctive are the flagellated *choanocytes* (pronounced 'koh-an-o-sites'), which normally beat synchronously, generating local water currents, and take in and digest tiny particles of microbes and detritus. A second type of cell secretes the skeleton as an extracellular matrix, and a third type forms the outer layer. Sponge cells can change form and move around: there are numerous cells called amoebocytes (because they resemble *Amoeba* in both structure and form of movement, see Ridge, 2001) that wander through this matrix, or may become gametes and break away from the parent. At any one time, however, the numbers of each type of cell are more or less constant.

Figure 1.1   (a) The dried skeleton of *Spongia*, the bath sponge, which lives in shallow, temperate seas including the Mediterranean. (b) The dried skeleton of *Euplectella*, Venus's flower basket, a glass sponge in which the skeleton consists mainly of regularly shaped spicules of silica. Both species can grow to tens of centimetres. (c) Diagram of a simple sponge with part of the wall enlarged to show the types of cell. The arrows indicate the direction of water currents. The diameter of the inflow pores is of the order of millimetres.

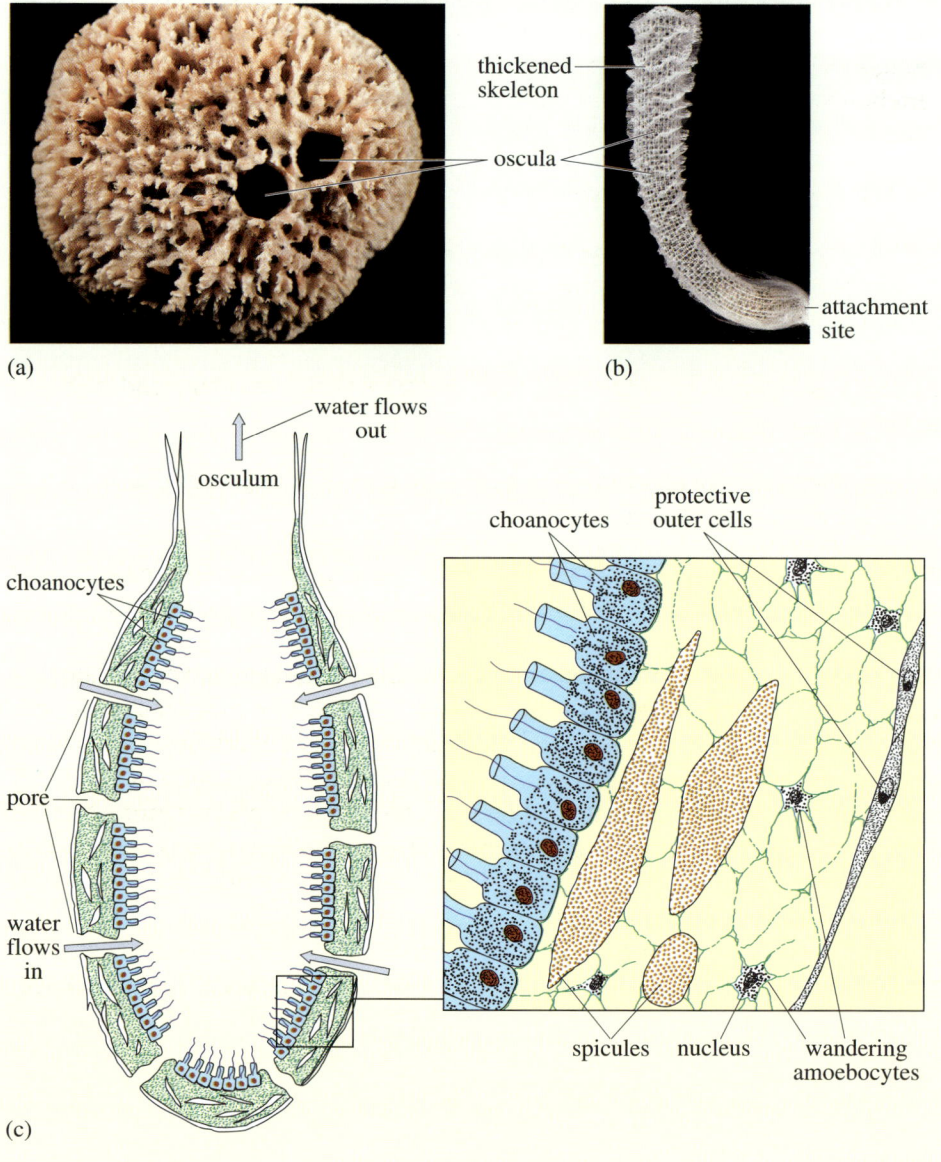

(a)

(b)

thickened skeleton

oscula

attachment site

water flows out

osculum

choanocytes

choanocytes

protective outer cells

pore

water flows in

spicules   nucleus   wandering amoebocytes

(c)

○   Are the sponge cells shown in Figure 1.1c tightly packed as in mammalian tissues (e.g. gut, lung, liver)? What lies between them?

●   No. Except for the layer of choanocytes the cells are loosely associated with spicules in the extracellular areas between them.

As with fungi and multicellular protoctists, the size and shape of sponges are very variable because they grow irregularly and often very slowly, depending upon the availability of food. A common feature, however, is the presence of holes through which water passes. Water usually enters through many small inhalant openings and is expelled through fewer (often only one) larger exhalant canals, called *oscula* (Figure 1.1a–c).

In the common bath sponge (Figure 1.1a) and many others, the skeleton consists mainly of a collagen-like fibrous protein called *spongin*. The gene that produces spongin is so similar to that for one of the many forms of collagen found in all vertebrates (including humans), that molecular biologists believe they must have a common ancestor.

○   What is the usual role of collagen in vertebrates such as mammals?

●   The mature protein is always extracellular and forms a framework, holding the cells in the tissue and contributing to the basement membrane that aligns epithelia.

Collagen is especially abundant in structural tissues such as skin, tendon, bone and cartilage.

○   Are the histological arrangement and function of collagen similar in sponges and mammalian tissues?

●   Yes. Spongin probably protects and supports the living sponge cells, though they are not organized into sheets or compact masses as mammalian cells are.

Both the molecular structure and anatomical role of collagen are essentially similar in all animals. Indeed, the ability to synthesize this chemically distinctive group of proteins, which is not found in plants or other eukaryotes, seems to be a key evolutionary advance that occurred very early in the evolution of multicellular animals and conferred on them their unique organization of tissues. Along with glycoproteins and glycosaminoglycans, collagen holds cells together, forming continuous epithelia sealed off from the outside world, and delimiting surfaces such as the lumen of the gut (Saffrey, 2001). Collagen also determines the mechanical rigidity of tissues, and maintains the spacing between cells. These gaps are essential to the transmission of chemical and electrical signals between cells (see Clements and Saffrey, 2001). Although sponges do not have a nervous system or muscles as such, they respond to touch or to certain chemicals by adjusting the beating of choanocytes near the site of the stimulus, and some have contractile cells that can slowly open or close the oscula.

The skeletons of some sponges, especially larger species living in deep, still waters such as glass sponges (Figure 1.1b), are reinforced with calcareous or siliceous spicules in various combinations. The spicules are by far the most consistent and durable components of sponges and their taxonomic classification is based mainly upon the spicules' form and chemical composition.

Many sponges live in areas of high water movements, in tidal pools or rivers, on actively moving animals or in the path of marine currents, so water and the particles on which they feed pass through them passively. Exposure to water movement is essential to obtaining food, but if too strong, waves and currents can crush or dislodge sessile organisms. The mechanical properties of the sponge skeleton determine what kind of water flow the animal can withstand and hence where it can live and the shapes it can assume. The spicules are stiff but brittle and tend to crack when bent, but these rigid structures are only about 0.1 mm long, which is very small compared with the size of the sponge. The spicules and collagen account for only a few percent of the total mass even in large, deep-sea sponges that grow to up to 2 m high.

Numerous tiny spicules embedded in flexible proteins enable the skeleton to deform in flowing water without breaking: bending occurs between, rather than within, the spicules, which thus avoid fracture. Sponge skeletons are 'spongy'; they bend with the flowing water, regardless of its direction, and recoil elastically to their original shape when released. Other sessile animals have similar skeletons consisting of many small, hard elements embedded in a flexible matrix.

○ How would skeletons of this structure be particularly efficient for animals that grow irregularly and often very slowly?

● Such skeletons are easily strengthened and/or extended as required, and would work equally well in specimens of widely different size and shape.

Sponge spicules also provide protection from predators. Many sponges also produce chemicals that, at least to terrestrial noses, smell obnoxious, and, in the case of a few species, irritate the skin if handled.

Sponges that live in stagnant waters, such as the deep sea, create their own feeding currents by active beating of the flagella on the choanocytes. Some remarkable recent discoveries about certain deep-sea and cave-dwelling sponges are illustrated in Box 1.1.

Like most sessile animals, sponges release their gametes into the sea, where they fertilize each other and form tiny ciliated larvae, which swim, and in many species feed, for some time before settling.

BOX 1.1   CARNIVOROUS SPONGES

Because sponges appeared to be immobile, early naturalists regarded them as plants. They were first recognized to be animals in the middle of the 18th century, following the observation that certain species actively generate the water currents that flow through the body. For more than two centuries, all sponges were assumed to be filter feeders, although many species were known only from dredged up remains and could not be studied alive. Indeed, most biologists believed that the simple poriferan body plan was incapable of any more elaborate style of feeding.

○ What structures lacking in sponges are present in most predatory animals? (If necessary refer to Pond, 2001b.)

● A proper gut with at least one large opening; movable jaws or stout feeding appendages; strong muscles capable of active, directed movement.

In 1994 two French scientists (Vacelet and Boury-Esnault, 1995) found evidence for non-filter feeding habits (Figure 1.2), challenging the long-established concept of the inherent limitations of the poriferan body plan.

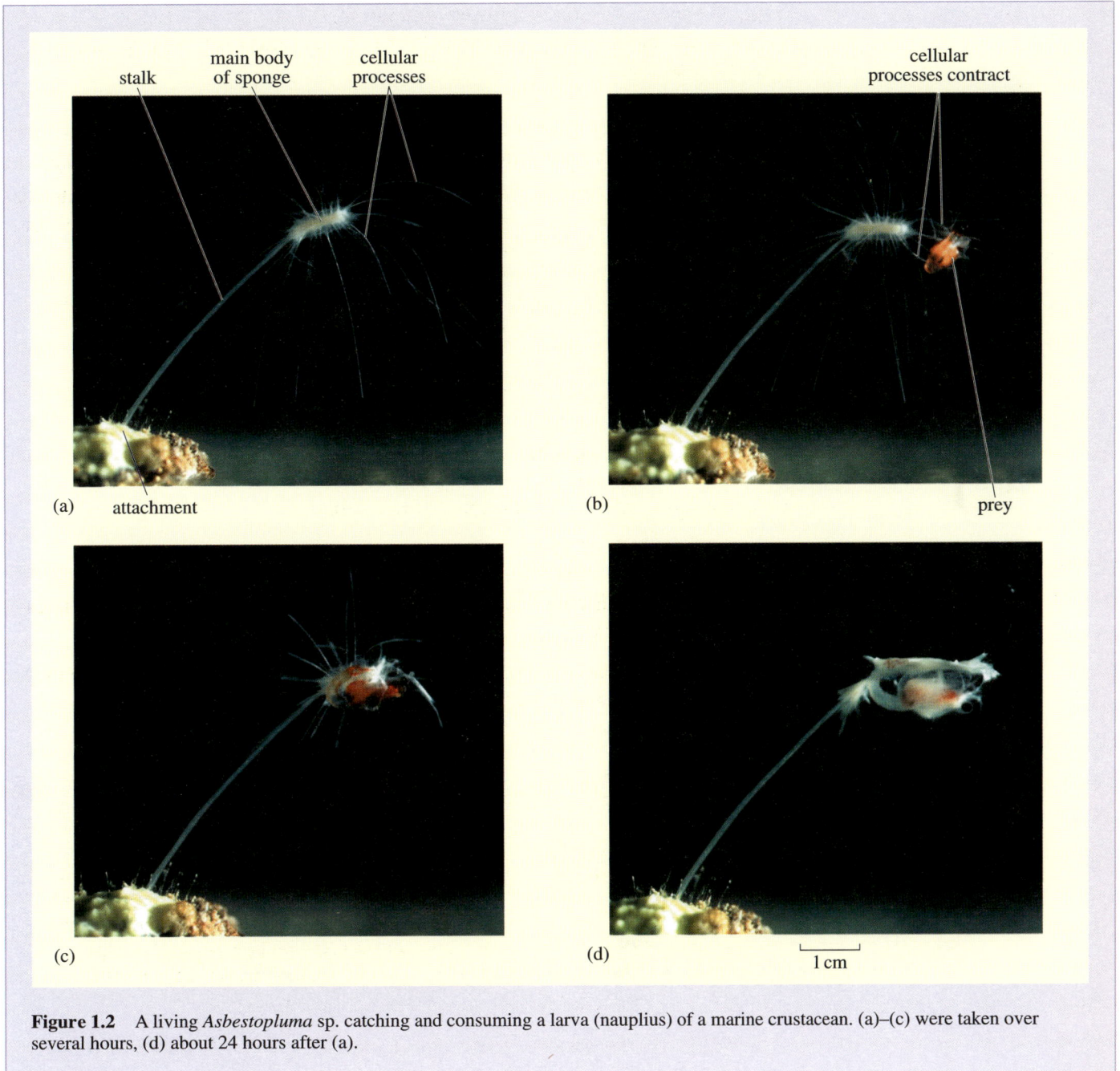

Figure 1.2   A living *Asbestopluma* sp. catching and consuming a larva (nauplius) of a marine crustacean. (a)–(c) were taken over several hours, (d) about 24 hours after (a).

1.2.2   CNIDARIANS

The internal structure of cnidarians is more highly organized than that of sponges. Cnidarians occur in two main forms united by a common body plan: cup-shaped **polyps** (Figure 1.3), including corals and sea anemones (class Anthozoa) and the anemone-shaped stage of hydrozoans (class Hydrozoa), e.g. *Hydra* itself (see Pond, 2001a), and bell-shaped **medusae**, the jellyfish (classes Scyphozoa, Cubomedusae and Hydrozoa). The former are nearly all sessile, living attached to the substratum or to other organisms often in dense colonies, while most medusae swim actively.

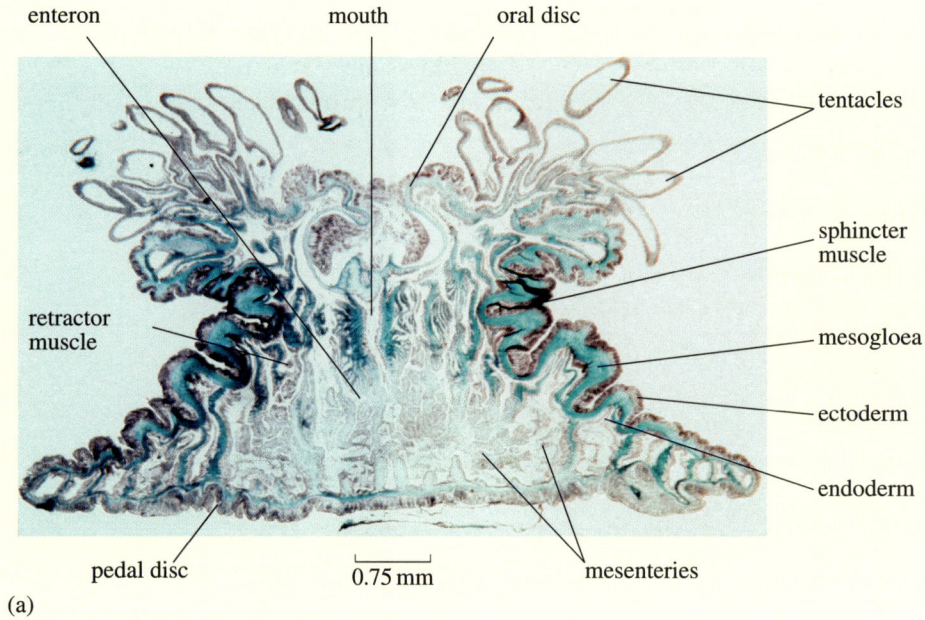

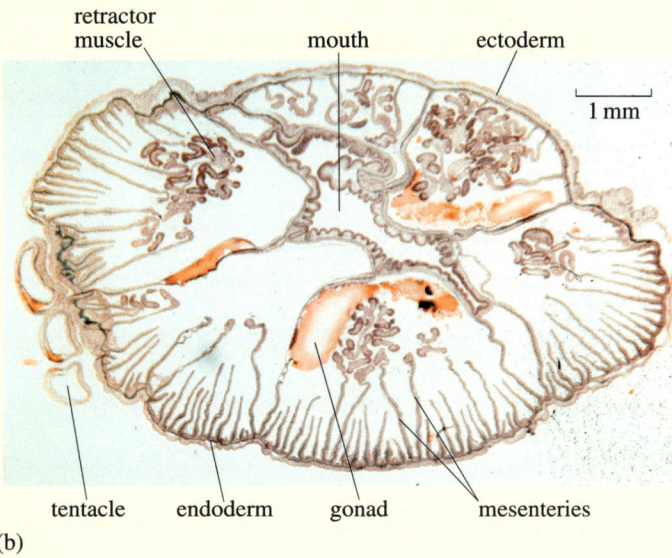

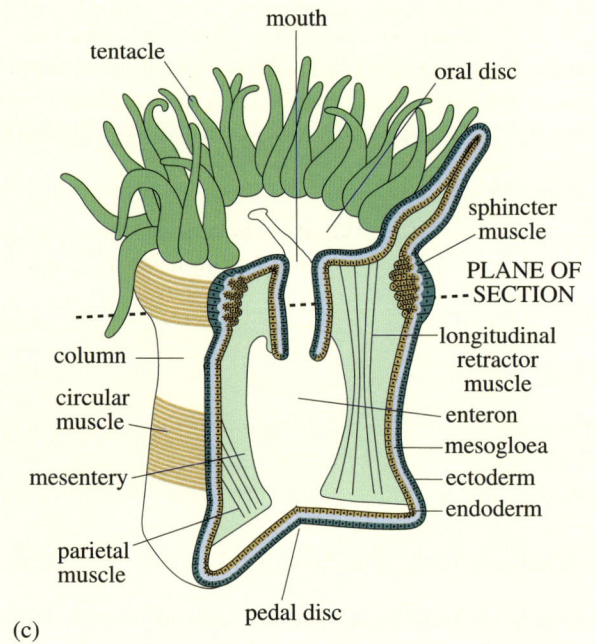

Figure 1.3   (a) Longitudinal section of a sea anemone near the middle of the body. Cells stain brown, extracellular collagenous material blue. (b) Horizontal section of a sea anemone at the level indicated in (c), cut just below the oral disc. The orange material is probably partially digested food. (c) Diagram of the anatomy of a sea anemone.

In the thin section of a small sea anemone shown in Figure 1.3a and b, the much-folded body wall stains most densely. The body wall forms the pedal disc (Figure 1.3a and c) that sticks the animal to other objects and the tentacles (seen here as oval loops of tissue) that surround the mouth, in the middle of the oral disc.

○ Are any cells visible in the regions of Figure 1.3a that stain turquoise blue? What could the blue-stained material consist of?

● No. A few cells are visible, but they are not blue. The blue-stained material could be extracellular matrix consisting of collagen and other proteins and glycoproteins forming an aqueous jelly.

The **mesogloea** (pronounced 'meso-gleer') between the cellular layers contains various fibrous macromolecules including collagen, that in sea anemones amount to about 8% protein and 1% polysaccharide, but very few cells. In some species, the mesogloea also has hard skeletal materials. This arrangement of cellular ectoderm and endoderm separated by an acellular layer, is sometimes called **diploblastic** (two layers of cells).

The *enteron* is the equivalent of a gut and is the site of extracellular digestion. The cells lining the enteron also take up small particles of food and digest them intracellularly. Since there is no anus, undigested remains are discharged through the mouth. The enteron cavity is partly divided by numerous vertical partitions called mesenteries, consisting of mesogloea with cells on both sides, seen in Figure 1.3b as lightly stained tissue. The structure of the mouth and mesenteries can be understood by comparing Figures 1.3a, b and c: the mouth is lined by a fold of the body wall and the mesenteries protrude from the body wall. In many sea anemones, mesenteries contain strong retractor muscles, which can tilt the mouth and tentacles in one direction or another. Synchronous contraction of the retractor muscles shortens the body, expelling the water in the enteron through the mouth (see Pond, 2001b).

○ What other differences between the basic body plans of cnidarians and poriferans can be seen by comparing Figures 1.1a and 1.3?

● The cnidarians have a clear symmetry, but the sponges are amorphous.

Cnidarians have a wider variety of types of cell than do sponges, including neurons and muscle cells that contract, waving or shortening the tentacles or changing the shape of the body. The cells are arranged in an inner (endodermal) and an outer (ectodermal) layer, which may be thickened by pleating or folding. In most cnidarians, only the inner layer of cells lining the enteron participates in digestion and absorption of food.

○ From Figure 1.3b, how is the area of cells involved in digestion and absorption increased?

● The enteron contains septa (mesenteries) that greatly extend the area of contact between its contents and the living cells.

Septa are found in the enterons of all but the smallest polyps, such as the hydrozoan *Hydra*.

The mouth of nearly all polyps is surrounded by tentacles that are often brilliantly coloured due to pigments in the cells. Concentrated in their outer surface are elaborate stinging cells called **cnidocytes** ('nettle cells', also known as nematocysts or cnidoblasts), which are unique to this phylum and contain long sticky threads, armed with barbs and toxins. Cnidocytes are discharged by rapid uptake of water into the cell, triggered by chemical or mechanical stimuli arising from contact with prey. The toxins immobilize animals, and the combined action of the cnidocyte threads and the tentacles deposit the prey in the enteron, where it is digested by secreted enzymes and by phagocytosis of small particles. The cnidocyte toxins of some jellyfish are, in high enough concentration, lethal even to very large animals, including humans.

○ In what ways is this predation strategy more efficient, and better suited to larger prey, than that of sponges shown in Figure 1.2?

● The toxins in cnidocytes enable cnidarians to subdue their prey much more quickly than carnivorous sponges can. Digestion in an enclosed enteron also avoids loss of secreted enzymes and may thus proceed faster.

Discharged cnidocytes die, and new ones form from the many undifferentiated 'interstitial' cells. Detailed examination of *Hydra* shows that all kinds of cells are continuously replaced: cell division is concentrated around the mouth, and newly formed cells migrate upwards along the tentacles and downwards to the base, where old cells are sloughed off. A fully-grown *Hydra* a few millimetres long is completely replaced in about a month, a process that may happen many times in what is often a long lifespan.

○ Does cell replacement occur in other kinds of animals?

● Yes, especially at high-wear sites such as the lining of the gut, where cells are replaced in a few days (see Pond, 2001b).

Cell replacement is easily demonstrated only when it occurs at a very high rate and/or starts from a distinct growth zone, but is believed to occur in the great majority of animals. Sponges, however, can grow at any point, determined by availability of food. Recent studies of the genes activated in and near cnidarian growth zones suggest some similarities to those that form anterior structures including the head and brain of more complex animals, as discussed in Chapter 5.

○ Does cell replacement promote longevity?

● Yes. Cells damaged by exposure to free radicals and other harmful agents are eliminated, enabling the organism itself to survive.

Animals such as the nematode, *Caenorhabditis,* in which there is no cell replacement, live only for about three weeks, while many cnidarians can live for decades, some probably for centuries (see Halliday and Pond, 2001).

Most of the animals known as corals are anthozoans. These 'stony' corals resemble sea anemones, but their outer layer secretes hard, calcareous material and many form colonies of numerous joined polyps. Some feed in a similar way to sea anemones, obtaining particles of food from water currents. Many, including almost all of those that build reefs, also obtain some or all of their nourishment from photosynthetic algae (mostly dinoflagellates, kingdom Protoctista) held in the cells lining the enteron. These symbiotic associations are broadly similar to that of *Hydra* (see Pond, 2001a). The cnidarian cells partially digest the cell walls of their internal symbionts, so up to 80% of all the sugars synthesized by the algae leak into their host's cells, and are used for energy production. The photosynthetic and other pigments of symbiotic algae also contribute to the brilliant colours of living coral.

The metabolism of the symbiont algae is crucial to the formation of the calcareous base as well as to the nutrition of the soft tissues. The removal of bicarbonate ions from the sea water surrounding the coral by algal photosynthesis favours the precipitation of calcium carbonate, so aiding the formation of the hard, calcareous material that over many years builds up to form huge reefs.

○   How would this symbiosis with algae limit where these corals could grow?

●   Algae need sunlight for photosynthesis, so corals are limited to living near the surface of clear waters.

Reef-building corals are also restricted to warm waters, though other species that do not have obligatory symbiosis with protoctists occur in deeper water and at high latitudes, including around Britain.

Under ideal conditions of light and temperature and with a full complement of healthy algae, corals deposit calcium carbonate fast enough for their reefs to grow upwards at up to $0.5\,\mu m\,h^{-1}$. But the symbiosis is easily disrupted by turbulence or sediment that make the water opaque and smother or break the coral, by a small change in sea temperature, or by excessive predation.

The variety of shapes and patterns of the colonies (Figure 1.4, overleaf) arise from the type of budding and the degree of separation of the resultant polyps. Individual polyps, each with a mouth surrounded by tentacles, are clearly visible in Figure 1.4a. In the 'brain coral' (*Meandrina* sp.; Figure 1.4c), the polyps do not separate, and thus form rows of small mouths surrounded by rows of tentacles; in life, the entire structure would be covered with living tissue with numerous small tentacles. A large colony consists of hundreds of intersecting rows that resemble the surface of a mammalian brain.

Sea-pens (Figure 1.4b) and precious coral *Corallium* (Figure 1.4d) belong to another group of corals in which the skeleton consists partially or entirely of collagen-like extracellular proteins impregnated with inorganic salts, which may confer brilliant colours. They are predators or filter feeders and do not harbour photosynthetic symbionts. Many therefore live in deep water and other dark habitats.

Most cnidarians take larger particles than do sponges, and some trap and kill living animals, but for them, as for sponges, flowing water is the main source of nourishment and a potential hazard.

○   In what ways can anemones avoid injury from waves, etc. that sponges cannot?

●   Cnidarians have muscles and can shorten the body and close the mouth tightly.

Figure 1.4 (a) Stony corals (*Goniopora* sp.), colonies of cnidarian polyps that secrete a calcareous skeleton. (b) A sea-pen. (c) The reef-building brain coral *Meandrina* sp. can grow to cover many square metres. (d) Precious red coral (*Corallium* sp.). Numerous tiny polyps cover the branches. The whole colony is variable in shape but can grow to tens of centimetres. The colour, which varies from almost white to deep red, depends upon iron salts and other inorganic materials impregnated in the proteinaceous skeleton.

(a)　　　　living polyps

(b)

skeleton　　living polyps

(c)

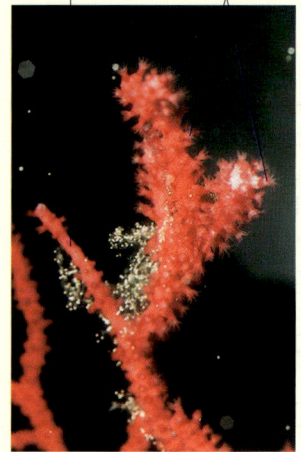

(d)

The mechanical properties of the mesogloea of sea anemones and hydroids determine what kind of water flow the animal can withstand and hence where it can live and what shapes it can assume. Short, squat sea anemones such as *Actinia equina* (Figure 1.5a) are common in shallow rock pools on beaches, usually in small crevices or under ledges. They remain upright and open with their stubby tentacles exposed even when pounded by waves and tidal surges, because their mesogloea is relatively rigid but elastic, storing the energy of deformation and recoiling when released, like a tennis ball. Tall sea anemones such as *Metridium* sp. (Figure 1.5b) are restricted to calm waters below the regions exposed by tides, where they can project their crown of long, delicate tentacles into slow-flowing currents. Slight differences in the composition and arrangement of the macromolecules make their mesogloea more plastic, enabling them to extend slowly from a compressed to a greatly elongated shape. Being on a beach in a storm would rip them to shreds.

○ How could radial symmetry be better than bilateral symmetry for sessile cnidarians?

● As well as enabling them to detect predators attacking from any direction, circular animals can withstand buffeting by water currents from many different directions.

Both the body form and the chemical composition and crystal structure of the secreted skeleton determine the corals' ability to withstand wave action and tides. Severe storms and tidal waves, or even just the wash from passing ships, often break up coral skeletons or smother the living tissues in sand and rubble.

Many corals collect particulate food and/or prey on small animals at night and during dull weather. Non-predatory species lack tentacles, but all corals have numerous cnidocytes on their surface, some of which, as divers have found, contain toxins that can 'sting' people quite badly.

○ What role could cnidocytes play in corals that never prey on animals?

● They act as a defence against predators.

Like plants, sea anemones and corals (and other sessile animals, including sponges) are vulnerable to grazing animals. Several kinds of echinoderms (see Section 1.4) and fish, including the parrot fish, eat the coral itself. Reefs support a rich variety of animals. Many fish that live in the open ocean as adults breed on reefs: their eggs and juveniles need the shelter and abundant food that reefs provide.

Most natural coral reefs are near coasts or surround islands and have been polluted and/or badly over-fished by the people who live there. However, underwater cables, offshore oilrigs and sunken ships and aircraft also make ideal substrata for sessile invertebrates including corals, sea anemones and sponges. Many such artificial reefs support thriving colonies of other kinds of invertebrates and fish. Harvests of many kinds of fish are noticeably better near old wrecks sunk in shallow water.

JELLYFISH

In jellyfish, the free-swimming medusae (Figure 1.6a, overleaf), especially the larger scyphozoan species, have more elaborate and extensive nerves, sense organs and muscles than most polypoid cnidarians, and surprisingly sophisticated behaviour. Some jellyfish are filter feeders, collecting detritus particles as well as small organisms on numerous fine tentacles (Figure 1.6a and b), and a few species harbour symbiotic algae in much the same way as do corals. The majority, however, are predators, ensnaring and poisoning a range of animals, including various kinds of shrimp, fish eggs and larvae and sometimes each other, in their long retractable tentacles (Figure 1.6b and c).

Some jellyfish, notably the scyphozoan *Chrysaora quinquecirrha,* commonly known as the sea nettle on account of its toxins (Figure 1.6b), have become so large and abundant around the Atlantic coast of North America as to constitute a threat to fish stocks as well as a hazard to bathers. Recent research (Purcell, 1997) has revealed that the various species select their food by the width and spacing of the predator's tentacles and properties of the cnidocytes, and the prey's size and

(a)

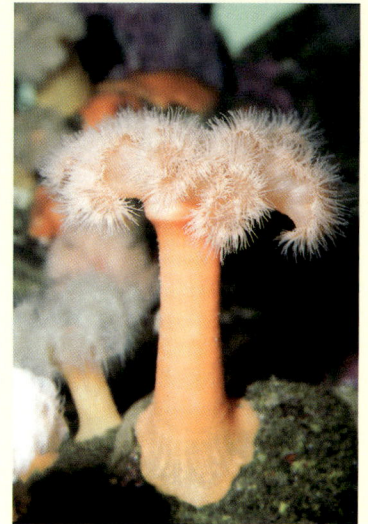

(b)

Figure 1.5 (a) The sea anemone *Actinia equina*, which is very common, attached to rocks in shallow tidal pools. (b) *Metridium* sp., found in deeper, calmer water on rocks and pilings, where thousands often form dense clumps. Both species can reach a diameter of several centimetres.

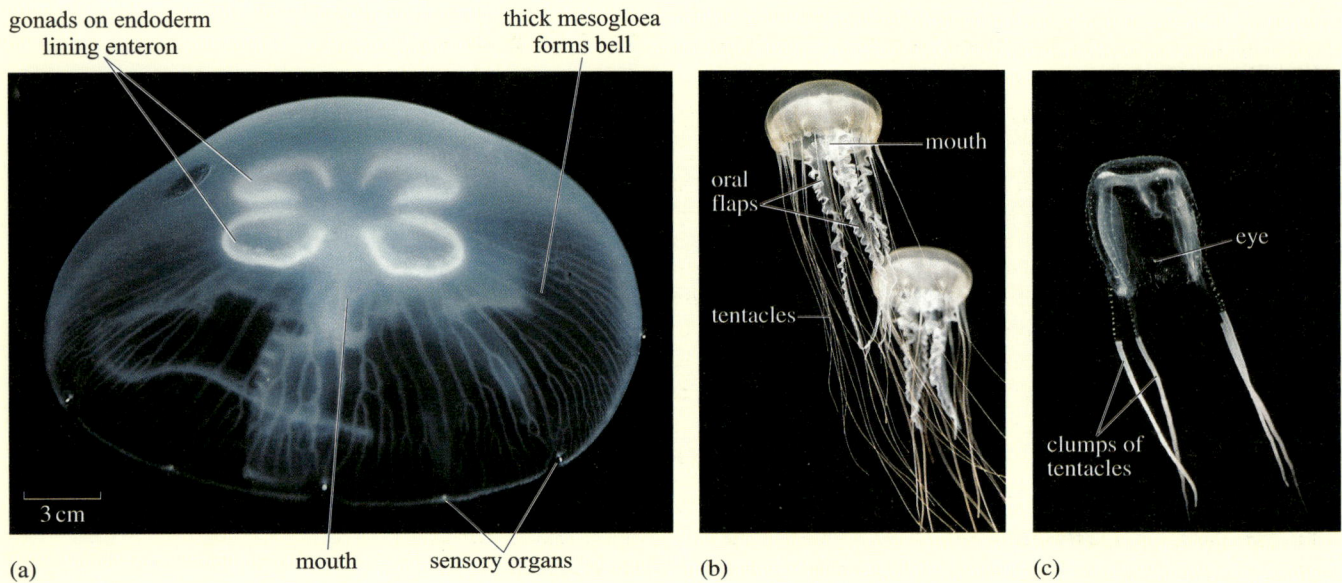

Figure 1.6 (a) The common jellyfish, *Aurelia*, seen from above, usually about 10 cm in diameter. The sense organs are clearly visible as bright spots on the margin of the bell. The four incomplete bright rings are gonads. This species is common in surface waters throughout the world. (b) The sea nettle, *Chrysaora quinquecirrha*, can grow to a diameter of more than half a metre. (c) The sea wasp *Chironex fleckeri* (Cubomedusae: box jellyfish), a small but deadly species that lives in the Indo–Pacific especially around northern Australia.

swimming speed. Jellyfish respond to touch by actively swimming away from obstacles and potential predators. Some, especially those that feed at night, have photophores (light-emitting structures) and can produce striking visual displays of flashing lights that probably startle predators. The remarkable proteins that produce the light are described in Box 1.2.

BOX 1.2   CNIDARIAN BIOLUMINESCENCE

The hydrozoan jellyfish *Aequorea* is almost transparent, but its outline is clearly visible as it swims near the surface at night because special cells around the margins of its bell emit a soft blue–green light. In the mid-1960s, biochemists examined the mechanism of its bioluminescence: a protein (which they named aequorin) combines with molecular oxygen and emits blue light, but only in the presence of calcium ions at micromolar concentrations.

○ What could be the natural role of regulation by calcium ions?

● Light production uses energy, and the light produced is much too faint to be useful during daylight. Regulation by calcium ions switches off the process when it is not required.

Physiologists realized that these properties meant that aequorin could be a useful tool, as it would function as a very sensitive indicator of free calcium ions in very small volumes and at concentrations too low to be measurable by chemical methods.

○ Why should physiologists be interested in where and when calcium ions appear?

● Local, transient release of calcium is important for many intracellular processes (see Clements and Saffrey, 2001).

Aequorin was first used to reveal the time course of the release of calcium ions in the coupling of membrane excitation to muscle contraction. The exceptionally large muscle fibres of giant barnacles were chosen for study because the aequorin had to be injected into the site of action, which is technically difficult with smaller cells.

In the 1990s this limitation, and the need to extract aequorin from wild jellyfish, were circumvented by cloning the gene (Kendall and Badminton, 1998), enabling aequorin to be produced much more cheaply and in larger quantities by engineering the cnidarian gene into bacteria. In a further step, the need to isolate the protein from the bacteria that synthesize it and to inject it into the cells to be studied was eliminated by inserting the gene into the genome of those cells (thus enabling them to synthesize aequorin), as has been done with other genes into yeast, and certain cultured plant and animal cells.

Further studies of the mechanism of bioluminescence in *Aequorea* revealed another protein that emits bright green light, using energy it acquires from oxidized aequorin bound to calcium ions. The combination of the two wavelengths accounts for the animals' blue–green appearance at night. A slightly different protein with similar properties is also found in an anthozoan coral known as the sea pansy, *Renilla reniformis*. This protein was named *green fluorescent protein* because it fluoresces if exposed to light of certain wavelengths, i.e. it re-emits the absorbed energy as light of a longer wavelength. The mechanism is so simple it works even in cells that do not naturally contain it: mice engineered to carry and activate the gene for this protein glow bright green when exposed to bright daylight. Biotechnologists cloned and manipulated the gene for green fluorescent protein in the same way as that for aequorin, but by 1996 they had gone further. Spontaneous and induced mutations of the cnidarian gene within *E. coli* have produced proteins that fluoresce red or yellow or blue, thus creating several easily distinguished marker molecules. They can be attached, for example, to other proteins or transcription factors and used to follow intracellular events such as gene expression and protein localization.

Eventually it may be possible to synthesize *de novo* genes that make proteins with such properties, but for the time being, molecular biologists have to rely upon jellyfish and corals to provide their raw material.

The Cubomedusae are tropical and subtropical medusae that differ from other jellyfish in several ways: the tentacles, which can be up to 3 m long, form four clumps giving the animals a square appearance, hence the name, box jellyfish. Cubomedusan cnidocytes are of unusual structure and, in most species, including the aptly named sea wasp (Figure 1.6c), contain powerful toxins that immobilize fish in seconds, and can be as dangerous to humans as many snake venoms.

Evenly spaced around the edge of the bell of Cubomedusae are eight surprisingly complex eyes, each complete with a lens and light-sensitive cells. Experiments show that the jellyfish can orient towards lights but cannot avoid hitting small obstacles, so the eye probably serves only for light gathering, not image formation. Some biochemical reasons why eye-like organs have evolved many times in the animal kingdom are discussed in Chapter 5, Sections 5.4.3 and 5.5.

Jellyfish swim by powerful, synchronous contractions of the circumferential muscles that bend the bell and deform the large quantities of jelly-like mesogloea, ejecting a jet of water from under the bell. As soon as the muscles relax, the mesogloea recoils, returning the bell to its original shape in time for the next cycle of contraction. Regular contraction of the swimming muscles produces the slow, pulsating motion that is characteristic of jellyfish.

Although graceful and energetically efficient, not least because it involves only one set of muscles, such swimming is slow and weak compared to fish or squid. At low speeds, streamlining saves very little energy, so the circular body shape and trailing tentacles are no handicap.

○ What is the advantage of an extracellular skeleton (such as mesogloea) over one consisting of living cells?

● Living cells, but not acellular materials, need access to oxygen and nutrients, and a means of eliminating waste. Once formed, the mesogloea imposes negligible metabolic cost.

The mesogloea of the medium-sized scyphozoan jellyfish *Aurelia aurita* (Figure 1.6a) is only 1% by weight collagen, glycoproteins and other organic matter.

○ How could these differing requirements affect the anatomical arrangement of cellular and acellular tissues?

● Living cells, especially those involved in metabolically demanding activities such as secretion, must be near the surface or in contact with body fluids such as blood. Acellular tissues can exist in large blocks without access to such nutrient-supplying media.

The diploblastic structure thus enables animals to grow large without the need for specialized respiratory organs or a blood system. Another advantage of the cnidarian body plan is mentioned in Pond (2001b): the nutrients in the cellular tissues and mesogloea can be withdrawn and broken down to produce energy that sustains the remainder of the body, thereby enabling the animal to withstand prolonged starvation.

○ How are slow growth, long lifespan and toxicity compatible with sedentary habits, sparse food supply and a large proportion of the animal's volume being non-cellular?

● All these features contribute to low metabolic rate.

○ What life-history problems do sedentary animals face?

● Arranging for gametes to meet in sexual reproduction, and dispersal of the offspring.

Like almost all sponges, cnidarians are hermaphrodite and shed large numbers of gametes into the water, where fertilization takes place. Thousands of individuals may release their gametes over a period of a few hours. Spawning is often synchronized with the phases of the Moon, but little is known about how these invertebrates perceive light, tidal movements or other coordinating stimuli. The zygote quickly develops into a ciliated larva that swims for some time before developing into the mature stages.

1.2.3 CTENOPHORES

Ctenophores, sometimes called sea-gooseberries, are often mistaken for jellyfish because they are marine predators, almost transparent and frequently bioluminescent. Their basic body plan resembles that of cnidarians: the gut has a single opening, two cellular layers are separated by thick mesogloea, and several aspects of the body are radially symmetrical. There are also some important contrasts, including the absence of cnidocytes, and experts still disagree on the exact relationship between the two groups. Like cnidarians, ctenophores have muscle cells, which expand the mouth and enteron enabling the animals to engulf prey as large as themselves. Many species catch prey on two long trailing tentacles, which, like those of cnidarians, are highly sensitive and capable of contracting far enough to bring the food near the mouth.

Eight rows of unusually long cilia called comb plates are the primary means of locomotion in nearly all ctenophores, with muscles and mesogloea having a minor role.

○ What kinds of organisms typically swim by means of cilia?

● Small organisms, including many protoctists and the larvae of poriferans and cnidarians (the larvae of many other groups of invertebrates are also ciliated).

Ctenophores are among the largest animals to swim using cilia. Those of each comb plate beat synchronously, making them clearly visible to the naked eye, especially when illuminated by bioluminescence.

Many species of ctenophore are abundant and widespread, and a few have become pests in certain areas. *Mnemiopsis leidyi* (Figure 1.7) is sometimes very numerous around the east coast of the USA where it preys on planktonic animals, especially small fish, including the juveniles of commercially important species. It was accidentally introduced into the Black Sea in the 1980s (probably with ballast water discharged from returning oil tankers) and has devastated the anchovy fisheries there.

Recent research (Bumann and Puls, 1997) suggests that *M. leidyi* owes its success to fast, thorough digestion. Although its gut has only one opening, there are three distinct phases of extracellular digestion that enable the animal to digest several sequentially caught prey simultaneously. The first phase is acidic, while the second and the third are alkaline, equipping the animal to digest a range of prey including those that are protected by shells or spicules. Food is moved between different regions of the flattened and elongated enteron by ciliary currents at a rate of $2-4 \, \text{mm} \, \text{min}^{-1}$. This system enables the animal to combine ejecting undigested material through the mouth with taking in more food, successfully evading the limitations of its simple body plan. *Mnemiopsis* can eat *Artemia* nauplii (tiny, shrimp-like crustaceans) at a rate of 20 an hour, and digests them

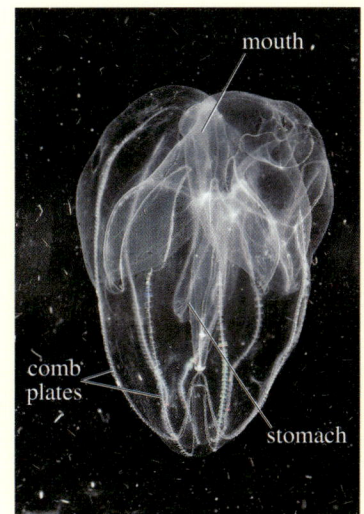

Figure 1.7 A comb jelly *Mnemiopsis leidyi* (phylum Ctenophora) from around the coast of northern Australia. The bands of cilia that form the comb plates produce interference colours. Comb jellies reach a few centimetres in length.

almost completely in less than an hour. Few animals indigenous to the Black Sea seem to kill *Mnemiopsis*, though in the western Atlantic, large jellyfish including *Chrysaora quinquecirrha* (Figure 1.6b) eat them.

In contrast to sponges and corals, ctenophores lack hard parts and so their fossil record is fragmentary. The few deposits in which they are found, notably the Burgess Shale in the Canadian Rocky Mountains, indicate that Cambrian ctenophores were similar in form and size to some modern species and accounted for about the same proportion of the total fauna as they do now. Nonetheless, with fewer than 100 living species, Ctenophora ranks as one of the least diverse animal phyla.

SUMMARY OF SECTION 1.2

1   Sponges have an irregular body form with many openings, and only three main types of cell, which are not rigidly organized into layers. They are long-lived and often abundant in aquatic habitats where all but a few are filter feeders.

2   Cnidarian cells are arranged in two layers separated by a largely acellular mesogloea, which is deformable but also resilient. The mouth opens into a sac-like gut. Most cnidarians are predators, catching prey with tentacles armed with cnidocytes that contain poison. They have organized muscles and medusae can swim strongly. Some have bioluminescent proteins that have recently been used for research purposes.

3   The body plan of ctenophores resembles that of cnidarians, but they lack cnidocytes. All are marine and some species often become very abundant in certain places.

1.3   WORMS

Worms are soft-bodied, elongated animals that move by wriggling or crawling, styles of locomotion that enable them to live in and on viscous media such as mud, sand, soil, and the tissues of other animals. Flatworms, roundworms, earthworms, many caterpillars, certain beetle and fly larvae, sea-cucumbers, eels, snakes and other legless vertebrates are all worm-like, and early classifications lumped them together as 'Vermes' (Latin for 'worms'). Comparative studies of their basic structures, and more recently biochemical analysis of genes and ribosomal RNA, indicate that these similar-looking animals are so fundamentally different that they should be classified in several phyla.

This section compares three of the commonest phyla of free-living worms that exemplify the diversity of internal organization and habits that can underlie superficial similarities of body form. The debate over how worm-like animals should be classified is far from concluded and new discoveries lead to new theories about their origins and relationships.

1.3.1   FLATWORMS

The phylum Platyhelminthes consists mostly of highly specialized parasites, that are discussed in Chapter 3, but one class, the Turbellaria, also known as

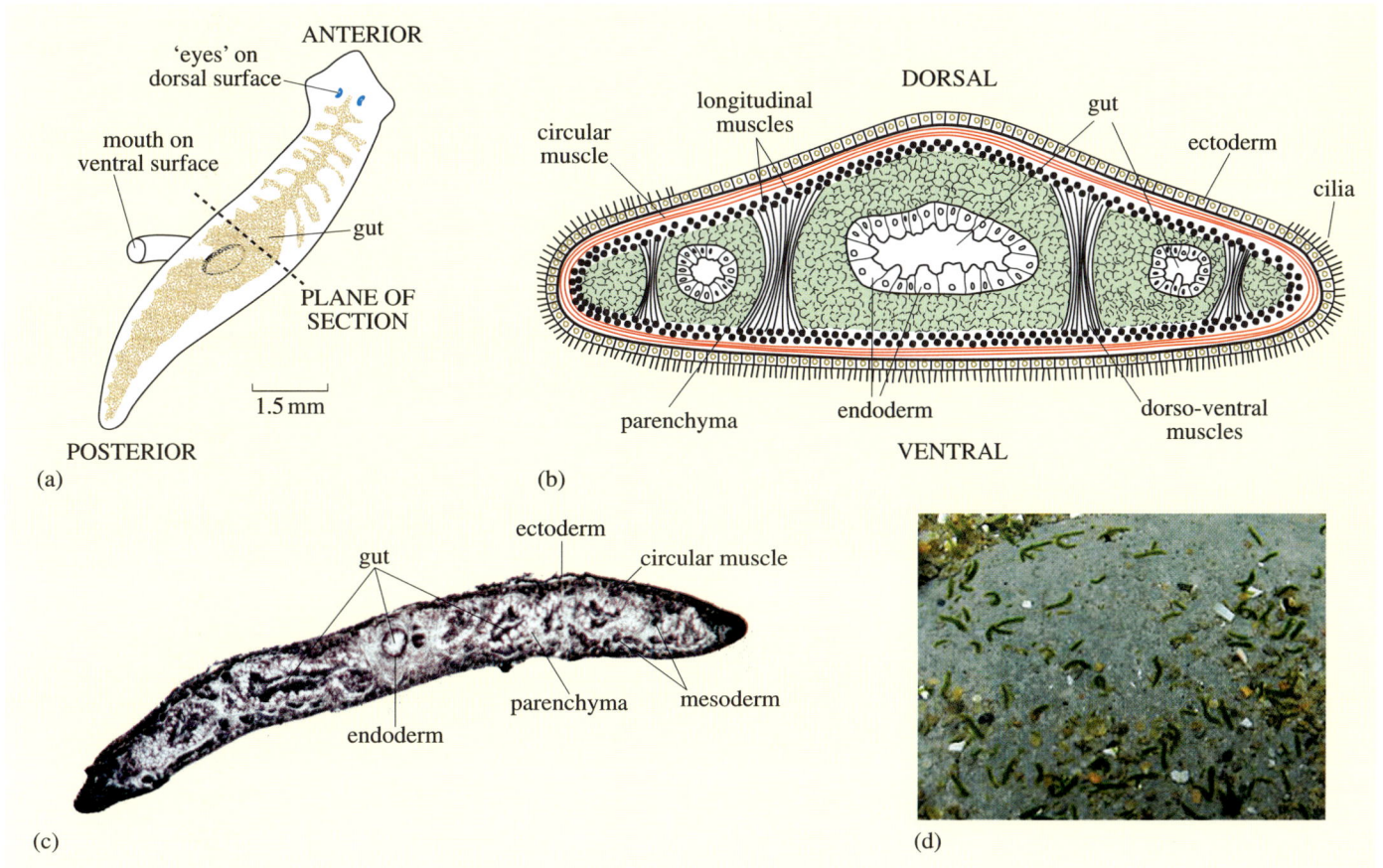

ANTERIOR

'eyes' on dorsal surface

mouth on ventral surface

gut

PLANE OF SECTION

1.5 mm

POSTERIOR

(a)

DORSAL

circular muscle

longitudinal muscles

gut

ectoderm

cilia

parenchyma

endoderm

dorso-ventral muscles

VENTRAL

(b)

gut

ectoderm

circular muscle

parenchyma

mesoderm

endoderm

(c)

(d)

flatworms or planarians (Figure 1.8), are common as free-living worms in fresh water, the sea, and in damp soil and leaf litter. As is often the case among invertebrates, the free-living groups have the most complex adult structure and simple life history (most aquatic planarians have a brief larval stage that swims in the plankton before settling), while their parasitic relatives (trematodes and cestodes) have complicated life histories but much modified adult structure.

Figure 1.8b shows the internal structure of a planarian. In place of the largely acellular mesogloea of simpler invertebrates, there is a cellular **mesoderm** between the layers of cells that form the outer covering and the lining of the gut. During development, the mesoderm gives rise to the majority of tissues, including muscles, excretory and reproductive structures. The mesoderm is a universal feature of structurally more complex animals, which are sometimes called **triploblastic** (i.e. cells form three layers, ectoderm, mesoderm and endoderm).

○ What features labelled in Figures 1.8a and b can also be seen in the section in Figure 1.8c?

● The outer layer of cells (ectoderm); the central part of the gut and several of its branches; cellular mesoderm filling the rest of the body.

Figure 1.8 (a) The general structure of a planarian. (b) A generalized diagram of a cross-section of a planarian. The mesodermal structures shown include the muscles and the parenchyma. (c) A cross-section of the planarian *Dugesia*, a common predator in ponds and slow-flowing rivers. (d) *Convoluta*, a marine planarian with its symbiotic algae, on Aberthaw beach in South Wales.

The planarian mesoderm contains numerous bands of muscle with which the worms can twist, flatten and shorten the body, enabling them to squeeze through crevices and to swim. They also crawl over dry surfaces by passing ripple-like waves of contraction along the ventral surface. These waves pass down the body as three or four muscular bulges with the body held slightly off the ground between each bulge. Cilia cover the entire body surface but are longer and more numerous in the head region and on the ventral side of the body. Planarians glide across damp surfaces such as vegetation by beating the cilia and secreting large quantities of mucus.

○ What are the main (a) similarity and (b) contrasts in the body plans of flatworms (Figure 1.8a) and cnidarians (Figure 1.3)?

● (a) In both flatworms and cnidarians, the gut is a sac with a single opening. (b) Flatworms are bilaterally symmetrical and triploblastic but cnidarians are radially symmetrical and have only two cellular layers, separated by mesogloea.

Platyhelminths are the simplest animals to have clear bilateral symmetry and a dorso-ventral axis. These features are shared by the majority of animals, including arthropods, molluscs and vertebrates (though some, including snails and flatfish, e.g. sole, plaice, have become secondarily asymmetrical), thus placing platyhelminths at the centre of the evolution of the more complex phyla.

The single-opening gut is a special feature of flatworms but, as in the case of cnidarians and ctenophores, the simple plan does not preclude catching and digesting large prey. As explained in Box 1.3, many planarians are highly efficient predators. The mouth is near the middle of the ventral surface, often on the end of a protrusible proboscis that enables the animal to engulf relatively large prey.

○ From Figure 1.8a, why is a single opening of the gut, the mouth, near the middle of the body a better arrangement than in the 'head'?

● The single opening enables the contents to move in several directions around the highly branched gut. Food is quickly dispersed from a central mouth through the whole gut and waste can be eliminated efficiently through the mouth.

As chordates, we are accustomed to expect the mouth to be intimately associated with the nervous system and special sense organs, but it does not have to be. The planarian 'head' is defined by a pair of simple eyespots, which differ in details of structure, as well as in anatomical arrangement, from those of Cubomedusae (Cnidaria), but which probably perform similar functions. As well as being sensitive to light, planarians also respond to touch and to many different chemicals, though little is known about how these sensory systems work.

Some planarians harbour symbiotic algae inside mesodermal cells as a source of nutrients, as well as or instead of eating other organisms, in much the same way as *Hydra* and corals do (Section 1.2.2). Such animals are green and live in brightly lit areas, including beaches and coral reefs. One of the first to be studied, *Convoluta*, is locally abundant (Figure 1.8d), but for unknown reasons occurs in only a very few sites around Britain.

○ How could you distinguish groups of symbiotic planarians such as that in Figure 1.8d from aggregates of free-living algae, without using any tools?

● Observe their reaction to abrupt disturbance of the water. Being animals, planarians have nerves and muscles and can move much faster in response to a 'predator' than algae can.

The *Convoluta* in Figure 1.8d crawl away immediately when disturbed by poking or shaking, or by a dark shadow passing over them, often disappearing into crevices or gravel.

BOX 1.3   IMMIGRANT PLANARIANS

Several species of planarians accidentally introduced from Australasia and Africa have become serious pests to agriculture in Scotland and other areas of northern Europe, and the USA. Travelling by air from New Zealand with imported ornamental plants, *Artioposthia triangulata* was first recorded in British gardens in the 1960s. Since then, it and several other exotic species have successfully established themselves in cool, well-cultivated soils over much of the country. Here, as in its native land, *Artioposthia* preys on other invertebrates, especially earthworms (phylum Annelida, see Section 1.3.2), but minor differences in the form and habits of European and antipodean annelids have enabled the immigrants to decimate the indigenous population. Earthworms benefit pasture and arable land by enhancing the availability of soil nutrients, increasing soil drainage and improving root penetration, and their extermination leads to compacted, unproductive soils that are susceptible to flooding.

The crisis has stimulated research (e.g. Blackshaw, 1992) into the habits and ecological relationships of planarians. One reason for their success seems to be their ability to attack and eat prey as large or larger than themselves, holding it in the proboscis and sucking it into the gut. *Artioposthia triangulata* weighs up to 1.5 g when fully grown, so it can tackle quite large earthworms, which it consumes at an average rate of 0.67 worms per week. Although a voracious eater under ideal conditions,

planarians can withstand prolonged starvation by 'degrowth' similar to that observed among cnidarians (see Pond, 2001b). In one experiment, all starved *A. triangulata* died within a week at 23 °C and within three weeks at 20 °C, but one individual survived for over a year at 5 °C. Its shrunken body regrew after feeding was resumed.

○ How could a temperature difference of only 3 °C enable planarians to survive three times longer? In view of this effect, what are the implications of transporting exotic plants in refrigerated containers for preventing the accidental introduction of similar pests?

● Longevity under starvation depends mainly upon the rate of energy utilization. Most enzymatic processes are slower at lower temperatures. Transporting plants in cooled containers greatly improves the chances that planarians (or other exotic soil animals) survive the journey.

Under ideal conditions the planarians breed prolifically and disperse efficiently, with the largest specimens able to crawl at up to 17 m h^{-1}, but they cannot burrow as well as their prey do, and are found in only in the top 0.2 m of fairly loose soil. In tropical areas, salamanders, frogs and juvenile snakes eat planarians, but northern Europe has relatively few amphibians and reptiles, and they prove to be reluctant to tackle these exotic species.

1.3.2 ANNELIDS

To most people, the term 'worm' means earthworms, fully terrestrial species of the phylum Annelida and class Oligochaeta, but this class also includes some freshwater worms (of which the most familiar is the blood worm, *Tubifex*), and a few marine species, some of which are found in mud on the floor of the deep sea. Of the two other classes in this phylum, the Polychaeta (ragworms, fan worms etc.) are almost entirely marine (a few on beaches and in brackish water) and are very diverse. Class Hirudinea (leeches) includes far fewer species but has representatives on land, in freshwater and in the sea.

Earthworms are found in all but the driest and coldest soils throughout the world; *Lumbricus terrestris* (Figure 1.9) is the commonest species in Europe.

○ What are the similarities and differences between the basic body plans of platyhelminths (Figures 1.8b and c) and annelids (Figure 1.9b)?

● Like platyhelminths, the annelid body is soft, without a stiff external covering, and the body wall and the lining of the central gut are highly cellular. But in annelids, the two layers of muscles in the body wall are much more massive than in planarians, and there are also muscles around the gut.

The unstained space between the outer layers of muscles and those around the gut is called a **coelom**, and is filled with a largely acellular, low-viscosity substance called coelomic fluid secreted by the epithelial cells that line it. Coelomic fluid differs from the blood in that it contains few cells and it is not actively pumped around the body by a specialized organ such as a heart. Coeloms have different functions in different groups of animal, and almost certainly evolved in parallel in several otherwise dissimilar lineages. Because it is easily recognizable by dissection and, as in Figure 1.9, in section, the anatomical arrangement of the coelom and its divisions have long been used as taxonomic characters for defining animal phyla.

Figure 1.9 The earthworm, *Lumbricus terrestris*. (a) A living worm. (b) A stained cross-section of a segment near the middle of the body; the muscles, gut and connective tissue and other cellular material stain brown and the external cuticle black. The blood in the vessel is naturally reddish.

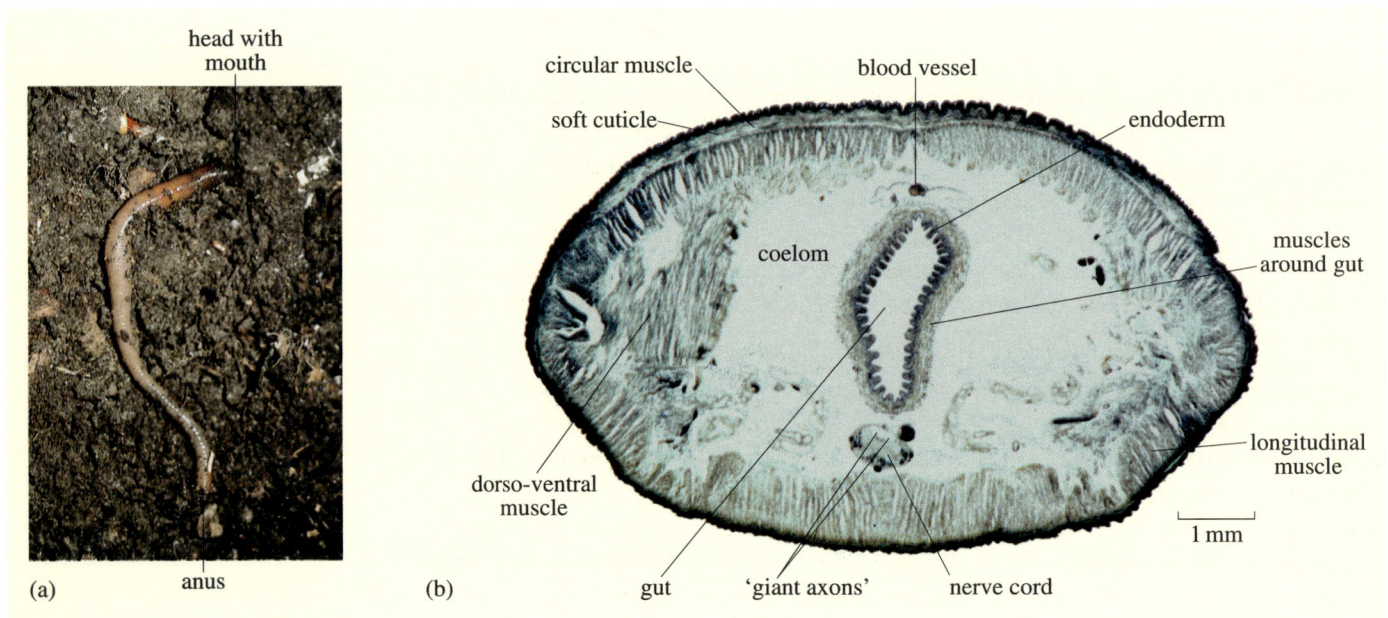

Other features visible in Figure 1.9b include a large dorsal blood vessel and a thick ventral nerve cord. The anterior region of this blood vessel is thicker and more muscular and actively pumps the blood through smaller vessels, and so acts as a heart (seen in the polychaete in Figure 1.11a). In many annelids, the blood and/or the coelomic fluid contain the respiratory pigment *haemoglobin* so most of their tissues, like those of vertebrates, appear red.

The haem parts of the molecule, which contain the iron atoms, have a strong affinity for oxygen, but their binding is modified by the much larger globin protein wrapped around them. Some worms have several respiratory pigments that differ in their affinity for oxygen because the globin components are of slightly different molecular structure. Those in internal, oxygen-using tissues have the highest affinity, and those in tissues near the gills (or body surface) the lowest. The pigments pick up oxygen at the low concentrations found in soil or mud and deliver it to the internal tissues. Some annelids even have a form of myoglobin that holds oxygen in the muscles (and confers a dark red colour).

○ In what habitats would annelids with such elaborate blood pigments be found?

● In burrows (where oxygen is often scarce), especially if, like earthworms, the annelids are large and active.

Other groups of annelids have quite different blood pigments and some smaller or less active polychaete worms that live in well-oxygenated habitats may have none at all.

The nerve cord consists of many small neurons that cannot be seen at the magnification in Figure 1.9b, and a few exceptionally large neurons. These fast-conducting 'giant' axons of earthworms (actually the processes of several neurons connected by electrical synapses, see Clements and Saffrey, 2001) mediate the fast shortening of the body that enables worms to retreat quickly into their burrows when attacked. Lugworms, fanworms and many other polychaetes have similarly rapid escape responses: they withdraw into their tubes or burrows at the slightest disturbance.

The nerve cord runs the whole length of the body. In each body segment the nerve cord forms clumps of nerve cells called **ganglia** that coordinate muscular activity within and between segments. In all annelids (and a great many other invertebrates), there is a thickened ring of several ganglia around the anterior region of the gut, just behind the mouth. Numerous nerves run from this structure, often called the brain, to sense organs that respond to touch, light and many different chemicals. The brain determines the direction of movement, and its close association with the mouth allows rapid, accurate control of feeding.

Compared to vertebrates, a nervous system consisting of comparatively few, relatively large neurons arranged in a constant pattern seems to be a basic feature common to all annelids (and animals built to an annelid-like body plan, see Chapter 2). The giant nerve fibres of earthworms were among the first invertebrate neurons to be studied intensively; their size and linear arrangement make them much more convenient to study using electrodes than the smaller, more diffusely arranged neurons of cnidarians and platyhelminths. Since the late 1960s the nervous system of the medicinal leech, *Hirudo medicinalis*, has been intensively studied. In spite of differences in overall plan, and in the size and shape of neurons, the neurotransmitters prove to be very similar to those of vertebrates.

Another feature of annelids, visible in Figure 1.9a, is the division of the body into many **segments**. Internally, the coelom is divided by 'curtains' of tissue called septa that partition each segment from its neighbours.

The annelid coelom serves as a **hydrostatic skeleton**, which confers rigidity on the body during burrowing, crawling, swimming and other forms of locomotion. Hydrostatic (or hydraulic) skeletons occur in many forms throughout the animal kingdom. They consist of a tension-resistant container (in annelids, the muscles and connective tissue of the body wall) inflated under pressure by a compression-resistant fluid (in this case, coelomic fluid, but in other groups including vertebrates, blood plays this role).

Shifted and compressed by the muscles as required, the coelomic fluid can be hard like a balloon filled with water, or compliant, as illustrated in Figure 1.10. Contraction of the circumferential muscles squeezes the coelom and the internal tissues (Figure 1.10a), producing pressures of up to 0.8 kPa and making the body thinner (Figure 1.10b). The volume of each segment is constant, since tissues and coelomic fluid do not normally escape, and the septa restrict the flow of fluid between adjacent segments, so the pressure rises, and the body becomes longer as it narrows. Conversely, if the longitudinal muscles contract, the segments become shorter but thicker and intrasegmental pressure falls to a minimum for a crawling worm, though still at least 1 kPa higher than a completely relaxed worm.

If the middle of the body is attached to or pressed against a surface, lengthening more anterior segments pushes the front forward as shown in Figure 1.10a. Waves of alternating contractions of longitudinal and circumferential muscles, called peristalsis, enable the worm to burrow, forcing its way through cracks in the soil, or to crawl. Collagen and other extracellular proteins between the muscle cells and in the body wall make the body flexible and, within limits, extensible.

○ From Figure 1.9b which muscles are more massive, longitudinal or circumferential? Does the relative size correspond to the pressure measurements in Figure 1.10b? Is there a functional explanation for your answer?

● The longitudinal muscles are much larger, although Figure 1.10b shows that higher internal pressure is generated by contraction of the circumferential muscles. This arrangement makes body shortening fast and powerful, most useful for avoiding predation from birds, moles, hedgehogs, tortoises, adult amphibians and the many other animals that find earthworms delicious.

Adhesion of the short, broad regions of the body to the substratum (or sides of the burrow) is aided by the bristle-like *chaetae* (pronounced 'kee-tee') that protrude from the body wall. (Easily felt if you let a worm crawl between your fingers.) Predators have to grip quickly and tightly, then pull hard to extract a worm from its burrow. Many burrowing annelids, including *Lumbricus*, can break themselves apart when molested in this way, in a process called autotomy. Bleeding is minimal and one or both fragments regenerates to form a smaller, but complete worm.

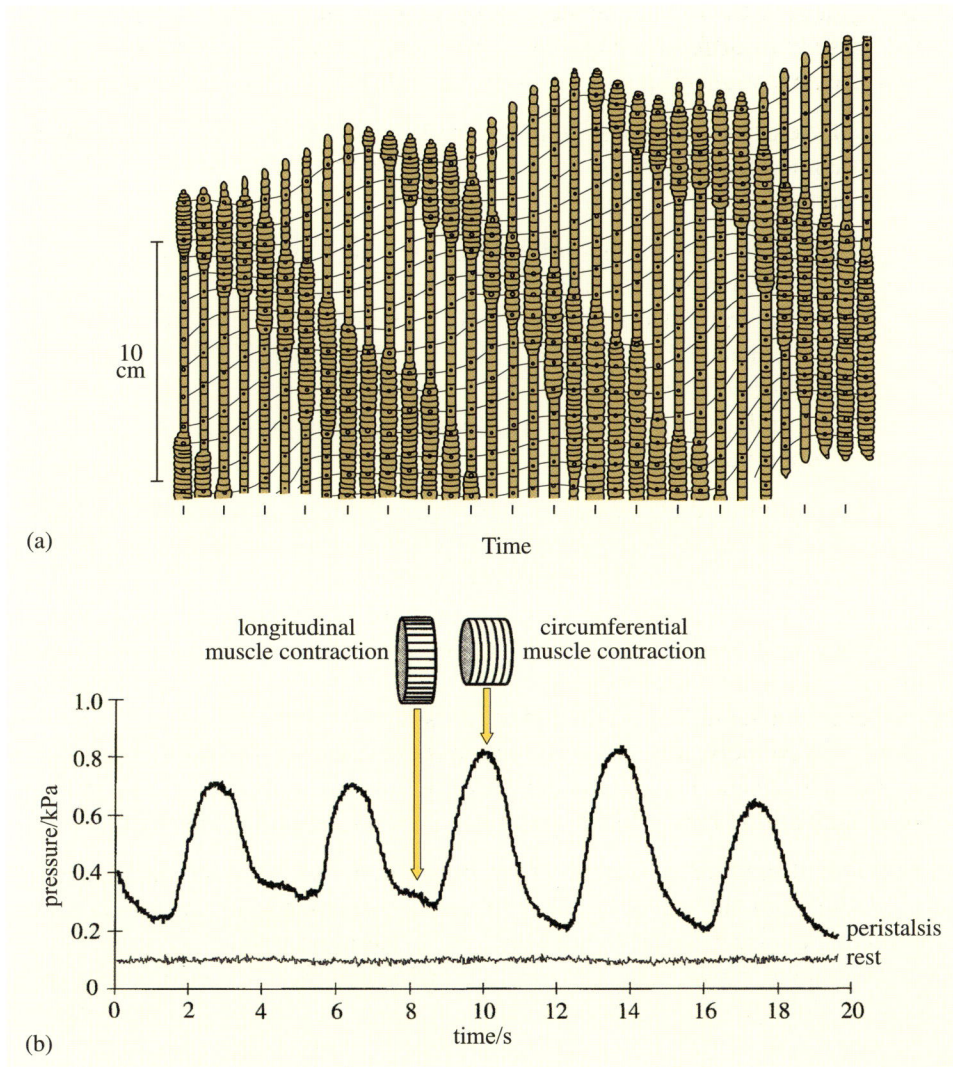

Figure 1.10 (a) General scheme for crawling of a soft-bodied, segmented worm, from Gray and Lissmann (1938). (b) Pressures recorded from inside a single segment of an earthworm while resting or crawling by peristalsis (from Quillin, 1998).

○ How does dividing the coelom into a series of cavities affect the mechanism of locomotion?

● In a non-partitioned coelom, muscular contraction anywhere could alter the shape of the whole body. If the coelom is divided by septa, contraction of the longitudinal or circumferential muscles alters the shape of each segment separately.

Dividing the coelom and musculature into segments makes locomotion using bands of muscle and a hydrostatic skeleton powerful and efficient. Although the septa are soft and deformable, allowing some of the pressure changes to be transmitted to adjoining segments, their presence permits much more precise movements than are possible for worms with an undivided hydrostatic skeleton.

○ How does a coelom differ from the cnidarian enteron and the annelid gut?

● The coelom is not directly open to the outside world as guts and enterons are, and it does not contain food being digested or absorbed.

In spite of these fundamental differences in structure, the enteron, and in some animals, the gut sometimes assumes some of the functions of a coelom by acting as a hydrostatic skeleton during burrowing or swimming.

Contraction of the longitudinal muscles on only one side of a segment produces side-to-side movements. Most polychaete annelids have a pair of paddle-like *parapodia* (singular, parapodium) on most or all segments that contain bands of strong muscles and terminate in stout chaetae, as in the ragworm *Nereis* (Figure 1.11a). Parapodia combine with changes in body shape similar to those of earthworms, and side-to-side movements produced by alternating contractions of the longitudinal muscles, to make swimming, crawling or burrowing fast and efficient, as shown in Figure 1.11b.

Polychaetes are most numerous and diverse on or near the seashore, including active predators such as the ragworm (Figures 1.11a and b) and the sea mouse (Figure 1.11c) which pursue prey through mud, sand and dense vegetation in shallow seas. Others feed on detritus, often filtering water through a crown of tentacles, as in fan worms (Figure 1.11d), or sucking sediment through a burrow,

Figure 1.11 Some common polychaetes living on beaches and shallow seas. (a) Ragworm (*Nereis pelagica*). (b) *Nereis* sp. swimming in shallow water. Ragworms grow to a length of up to 10 cm. (c) Sea mouse (*Aphrodite* sp.), which can be several centimetres long. (d) Fan worms (*Sabella* sp.) protect the body with a 'case' of sand particles from which filtering tentacles can be protruded.

parapodium | large dorsal blood vessel

eye

(a)

(b)

eye

(c)

parapodia

(d)

as in lugworms (*Arenicola*). These large burrowing polychaetes are often extremely abundant in shallow muddy bays and estuaries where the water is rich in microbes and tiny particles of detritus. Their presence is indicated by little piles of defaecated sand. Together with burrowing molluscs, they are a major source of food for many kinds of wading birds as well as for fish, crabs etc.

Earthworms burrow to eat rotting plant material and tiny soil animals and microbes, so many species become abundant in and near compost heaps and well manured soils.

○ In what ways would the basic organization of the annelid gut be suited to the earthworms' diet?

● Plant material is bulky, especially when diluted with the mineral components of soil, and its digestion is slow. A gut surrounded by muscles with 'one-way' traffic from mouth to anus would be able to move the large mass.

Digestion is quite thorough: earthworm droppings, often seen on the soil surface in the early morning, consist mainly of fine mineral fragments with almost all the plant and microbial matter extracted.

○ Compare Figure 1.9b with Figure 1.8a–c; how do nutrients reach the mesodermal and superficial tissues of flatworms in the absence of a blood system?

● Platyhelminths do not need a vascular system: their gut is highly branched and, in the absence of a coelom, extends throughout the mesoderm, so no cells are very far from where nutrients are absorbed, and they diffuse to where they are utilized.

The presence of a coelom thus entails the formation of a circulatory system, at least for larger, more active worms. As well as functioning as a hydrostatic skeleton, the coelom, at least that of certain segments, accommodates tissues of variable size.

○ Which tissues undergo large changes in volume as an integral part of their normal role?

● Gonads, especially ovaries, and storage tissues such as adipose tissue, normally undergo large changes in mass, particularly in animals that breed seasonally.

The coelom can expand as required to accommodate massive reproductive and storage tissues. In all coelomate animals including ourselves, gonads form from cells in the lining of the coelom. In annelids mature gametes accumulate in the coelom of many, in some species almost all, segments. Some marine polychaetes rupture (and die) when releasing eggs and sperm into the sea, where fertilization occurs. In other polychaetes, especially free-living predators rather than burrowers, eggs fill a portion of the body called an epitoke, which is shed and swims by itself for a few hours. Sometimes thousands of worms spawn simultaneously at certain times of year, releasing epitokes that contain millions of nutrient-rich gametes. The transient abundance attracts many predators, including humans in some areas, who relish the yolky eggs. The zygote develops into a tiny, ciliated larva called a *trochophore*, which swims in surface waters before settling and developing into a worm.

Earthworms (and most other oligochaetes) have more elaborate reproductive structures restricted to a few anterior segments, which enable the ova to be fertilized internally with sperm exchanged between two hermaphrodite adults. The relatively large, white eggs laid in clumps of about a dozen are a common sight in the upper layers of well-cultivated soil. There is no larval stage: the eggs hatch into tiny worms that have the same diet and habits as their parents.

Leeches (class Hirudinea) are highly distinctive annelids that have a constant number of segments and strong suckers at each end of the body.

○ Comparing Figure 1.12b with the section of an oligochaete (Figure 1.9b), what features unite them in a single phylum? What are some major contrasts between leeches and other annelids?

● Both animals have a central gut with muscles around it, a ventral nerve cord and layers of body musculature. The leech differs from the earthworm in that there are bands of muscle across the coelom, which is largely obscured by cellular material, and there is no external cuticle.

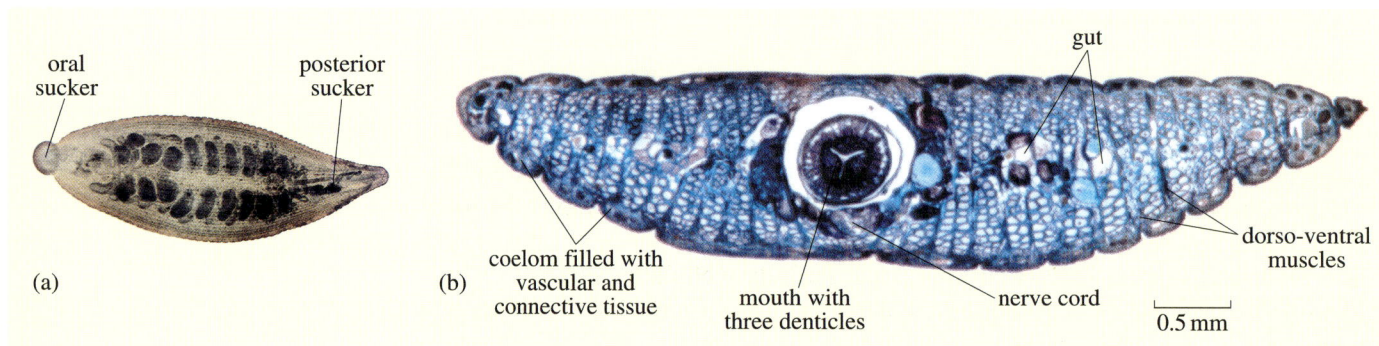

Figure 1.12 (a) A whole leech (*Glossiphonia* sp.) up to 1 cm long. (b) Cross-section of a leech near its anterior end; cells stain brown, extracellular collagenous material blue. Traces of red blood from its prey can also be seen in branches of the large gut.

Without much coelomic fluid, the septa are also greatly reduced. Leeches do not burrow but 'loop' over damp surfaces using their suckers, or flatten the body with their dorso–ventral muscles and swim by up-and-down undulations.

Like oligochaetes, leeches occur in freshwater, in the sea, and on land, but are most abundant and diverse in ponds, lakes and slow-flowing rivers and in moist terrestrial habitats such as soil and rainforest. Some are predators on other soft-bodied invertebrates, but the majority are bloodsuckers, attacking vertebrates such as fishes, frogs and mammals and large invertebrates, including snails and crustaceans. Some pounce on passing hosts from leaves and branches in swamps and rainforests, but the majority specialize on aquatic prey.

Hirudo medicinalis feeds only on vertebrates, which it detects by vibration and by smell. Leeches grow faster when feeding on mammals or birds, whose blood contains a higher proportion of cells than that of amphibians or fish. Once close to its prey, the leech selects a warm patch of skin (possibly because warmth means blood vessels near the surface of thinner skin) and cuts a small wound with its three, hard denticles that give the mouth its Y-shape, visible in Figure 1.12b.

Any objections from the host are suppressed early by the secretion of copious amounts of saliva containing pain-killing and anticoagulant proteins that are small enough to avoid eliciting an immune response. These properties have been exploited for medicinal purposes, as described in Box 1.4.

○ How is a local immune response to blood-sucking animals recognized?

● Mosquitoes, horseflies and most other blood-sucking arthropods leave proteins in their hosts' wounds that provoke swelling and itching for hours, often days, afterwards.

Bites from most kinds of leech rarely cause such responses. Prolonged bleeding and the mechanical presence of the leech itself is still unpleasant and heavy infestations can lead to severe blood loss. *Hirudo* is one of the few species that can bore through tough skin: most smaller leeches select easier sites such as the lining of the mouth, nose and throat, reached as the animal drinks, thereby causing great distress to their host.

In a single meal lasting 10–40 min, leeches ingest up to ten times their fasting body mass, or 5–15 ml of blood in the case of fully grown *Hirudo*. The engorged leeches then drop off their host and, often so swollen that they are unable to swim properly, retreat into a damp place.

○ What physiological problems arise from consuming blood?

● Blood contains too much water for efficient digestion (invertebrate blood is even more dilute than that of vertebrates), so the excess must be excreted. The large quantity of iron can also be toxic if it is all absorbed (see Pond, 2001b).

The excretory system (called nephridia) is particularly well developed in leeches and eliminates the excess water quite quickly. Digestion is slow but thorough, and a large, well-sated leech may not feed again for 12–18 months. Juveniles eat more frequently, but at 20 °C, *H. medicinalis* needs only about nine meals spread over a period of about 10 months to grow from hatching to sexual maturity.

BOX 1.4   THE MEDICINAL ROLES OF *HIRUDO MEDICINALIS*

Therapeutic bleeding was a respected treatment for a variety of disorders from the 5th century BC in Europe, China and India. Being less painful and leaving smaller scars than lancing, leeches were a preferred method. Linnaeus recognized their importance in this role when he named the European leech *Hirudo medicinalis*. Sometimes several dozen animals were applied simultaneously to a single patient and treatment was repeated every few weeks. In 1832 alone, St Bartholomew's Hospital in the City of London used over 97 000 leeches. Such large numbers could not be stored for over a year until they were hungry enough to be re-used, so, expensive though they were, leeches were disposed of. Overcollecting made the animal rare over much of its natural range (as they still are in most places), and by the early 19th century, medicinal leeches were imported into France and Britain from as far away as Russia.

With the development of new drugs and safer surgery in the mid-19th century, the medicinal use of leeches went out of fashion. Fortunately, biologists continued to study them both for their scientific value and with a view to finding therapeutically useful materials. The principal anticoagulant in leech saliva, named hirudin, was identified in 1884 and has been used to prevent clotting during blood transfusions and surgical operations ever since. Hirudin binds irreversibly to thrombin, a key enzyme in the process of blood clotting, thereby inactivating it. As well as being highly specific and non-allergenic, hirudin is such a small peptide that any excess is quickly eliminated through the kidney. Large doses can therefore safely be administered during surgery but hirudin is soon cleared from the blood after the treatment ends, thus restoring the clotting mechanism to normal.

For the first 70 years of its use, the only source of hirudin was live leeches, but in 1984, the protein was sequenced and its gene cloned. Hirudin is now produced commercially by genetically engineered *E. coli* bacteria. However, such progress was not the end of the medical uses of annelids.

More than a dozen other proteins have been isolated from the saliva and intestine of *Hirudo medicinalis* and other species of leech, including the enzymes hyaluronidase and collagenase, and different anticoagulant proteins that prove useful for open-heart surgery where prevention of blood clotting is essential. Some of these proteins also occur in blood-sucking arthropods such as ticks, bedbugs and mosquitoes, but others seem to be unique to annelids.

The leeches' special style of drawing blood works particularly well on very small vessels. When re-attaching severed fingers, ears etc., plastic surgeons are usually more successful in uniting the tougher and more muscular arteries and arterioles that carry blood into the extremity than the veins and venules that carry it away, so blood accumulates. Leeches prove to be the ideal treatment: as well as removing the unwanted blood and so relieving swelling, the enzymes and anticoagulants 'open up' the vessels and encourage the restoration of normal circulation. After more than a century in disfavour, leeches and some of their unique proteins are back at the forefront of medicine and surgery (Eldor *et al.*, 1996). *H. medicinalis* is now bred for research and therapeutic use in authorized leech farms in France, Wales and elsewhere.

1.3.3 ROUNDWORMS

The phylum Nematoda, commonly known as roundworms, threadworms or eel-worms, is second only to arthropods in numbers of species, and probably also in numbers of individuals. But most people rarely notice nematodes, which live inconspicuously in soil and mud, and among stones and encrusting organisms of both freshwaters and the sea (especially the bed of deep seas) and above all as parasites of many different kinds of plant and animal.

○ Why were nematodes chosen for use in laboratory studies of cell biology?

● The soil worm, *Caenorhabditis elegans*, was chosen for study because it has a short life cycle and consists of a fixed number of cells (see Pond, 2001a).

Like *Caenorhabditis,* most free-living nematodes, and many parasitic species, are only a few millimetres long, but some that are gut parasites of large animals such as horses, pythons and whales are much larger. One nematode parasite of sperm whales is said to grow to 9 m long.

The nematode body form is very constant: long and slender with round cross-section (Figure 1.13), entirely or partly covered by a stiff, impermeable cuticle composed of various proteins and other macromolecules, including several distinct forms of collagen.

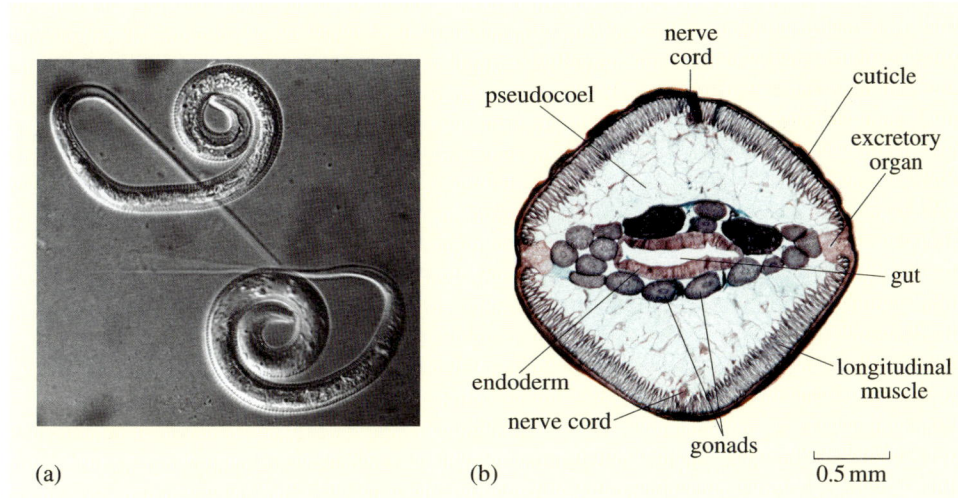

(a)   (b)   0.5 mm

Figure 1.13 (a) A whole nematode. (b) Cross-section of a roundworm; cells stain brownish red, extracellular material blue, cuticle black.

○ How do (a) the muscles, (b) the nervous system and (c) the coelom differ in nematodes (Figure 1.13b) compared with those in annelids (Figure 1.9b)?

● (a) As in annelids, the longitudinal muscles are clearly visible in the nematode but there are no circular muscles. In this section, no muscles can be seen associated with the nematode gut. (b) Instead of a single large ventral nerve cord, nematodes have several much smaller cords (two visible in Figure 1.13b). (c) The body wall and the gut form solid rings of tissue but the space between contains few or no cells.

These fundamental contrasts are the bases for classifying the two kinds of worm in different phyla. The nematode gut has mouth and anus, but its walls are muscular only in the pharynx. The types of cell that form the boundaries of the coelom are different in nematodes and annelids, so that of the former is called a **pseudocoel**. Nematodes do not have septa or segments dividing the pseudocoel into compartments. In contrast to the soft, deformable bodies of many other kinds of worm, the body is covered with fairly stiff cuticle that allows little expansion of the nematode body. The cuticle is shed periodically by a process called ecdysis. The soft tissues grow rapidly before the newly secreted, larger outer covering hardens.

○ Would nematodes be able to crawl or burrow like annelids (see Figure 1.10)?

● No. Worms with these anatomical structures are unable to constrict or lengthen the body as annelids do.

Nematodes move by a distinctive, whip-like thrashing, hence their common name, eel-worms: muscles on one side of the body contract, bending the cuticle, which recoils when released. Movement is coordinated by a sparse nervous system (only 302 neurons in the case of *Caenorhabditis* (see Pond, 2001a)) assembled into several nerve cords, of which the largest are ventral, that meet at a ring of neural tissue surrounding the gut just behind the mouth.

Most nematodes feed by sucking in nutrients such as soil bacteria and fungi, or the blood or gut contents of other animals. Several groups of nematodes pierce hard external tissues before sucking out the soft contents. This technique is used by some predatory species and also by many herbivores that feed on plant roots in the soil.

○ Which other animals mentioned so far in this chapter are terrestrial herbivores?

● None. The ability to eat higher plants in significant quantities is limited to just a few animal phyla, of which nematodes are the simplest.

The sexes are separate in most nematodes and fertilization is internal. The juveniles are very similar to the adults and usually share the same habitat. As they grow, the cuticle is shed in a process called *ecdysis*, and replaced with a new, larger version three or four times before reaching maturity. The most thoroughly studied nematode, *Caenorhabditis elegans,* and related soil-living worms, are unusual in being mainly hermaphrodite (see Dyson, 2001).

○ What aspects of its biology are consistent with this reproductive strategy?

● *Caenorhabditis* lives in soil, where its main food, bacteria, occurs in patches at the sites of decaying plant or animal matter. This tiny worm breeds and matures very rapidly near temporarily abundant sources of food.

Hermaphroditism facilitates very rapid, opportunistic breeding from just one or a few founders. Many parasitic nematodes have more complicated life cycles that are discussed in Chapter 3.

SUMMARY OF SECTION 1.3

1   Worm-like animals with soft, extensible bodies are classified into several different phyla.

2   Platyhelminths lack a coelom and the gut has only one opening. They creep using cilia as well as muscles. Only one group, the planarians, are free-living, but some are very successful predators.

3   Annelids have an anus as well as a mouth, paired appendages and a coelom divided into segments, that serves as a hydrostatic skeleton for burrowing and crawling. Polychaetes are marine predators or filter feeders and have small ciliated larvae. Oligochaetes and leeches are mainly terrestrial or freshwater; the former are mostly detritus feeders while leeches are predators or bloodsuckers.

4   Nematodes secrete a stiff cuticle, which is replaced by moulting during growth. Although structurally uniform, they live in a wide range of habitats and are often abundant.

1.4 INVERTEBRATES WITH HARD SKELETONS

Hard skeletons have evolved in several quite different lineages of invertebrates, in each case enabling the animals concerned to become larger, more powerful and often more diverse. The most abundant invertebrates with hard skeletons are the arthropods, discussed in Chapter 2. The skeletons of echinoderms and molluscs are chemically quite similar to each other, but while that of echinoderms is internal, the skeleton of molluscs is secreted externally. The body plans of the two phyla are also quite different, so they probably arose from different worm-like coelomate ancestors that followed contrasting paths through evolution into complex and powerful invertebrates.

1.4.1 MOLLUSCS

A secreted external skeleton, the shell, is a fundamental character of molluscs, but while in nematodes (and arthropods), almost the entire body surface secretes cuticle, molluscan shells are formed only by the **mantle** (Figure 1.14a–c). The outer surface of this uniquely molluscan structure synthesizes and secretes the shell. The inner surface forms a space outside the body that contains the gills, thin, permeable and usually highly frilled structures (Figure 1.14b) or, in terrestrial species, a lung (Figure 1.14c).

Figure 1.14 Essential features of molluscs. (a) Generalized mollusc showing the shell, mantle and other features of the body plan. (b) Ventral view of a marine mollusc showing the foot, gills and underside of the mantle. (c) A terrestrial slug showing the head and the mantle, which lacks a shell. (d) The radula of the abalone limpet (*Haliotis* sp.), a grazing gastropod.

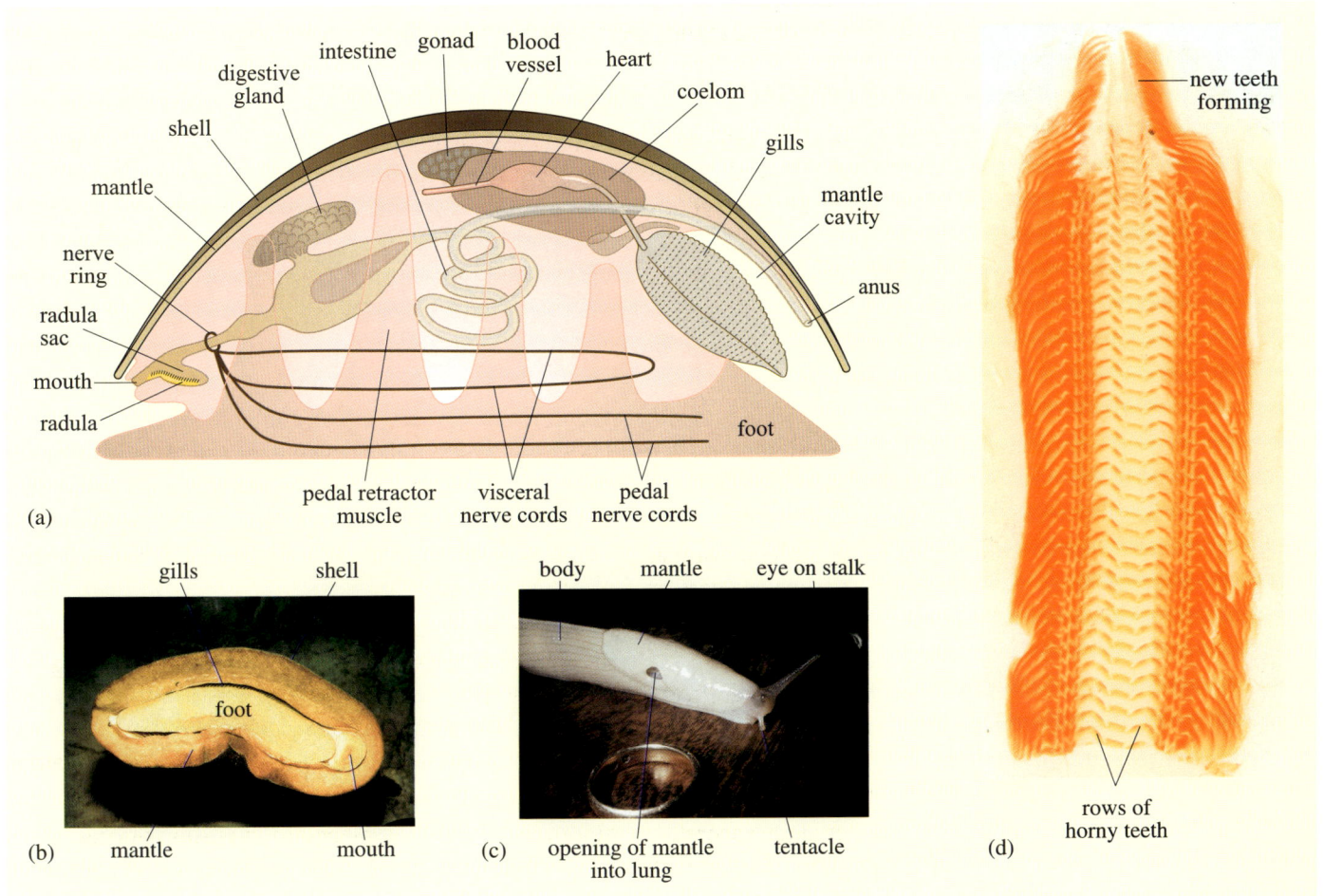

○ How does the arrangement of the musculature and viscera in molluscs (Figure 1.14a) resemble and differ from that of platyhelminths (Figure 1.8b and c) and annelids (Figure 1.9b)?

● As in annelids but not platyhelminths, molluscs have an anus as well as a mouth, a nerve ring around the anterior end of the gut, and a coelom that separates the viscera from the body wall. In the worms the muscles surround the gut and associated glands, reproductive organs and nervous system, but in the mollusc, the viscera are dorsal while the muscles are concentrated on the ventral side. In contrast to annelids, but like platyhelminths, segmentation is not evident.

The muscular region is called the foot and, as in soft-bodied worms, is usually expanded by the hydrostatic pressure of the fluid in the **haemocoel**, which combines the functions of the blood system and coelom. In all but the most primitive molluscs, the locomotory structures are not divided into segments.

Many living molluscs (and probably the earliest molluscan ancestors) have a **radula** (Figure 1.14a and d), a tough, horny 'tongue' supported by muscle, cartilage and collagenous connective tissue that rasps against firm surfaces such as coral, encrusting algae or plants. Its rhythmical protrusion and scraping are easily seen when a pond snail rasps algae off the glass wall of an aquarium, leaving a distinctive pattern in its wake. Constant abrasion soon blunts the teeth, so the radula is continuously replaced with new rows of teeth secreted by the radula sac.

Differences in the shape and relative size of these body components create the huge diversity of molluscs. Seven living orders are recognized, of which only the three most abundant and diverse, Gastropoda, Bivalvia and Cephalopoda, are discussed here.

Gastropod and bivalve shells grow only from the edge of the mantle, although wounds and breakages anywhere are patched with newly formed shell. Shell-forming epithelial cells secrete a thin outer layer of a horn-like protein, and calcium and carbonate ions, which form the thicker middle layer, together with some highly acidic proteins that promote the precipitation of the inorganic ions in ordered crystals (mostly aragonite). The crystals are arranged in layers of alternating orientation, like plywood, which makes even a thin sheet very strong and resistant to cracking. On the right of Figure 1.15a, a crack has passed

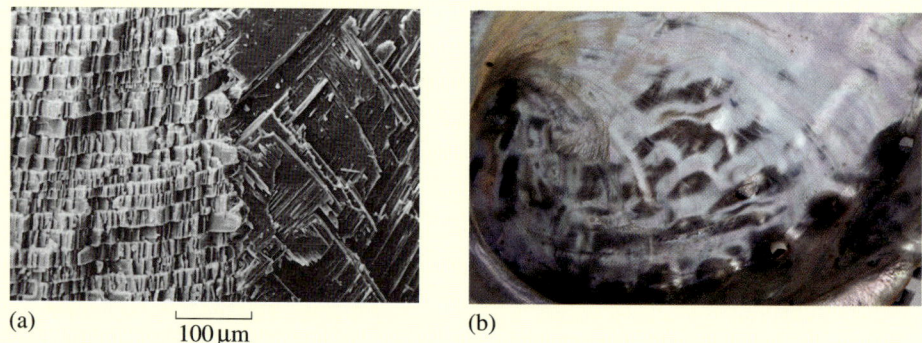

Figure 1.15 The structure of molluscan shell. (a) Scanning EM of the fracture surface, formed by cracking, of the common cone snail (*Conus litteratus*), a robust predator. (b) Nacre (mother-of-pearl) on the inner surface of a limpet (*Haliotis* sp.).

(a) 100 µm   (b)

relatively easily between layers of crystals, leaving a smooth surface, but on the left, the orientation of the crystals is turned by about 90°, and much more energy is dissipated in forming a rough fracture surface. Cracks started by all but the hardest blows would have stopped at the interface.

The protein matrix comprises only a few per cent by weight but makes the shell up to 3000 times tougher than pure aragonite crystals. As well as being hard, molluscan shells are resistant to cracking when smashed against rocks or in predators' jaws, or trodden on. The epithelial cells also produce pigments, which may be incorporated into the horny or calcareous layers, and may form bright colours and elaborate patterns (Figure 1.16).

The smooth inner layer called **nacre** (Figure 1.15b) is even tougher. It consists of sheets of aragonite crystals only 0.5–1.0 µm thick, separated by even thinner layers of protein that together produce colours by interference between light beams reflected from them.

○ Would these interference colours be functional in the living mollusc?

● No, usually not, because, being the inner layer of the shell, nacre is rarely exposed to light in the living animal.

Nacre is 98% mineral crystals and only 1% each protein and water, while cow bone is 65% mineral and 2% organic material (the rest being water). Nonetheless, mechanical tests show that nacre is at least as tough and resistant to fracture as most kinds of bone. Being tough as well as lustrous makes pearls one of the most valuable biological materials (see Box 1.5).

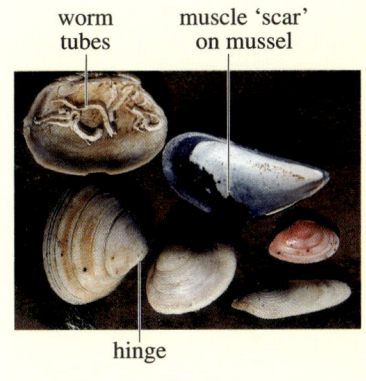

Figure 1.16 Various bivalves common on sandy shores around Britain. The worm tubes are of a polychaete *Spirorbis,* a fan worm that attaches itself to plants and other animals, both living and dead, and secretes a snail-like tube.

BOX 1.5 PEARLS

The nacreous inner layer of molluscan shells, sometimes called flat pearl or mother-of-pearl, forms nodules of gem pearls around exogenous objects, such as a sand grain or a small parasite, that find their way between the epithelium and the shell. The secretions stimulated by the foreign material contain more protein, and the embedded inorganic crystals are finer than in normal shell, making pearls lustrous and mechanically tough enough to be ground and polished. Gem pearls are believed to glisten more brightly when worn continuously, perhaps because the salts, water and organic molecules in sweat secreted from the skin interact with the charged proteins, maintaining an appropriate crystal structure.

Such accretions are found naturally in many species of marine and freshwater bivalves and a few gastropods, and can be created artificially by 'seeding' an irritant material under the shell. Marketable pearls form in a few years. The choice of species for commercial production is determined by ease of maintenance and growth rate. Species that naturally assemble in quite large groups are usually chosen because they are more resistant to diseases and to ectoparasites such as polychaete worms that grow in and on their shells. Oysters of the genus *Pinctada* are widely cultivated in warm shallow lagoons around the Indian and Pacific Oceans. They eat mostly algae in the wild, but in captivity they can be raised on yeast or just detritus. In common with many marine invertebrates, *Pinctada* and other tropical bivalves grow much faster than cold-water species. The freshwater pearl mussel, *Margaritifera margaritifera*, is found in unpolluted rivers throughout Europe and western Asia. Its age can be estimated from the annual growth rings in the shell. Spanish mussels were found to live up to 28–40 years, but members of the same species collected in northern Sweden and Russia grow much more slowly and live as long as 114–190 years.

In many molluscs, the shell is extensive, covering most of the body. While this arrangement protects the body from predation and desiccation, an impermeable covering can limit other physiological functions.

○ How do animals that live in conditions where dissolved oxygen is scarce, such as in tubes, burrows and under impermeable shells, obtain sufficient oxygen?

● The animals that live under such conditions often have blood pigments, which bind oxygen, facilitating its uptake and delivery to tissues (Section 1.3.2).

The blood of earthworms, lugworms and many other annelids contains the iron-based pigment haemoglobin, like that of vertebrates. The blood of large molluscs contains the pigment **haemocyanin**, which is pale blue when oxygenated but colourless when deoxygenated. Like haemoglobin, haemocyanin is a very large and complex molecule, which consists of seven or eight distinct but linked subunits, each containing a pair of copper atoms to which the oxygen binds.

BIVALVES

In bivalves, the mantle and shell form two usually symmetrical hinged lobes that protect and support the extensive and elaborate gills that serve as food-collecting surfaces as well as for respiration. Except for a few species that live as ectoparasites or in symbiotic associations, all other bivalves are detritus feeders and most are sedentary, living in or on sand, rock or other substrate. The main bases for diversity are the manner of attachment or burrowing, and the mechanism of food collection and selection. Most bivalves burrow, or live attached to a rock surface from which they filter fine particles from the water in various ways.

The muscular foot is expanded by body fluids acting as a hydrostatic skeleton, and can probe and obtain a grip by changing shape, for burrowing into sand or mud. The buried mollusc then extends its siphons to the surface and creates a water current that passes over its many gills. Most species can rebury themselves if their burrow is washed away, and a few, including the familiar scallop shells (*Pecten* sp.), can swim short distances by clapping the valves together. Most bivalves can withdraw the delicate siphons and shut the shell when attacked, or exposed to air, or if the water current becomes toxic, and hold it tightly closed for many hours.

○ Can typical muscles maintain tension for hours at a time?

● No. Muscle contraction usually dies away after a few seconds or minutes and cannot sustain forces except by repeated stimulation, which would use a lot of energy.

○ Would bivalves be able to expend energy at a high rate for several hours?

● No. They cannot take up oxygen from the water when closed, so would quickly become anoxic.

Physiological measurements show that bivalves use hardly any energy when closed up, but, as anyone who has tried to open a living mussel or oyster knows, considerable force is needed to prise the valves apart. The muscle seems to 'catch' in a shortened position, which can be maintained indefinitely with minimal ATP breakdown. The biochemistry of the shell-closing ('adductor') muscle (whose attachment to the shell forms a 'scar' seen in Figure 1.16) has been extensively studied to discover the mechanism of this unique property.

Several families of bivalves burrow into rock, coral, wood or other hard substrata: gripping the side of their tunnel with the foot, they rock the shell, which is equipped with hard, cutting 'teeth', from side to side (Figure 1.17). As the animal burrows deeper the siphons grow longer so their tips remain at the surface (Figure 1.17b). Beating cilia draw a current of water in one siphon, over the gills where food particles are extracted and out through the other siphon. Once established, the animal cannot leave or turn round in its tunnel but it is almost completely protected from predators. Drilling rates depend upon the hardness of the rock but can exceed 11 mm per month.

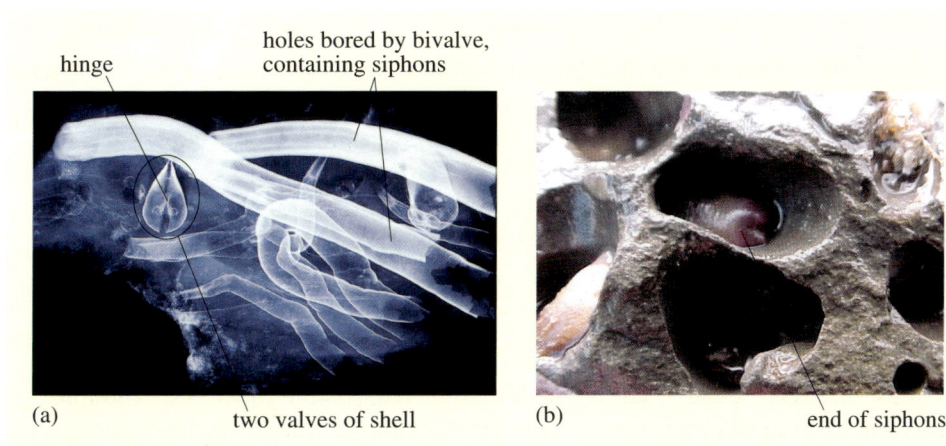

(a) hinge — holes bored by bivalve, containing siphons — two valves of shell

(b) end of siphons

Figure 1.17 (a) X-ray of a rock-boring bivalve. Each animal has its own burrow that it grows to fit. The burrows, which can reach several centimetres in diameter, rarely intersect, indicating that animals are aware of each other. (b) *Hiatella* sp., a bivalve that bores into soft limestone. Only the tips of the siphons are visible; the rest of the body is wedged into its burrow.

Several genera bore into wood, of which the most specialized are the 60 species of *Teredo*, the shipworm, which can drill deep into all but the hardest timbers, weakening their structure and exposing them to further attack from fungi and other microbes. With the help of symbiotic bacteria in their enlarged stomach, shipworms eat the 'sawdust' that their excavations produce, as well as or instead of filter feeding.

○ How could human activities have extended the range and abundance of *Teredo* during the past 20 000 years?

● By building sea-going wooden ships and timber piers, thus greatly increasing the amount of wood in the oceans.

Before humans took to fishing and travelling by boat, wood-boring bivalves were confined to the small amount of naturally occurring driftwood, and mangroves, which are among the very few woody plants to grow in the sea.

In mussels, the foot is reduced and secretes byssus threads: tough, collagenous strands tipped with glue which stick to rock so efficiently that the bivalves can take advantage of the water currents in and around the surf zone (thus saving the energy that would be needed to create their own feeding currents) and are rarely dislodged by even the fiercest storm. Similar glues enable oysters to stick one of their valves to the substratum.

GASTROPODS

Gastropods are much more mobile than bivalves, using their muscular foot in a wide variety of ways. On land and in intertidal habitats they creep over a trail of mucus secreted from glands in the foot (Figure 1.18a). Mucus consists of highly hydrated polysaccharides and proteins, so glucose, amino acids and large quantities of water are needed to produce it.

Measurements show that secreting mucus is metabolically expensive, requiring more than 30 times as much energy as the work done by the muscles in locomotion. Around 70% of slugs' total energy expenditure is devoted to slime production. Gastropods are estimated to use about 10 times as much energy per unit of distance travelled as a legged arthropod or vertebrate of similar body size.

○ What cellular processes would use all this energy for making mucous trails?

● Energy-consuming processes include: the synthesis of the proteins themselves and enzymes for modifying and polymerizing simple sugars to make polysaccharides; moving the molecules through the Golgi apparatus and secretion through the cell membrane; ion pumps create the gradients that allow the uptake of large quantities of water for hydrating the macromolecules.

This apparently extravagant form of locomotion has its advantages: the slime trails enable molluscs to 'home', returning to a safe resting place after foraging excursions. The common limpet *Patella vulgata* is famous for its ability to find its way back after grazing on algae under water to its own patch of rock just before it is exposed by the receding tide (Figure 1.18b). Its shell forms a tight seal, having grown into a shape that exactly fits its 'home', enabling it to withstand wave action, hot weather and desiccation. The mucous trail is exceptionally durable, having a half-life of around 40 days. Thicker mucus is sticky, which enables smaller snails and slugs to hang upside down under rocks and leaves.

Figure 1.18 Gastropods. (a) A terrestrial snail (*Helix pomatia*) laying a trail of mucus. (b) Limpets (*Patella vulgata*) at low tide. (c) A predatory marine snail (*Buccinum* sp.).

(a) coiled shell   eye on stalk   foot   tentacle
(b)
(c) eyes   mouth

Several kinds of marine gastropod are filter feeders, collecting tiny particles from water currents on their gills, mucous threads or other parts, rather like bivalves. But the majority, including many of the most primitive such as *Patella*, use their large radula to rasp encrusting algae off rocks, living coral tissue, etc. The patterns of radula 'teeth' (Figure 1.14d), their hardness and growth rate, determine the kinds of food each species can collect.

Many of the more advanced gastropods are predators (Figure 1.18c), killing invertebrates including other molluscs and in some cases small fish. The radula is greatly modified with barbs that inject toxins. Many of these predatory snails and shell-less gastropods are capable of brief bursts of quite rapid movement. Their eyes, situated on retractable stalks just behind the tentacles, are also larger and probably more efficient. For example, *Conus* (Figure 1.19a) ambushes passing prey and stabs it, injecting nerve toxins through a hollow radula tooth before engulfing its meal whole with its large muscular proboscis. A few tropical species are dangerous to humans.

Gastropods are the only class of mollusc in which a major order has become fully terrestrial.

○ Which organs would be radically different in aquatic and terrestrial molluscs?

● The respiration system; gills are efficient only in water.

Terrestrial slugs and snails, such as that in Figure 1.18a, are pulmonate gastropods. As their name implies, pulmonates have a lung in place of gills, formed from the inner surface of the mantle, as shown in Figure 1.14c. The mantle cavity and its permeable walls are ventilated at a single hole, which can be sealed up completely to prevent water loss in dry conditions. Like their aquatic counterparts, many terrestrial snails are herbivores, eating algae, bryophytes and flowering plants.

○ With which major group of invertebrates would terrestrial, leaf-eating molluscs be in direct competition?

● Insects (especially caterpillars, bugs and beetles), which also eat terrestrial plants.

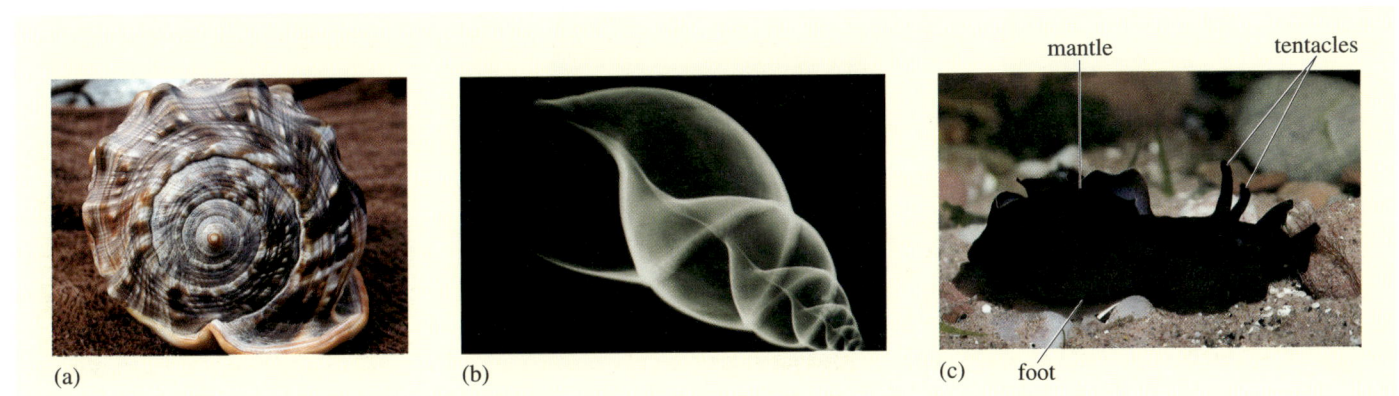

(a)   (b)   (c)   foot

Figure 1.19   (a) A *Conus* shell showing the coiled structure. (b) X-ray of a snail to show the internal structure of the shell. (c) A sea hare (*Aplysia* sp.), which can grow to a length of several centimetres, is one of several kinds of shell-less marine gastropod.

The most distinctive feature of gastropods is the shell, which is similar to that of bivalves in basic structure (Box 1.5). The coiled shape (Figures 1.18a and 1.19a and b) arises from continuous, asymmetrical growth at the edge of the mantle; the oldest and smallest whorls are at the apex, and may become hollow as larger whorls grow around the soft tissues. In most species, the aperture is to the right of the shell, but some are 'left-handed' and a few species include right- and left-handed snails. Most of the viscera and reproductive organs are permanently encased in the shell, and, especially in terrestrial and littoral species, all or part of the foot can be retracted into it for protection from predators and exposure. In several lineages of marine gastropods (Figure 1.19c), and of course terrestrial slugs (Figure 1.14c), the shell is secondarily reduced or absent.

CEPHALOPODS

Cephalopods are the most complex molluscs and some are very large, up to 16 m in total length, by far the largest free-swimming invertebrates. They rival fish for size, power and complexity of behaviour, but in the modern fauna cephalopods are much less diverse, totalling about 600 species compared to thousands of marine fish.

Their most distinctive features are the prehensile 'arms' around the mouth (sometimes called tentacles), often equipped with rows of circular suckers (Figures 1.20 and 1.21). The arms, which are often much longer than the rest of the body, are used for slower movement, food-gathering and, not least, mating. Cephalopod arms and mantle contain several distinct muscles intricately arranged around layers of fibres in connective tissue, which recoil elastically after being deformed by muscle contraction, in much the same way as jellyfish mesogloea.

Figure 1.20 Cephalopods. (a) *Nautilus*, one of the few surviving shelled cephalopods that grows to a body mass of a kilogram or more. (b) A small squid. These elongated cephalopods range in size from a few centimetres to many metres in length and have eight arms and two highly extensible tentacles around the central mouth. Most are powerful swimmers, using the lateral fins on the mantle for cruising and the siphon for jet propulsion. (c) *Octopus* lives near the bottom, usually in shallow coastal waters and can grow to several kilograms. Octopods resemble squids but lack the two tentacles. They are less strong swimmers and often have much reduced fins, rounder bodies and very long flexible arms equipped with numerous suckers.

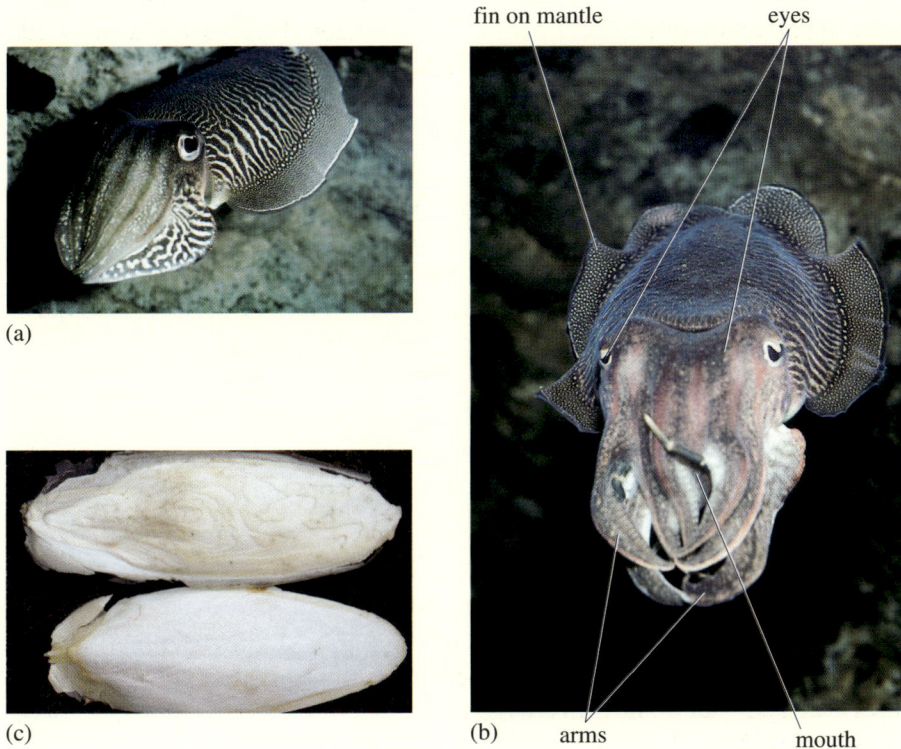

fin on mantle   eyes

(a)

(c)   (b)   arms   mouth

Figure 1.21 (a) and (b) Cuttlefish, *Sepia*. Note the large eye, short but highly mobile arms, the extensive fin along the edge of the mantle and the pattern of pigmented cells. These shallow water species grow to around half a metre long, but many deep sea forms are much bigger. (c) Cuttle 'bone', showing upper and lower surfaces, shown at about one-third actual size.

The shell is large and external in most extinct cephalopods and a few living species notably *Nautilus* (Figure 1.20a), but in squid (Figure 1.20b), cuttlefish (Figure 1.21a and b) and octopus (Figure 1.20c), it is reduced and has become internal, assuming various forms including cuttlebone (Figure 1.21c). The cuttlebone secretes gases that confer buoyancy on free-swimming cephalopods, enabling them to control their depth nearly as well as fish can.

As well as enclosing the gills, the very muscular mantle powers the 'jet propulsion' system that is particularly impressive in small squids and cuttlefish (Figure 1.20b, c and Figure 1.21a, b). Synchronous contraction of the mantle muscles, coordinated by the giant neurons (see Clements and Saffrey, 2001), expels water from the mantle cavity through the narrow funnel, producing backwards movement at speeds of up to $1 \, \text{m s}^{-1}$. Measurements of oxygen consumption show that such jet propulsion is energetically expensive and probably only used for catching prey or escape.

Cuttlefish and squid can also hover or swim backwards or forwards by undulating the long fins on both sides of the mantle (Figures 1.20b and 1.21a, b). This form of movement, combined with submaximal use of the mantle muscles, enables medium-size migratory squid (*Illex* spp.) to travel about $20 \, \text{km day}^{-1}$. Bottom-living species such as *Octopus* spp. (Figure 1.20c) have very long, strong tentacles with which they propel themselves, often hauling their very flexible body through surprisingly narrow crevices.

○ Which physiological systems are essential to support fast, sustained and powerful swimming?

● The circulatory and respiratory systems, i.e. the heart, blood vessels and gills.

The haemocyanin pigments in cephalopod blood are more abundant and more efficient than those of gastropods.

In sedentary and slow-moving molluscs, as in worms, blood pressure is locally variable but generally very low. The body muscles exert much higher pressures than the small heart can; its pumping maintains only a sluggish flow. The circulatory system of cephalopods is much more sophisticated. As well as the main heart, the blood is pumped by two supplementary hearts near the gills, through a vascular system that consists of muscular arteries and veins and capillaries, similar to that of vertebrates.

○ What function of the haemocoel in gastropods and bivalves is incompatible with the high blood pressure of cephalopods?

● The haemocoel can no longer serve as a hydrostatic skeleton.

The biggest and most powerful cephalopods live in middle or deep waters. Wounds caused by suckers and beaks suggest that large species put up formidable resistance to sperm whales (*Physeter catodon*), one of their best-known predators. An adult male *Physeter* 13.5 m long was found stranded on a beach near Penzance in Cornwall after a severe storm in February 1990. Its stomach was found to contain the remains of over a hundred squids, of which almost half were *Architeuthis*, the largest known cephalopod. This giant squid grows to a body mass of up to 200 kg, although nearly all of those eaten by the whale were smaller, presumably younger specimens.

We know very little about the physiology or habits of giant squid because the few that have been found near the surface were dead or dying from the decrease in pressure or from exposure to light and heat. Comparative studies of the fine structure and properties of the tissues enable biologists to propose some plausible hypotheses, as indicated in Box 1.6.

BOX 1.6   INDIRECT METHODS FOR ASSESSING PHYSIOLOGICAL CAPACITIES OF DEEP-SEA SQUID

A. ANATOMICAL EVIDENCE FOR HIGH BLOOD PRESSURE IN GIANT SQUID

The heart of *Octopus* beats at about 0.14–0.23 Hz at 10 °C (compared to about 1.7 Hz in adult humans). The resulting blood pressure ranges from 2 to 6 kPa at rest to a maximum of 8 kPa during spontaneous exercise, much higher than most other invertebrates, but about a third that of similar-sized mammals such as a small dog. Examination of the structure of the arteries (Shadwick, 1994, 1999) reveals some similarities between the closed circulatory system of cephalopods and that of vertebrates. As in vertebrates, the artery walls expand easily to accommodate the blood at low to moderate pressures, but

are burst-resistant, increasing in stiffness when greatly extended by high pressures (see Section 4.2.1). However, the molecular structure of the proteins that confer these properties was found to be substantially different in octopus and cuttlefish compared with rats and other vertebrates, implying that high blood pressure and burst-resistant arteries evolved independently in molluscs and vertebrates.

Within the major classes of vertebrate, larger species have higher blood pressures than smaller ones with similar habits (e.g. weasels compared with bears within the order Carnivora). The shift in function is matched by changes in the fine structure and mechanical properties of the large arteries, making it possible to deduce an animal's blood pressure from inspection of a piece of a major artery. Applying the same reasoning to observations and measurements on samples of arteries taken from washed-up giant squids (*Architeuthis*), cephalopod physiologists have concluded that the large adults may have blood pressure, and presumably athletic performance, within the same range as our own. No wonder sperm whales and other deep-diving whales are injured by their prey!

B. BIOCHEMICAL EVIDENCE FOR METABOLIC PARSIMONY

Many enzymes, including citrate synthase, a key enzyme in the TCA cycle, are robust enough to remain active for some time after death, and can be extracted and assayed *in vitro*. As shown in Figure 1.22a, the activity of citrate synthase in squid muscles correlates strongly with the total oxygen consumption of the whole animal.

○ What do the data in Figure 1.22b show? What can you conclude from them about the habits of deep-sea cephalopods?

● The citrate synthase activity is variable, but in general, it is lower in animals living in deeper water. Combining this conclusion with that from Figure 1.22a indicates that the capacity for aerobic metabolism of squid found in deep water is only about a tenth of that of specimens caught within 100 m of the surface. The deep-water squid probably move mainly by means of the fins and/or the arms and only rarely use energetically expensive jet propulsion.

Such economies of physiological effort are common in the deep sea: seawater is usually colder and contains less oxygen far from the surface, and food is scarce. *Architeuthis* itself has quite high aerobic capacity, consistent with its adaptations to high blood pressure. But most biologists believe it uses these abilities only occasionally for brief bursts of vigorous exercise; most of the time it probably conserves energy by swimming slowly using its fins. Deep-water submersibles are improving rapidly; one day, people may witness an encounter between a whale and a giant squid 0.5–1.0 km down, and confirm or refute these deductions.

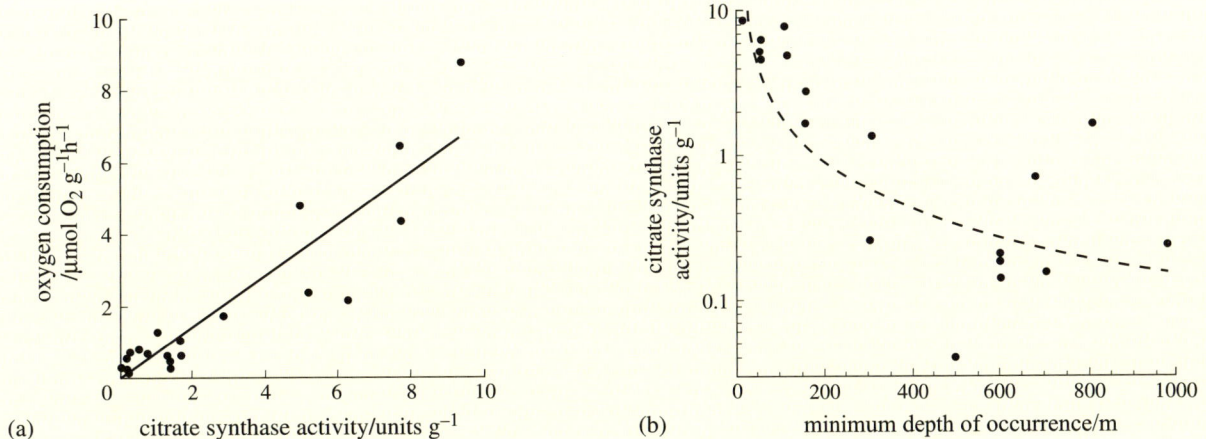

Figure 1.22   (a) The rate of oxygen consumption and the activity of citrate synthase in swimming muscles of various cephalopods caught off California and Hawaii. (b) Citrate synthase activity and depth at which the species are found. The data for enzyme activity are presented on a log scale on the vertical axis (data from Seibel *et al.*, 2000).

Most cephalopods are predators of fish and large invertebrates, which they stalk and catch with finely controlled movements of the mantle and tentacles, often immobilizing the prey with venom before eating it. The mouth is equipped with a horny beak as well as a radula, and the nervous system and sense organs are much more sophisticated than in other molluscs. At least one species uses one kind of venom for prey capture and another for defending itself against predators. The unusual habit of squirting a large volume of secretions from a gland near the anus when disturbed is also probably a means of escaping predators. In *Sepia* (Figure 1.21) and other species that live in brightly-lit surface waters, this 'ink' * contains the pigment melanin and forms a dark cloud, but the secretion is luminescent in deep-sea species.

Serious laboratory study of cephalopod behaviour began in the 1950s using *Octopus* (Figure 1.20c), which is found on the sea-bed in shallow waters, where it crawls around rocks and into crevices to find crabs and carrion, and rarely swims. It adapts readily to life in an aquarium and can be trained using rewards of food. Everyone was surprised when its memory, visual and tactile discrimination and problem-solving ability proved to be as good as that of many fish and advanced arthropods such as crabs.

Keeping open-water cephalopods proved much more difficult: they swim so fast that a huge tank is necessary to keep them healthy. When scientists began to use diving equipment and underwater cameras to investigate such animals in their natural habitat, other aspects of cephalopod behaviour were revealed. Cuttlefish (Figure 1.21a and b) and some squids that live in brightly lit waters undergo very rapid and complex changes of colour, produced by pigmented cells called chromatophores that shrink or spread out by the action of tiny muscles under neural control. As well as camouflaging themselves instantly and accurately against sand, rocks or other background materials, many cephalopods use colour change for elaborate courtship display and other forms of interspecific communication. In some species, reflective particles in the skin add to the brightness and intricate patterns of the displays.

○   What does such behaviour suggest about the quality of cephalopod vision?

●   The eye must be able to detect colour and shape.

Cephalopod eyes (Figure 1.23) are among the largest and most sophisticated in the animal kingdom. In general plan and probably also in the biochemistry of the

Figure 1.23 Section of the head of a small cephalopod to show the large eyes and brain.

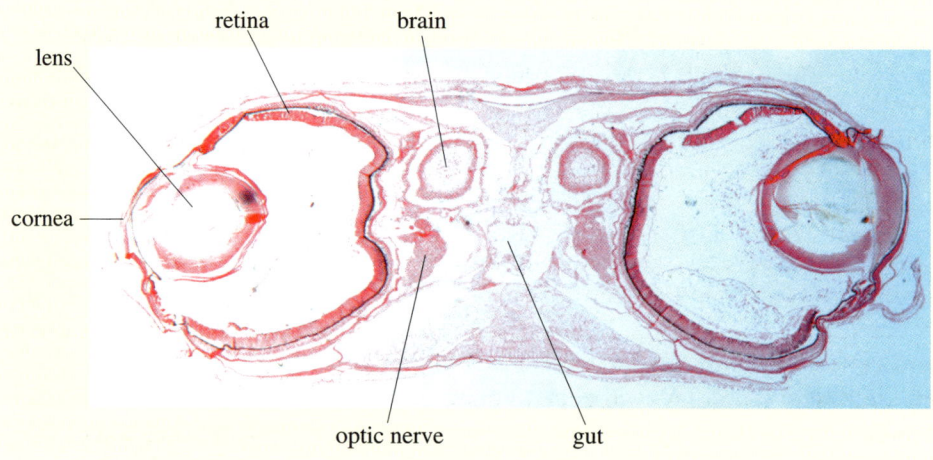

* Ancient Mediterranean people collected the secretion for use as a writing material; in both Greek and Latin *sepia* means 'ink' as well as 'cuttlefish'.

lens and light-sensitive cells, they resemble those of vertebrates (see Sections 4.5.2 and 5.4.2) to a remarkable extent.

Many of the cephalopods that live at depths beyond the reach of sunlight or are active at night have elaborate light organs: special pockets of tissue that harbour bioluminescent bacteria. Some, including *Euprymna scolopes,* which forms an intricate symbiosis with *Vibrio fischeri* (Walker, 2001), have supplementary photoreceptors situated just behind the paired eyes. Perhaps these animals use the light produced by their symbionts to find their way around, as well as for communication and predator evasion.

1.4.2 ECHINODERMS

Animals of the phylum Echinodermata have been numerous and diverse in the sea almost continuously since animals first appeared in the fossil record. Their most striking feature is their *pentamerous* (five-fold) symmetry (Figure 1.24a and b), usually with five 'arms' equally spaced around the body. The mouth and anus are quite close together near the centre of the body, the former normally on the ventral surface and the latter on the dorsal surface. The gut is often coiled or branched, and in many echinoderm species extends into the 'arms', which thus cannot strictly be called limbs. The nervous system consists of a network of nerve cells concentrated around the mouth, and radial nerves extending into the arms, but there are no large ganglia as in annelids, arthropods or molluscs.

Echinoderms do not have a shell or cuticle like that of molluscs or nematodes: the hard skeleton is internal, and thus classified as an endoskeleton, although most of it is close to the surface, and in many species, spines protrude through the soft tissues. The spines of echinoids (Figure 1.25a, overleaf) are particularly conspicuous, and are often very long, with sharp tips armed with powerful venoms that deter all but the most determined predators.

Adjacent plates are linked by sutures of flexible connective tissues, as shown in Figure 1.25b and c, and are the main sites at which growth takes place. On a diet of 5 g of seaweed a day, *Strongylocentrotus* grows well, reaching a diameter of about 8 cm (Ellers *et al.,* 1998). Their sutures appear as a gap in dried skeletons (Figure 1.25b). If given very little food for a year, growth stops and the sutures close to form a rigid 'shell', as shown in Figure 1.25c. Because most of the plates interlock tightly, they stay together after death, so the 'shells' of sea urchins or starfish are often found almost entire.

The **tube-feet** (Figure 1.26) are unique to echinoderms. Adults may have from dozens to several thousands of tube-feet, each consisting of an internal ampulla and a portion that protrudes through the skeletal plates into the exterior (Figure 1.26b). As shown in Figures 1.26c and d, the shaft of the tube-foot can be shortened by contraction of the muscles along its walls, and can be extended, sometimes to a length of several centimetres, by forcing fluid from the muscular ampulla into the shaft, by mechanisms similar to the hydrostatic skeleton of coelomate worms (see Section 1.3.2). The tip is a disc (Figure 1.26d), often in the form of a suction pad or sticky with mucus, and/or equipped with sense organs sensitive to touch and chemicals.

(a)

(b)

Figure 1.24 Various starfish viewed from above. (a) A tropical starfish (*Oreaster*) with five arms. (b) The crown-of-thorns starfish (*Acanthaster*) with numerous arms (usually multiples of five). These species can reach a diameter of more than half a metre.

(a)

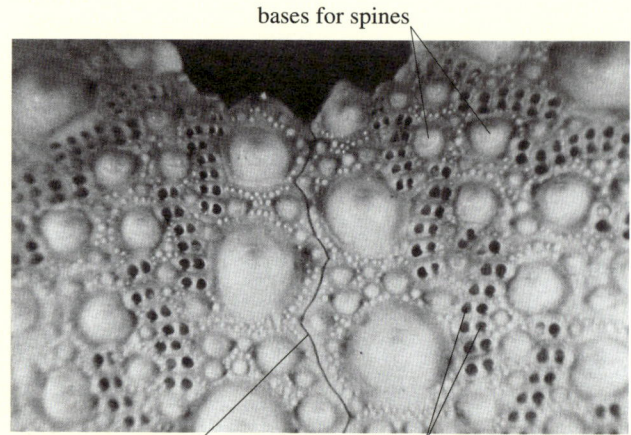

bases for spines

(b)   suture   sockets for tube-feet

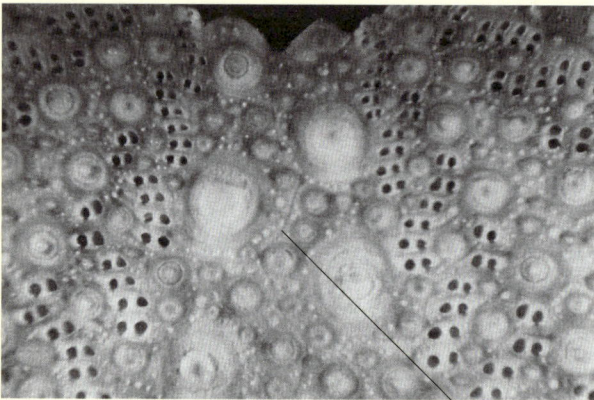

(c)   closed suture

Figure 1.25 Echinoids. (a) A well-armoured grazing echinoid, spines are black and white striped, the test is red. Sutures in the skeleton of (b) well fed, and (c) starved *Strongylocentrotus purpuratus* from the coast of California. Each photograph shows an area about 1 cm across.

The water pressure inside individual tube-feet is adjusted by the flow of fluid along the narrow tubes of the **water vascular system** that links rows of tube-feet together (Figure 1.26b). The muscles of the shaft can also bend the tube-foot from side to side. Waves of attachment and detachment of many tube-feet to the substratum are the principal means of locomotion in echinoids such as sea urchins, sand-dollars and heart urchins (sometimes called sea potatoes), most starfish (asteroids), and sea-cucumbers (holothurians).

The water vascular system is anatomically part of the coelom. The fluid it contains is similar in composition to seawater, except that potassium ions are enriched and there are a few large cells called *coelomocytes* that creep by means of amoeboid movement.

○ What roles could cells with these properties play?

● Like phagocytes in vertebrates, coelomocytes may protect the tissues from disease by engulfing foreign particles, dead tissue and pathogenic microbes, and may prevent leakage of coelomic fluid by sealing off wounds.

Larger echinoderms also have other circulatory channels that form a rudimentary 'blood' system.

(a)

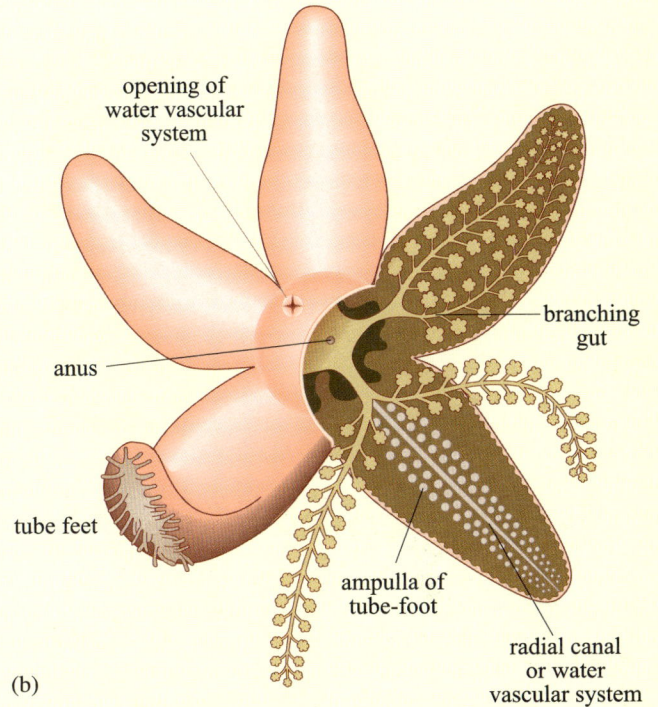

opening of
water vascular
system

anus

tube feet

branching
gut

ampulla of
tube-foot

radial canal
or water
vascular system

(b)

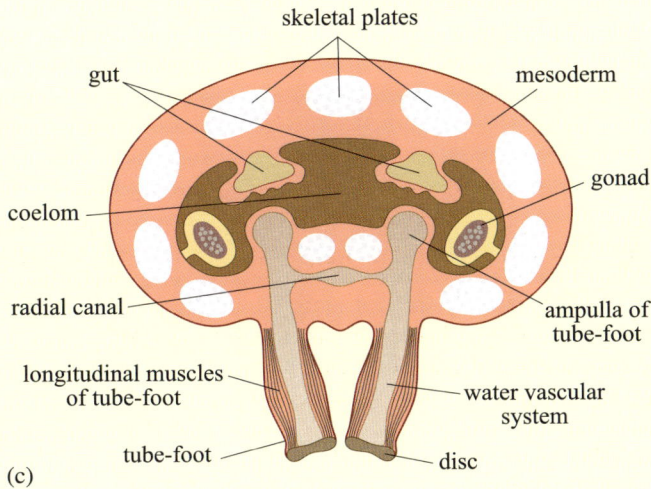

skeletal plates

gut

mesoderm

coelom

gonad

radial canal

ampulla of
tube-foot

longitudinal muscles
of tube-foot

water vascular
system

tube-foot

disc

(c)

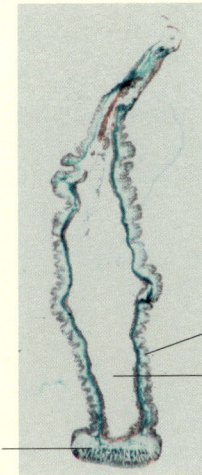

muscle

water
vascular
system

disc

(d)

Figure 1.26   (a) A living starfish turned upside down, showing the large ventral tube-feet, many extended far enough to form shadows. (b) Diagram of a starfish seen from above, with two arms dissected to reveal parts of the gut, ampullae of tube-feet and radial canals. (c) Transverse section through an arm of a starfish to show the muscles, skeleton, gut, coelom and ampullae of tube-feet. (d) A longitudinal section of a single tube-foot of a sea urchin, *Echinus*.

Asteroids (starfish) usually have five arms that can bend and twist, with each supporting a double row of stout tube-feet, but in some larger species (a few reach 1 m in diameter), the arms are branched. Others have up to 40 arms, with multiples of five being the most common numbers (Figure 1.24b). In 'sea-cushions', the arms are reduced, giving the animal a pentagonal outline.

Starfish are common on rocks, sand or mud, especially in coastal waters and around reefs. Many are active predators: they fold the arms around mussels and other bivalves, gripping the shell using the tube-feet as suckers, then gradually extend, prising the two valves apart and exposing the soft tissues inside. The starfish everts its stomach through its mouth and into the gap so formed, and maintains its grip while the prey's muscles are weakened by the predator's digestive enzymes. Others 'graze' on encrusting animals such as sponges and corals, including the large, spiny crown-of-thorns starfish (*Acanthaster planci*, Figure 1.24b), that became extremely abundant during the last third of the 20th century, destroying huge areas of coral in the Australian barrier reef, the Red Sea and elsewhere. Smaller and more sedentary asteroids collect small animals that fall on their body surface with their cilia and modified tube-feet.

Ophiuroids (brittle-stars) derive their name, which means 'snake-like', from their long, highly mobile arms. As Figure 1.27a shows, the central oral disc is distinct from the flexible arms, which are composed of numerous 'ossicles' between which the tube-feet protrude. Ophiuroids are generally smaller than asteroids and most are filter feeders or scavengers that live on or in the sea bed. Tiny particles are collected in mucus suspended from the long, flexible arms, which are also

(a) (b)

Figure 1.27 (a) A brittle-star *Ophiothrix fragilis* seen from above. On parts of the sea-floor of the North Atlantic these small ophiuroids form dense colonies, as many as 2000 per m². The central disc, a few centimetres across, is pink. Spines and tube-feet project from the long, flexible arms, in which the skeleton is divided into articulating ossicles. (b) A filter-feeding basket-star, *Conocladus,* on a sponge. Four of the five branching arms can be seen protruding from the central disc. These ophiuroids are mainly sessile and trap particles from the 'basket' formed by the five branched arms and their tube-feet. Some grow to more than a metre in diameter.

used for swimming and crawling. In many areas, the sandy bed of the deep sea is covered with brittle-stars and the more sedentary basket-stars, whose highly branched arms trap small prey (Figure 1.27a).

The echinoid body is not drawn out into 'arms' as in asteroids and ophiuroids, but the pentamerous symmetry is still clearly visible in the arrangement of rows of tube-feet, which are long and muscular, often reaching out beyond the spines (Figure 1.28a). Most echinoids live buried in sand or in sheltered waters on coral reefs or intertidal rocks, where they scrape off encrusting organisms, such as algae and corals, using the sharp teeth that form part of a unique, tooth-bearing structure called *Aristotle's lantern* (Figure 1.28a).

The size, shape and composition of the Aristotle's lantern correlate with species differences in diet and feeding habits. Many echinoids including *Echinometra* (Figure 1.28b) 'graze' on coral, and although they eat only a little of the skeleton along with the living tissues, their mode of feeding damages the reef itself. In some places, echinoids grind as much as 29% of the corals' total accretion of calcium carbonate (see Section 1.2.2) into a fine powder, which contributes to the beautiful white sand characteristic of beaches near coral reefs. Table 1.1 shows some data for two species of sea urchins in the Red Sea near the Israeli resort of Eilat.

Figure 1.28   (a) The ventral surface of a sea urchin showing the external portion of Aristotle's lantern. (b) *Echinometra*, a warm-water sea urchin found on coral reefs. Its stout spines resist fish, turtles and many other predators that attack slow-moving grazers of coral and encrusting algae. (c) *Diadema*. The tips of the long black spines of this grazing urchin are barbed and easily embed in the flesh of potential predators. (d) Ventral and (e) dorsal views of a sand-dollar skeleton without spines. The five 'arms' are represented by double rows of tube-feet.

Table 1.1 Size and population density of two common echinoids in two different areas of the Eilat Nature Reserve. Data from Mokady *et al.* (1996).

| | Diameter of body/mm | Aristotle's lantern | | Urchin density/m^2 | |
		Diameter/mm	Height/mm	Reef slope	Flat reef
Diadema	30	11	15	6.4	0.1
Echinometra	23	7	9	3.7	10.5

○ Which of the species listed in Table 1.1 has a proportionately larger Aristotle's lantern? Assume that the body is spherical (volume of a sphere = 4/3π radius^3) and the lantern is a cylinder (π radius^2 × height).

● *Diadema*, whose Aristotle's lantern accounts for 10% of its volume: $[\pi(11/2)^2 \times 15] / [(4/3) \times \pi(30/2)^3] = 1426/14\,137 = 0.101$.

That of *Echinometra* is only 5% of its volume: $[\pi(7/2)^2 \times 9] / [(4/3) \times \pi(23/2)^3] = 346/6371 = 0.054$.

Even in the smaller species, the feeding organ is relatively massive, and the animals spend a very large proportion of their time grazing. Much of the rest of the body is occupied by the gut and, in the breeding season, by eggs or sperm, which are produced in huge numbers by large adults.

○ In view of the structure and composition of corals (see Section 1.2.2), would they constitute nutritious food for other animals?

● No. Corals consist of thin cellular layers that in many species include symbiotic algae, but may also contain toxins. The mesogloea provides few nutrients since it contains only a little protein and even less fat or carbohydrate.

Animals must eat a lot of coral to sustain themselves. In this Red Sea study (Mokady *et al.*, 1996), *Echinometra* was found to eat less coral per day than *Diadema*. Although it comprised nearly 40% of the urchin population, *Echinometra* was estimated to destroy only 18% of the coral broken down by echinoids on the reef slopes. However, this smaller species accounted for 97% of the erosion of coral skeletons in the flat areas of the reef, where *Diadema* (Figure 1.28b) cannot survive the much stronger wave action and currents.

In many kinds of sea urchin, for example Figure 1.25a and 1.28a, b and c, the spines can be moved independently by muscles at the base. They vary in shape from short and stubby (Figure 1.28a), to long and razor-sharp (Figures 1.25a and 1.28c), sometimes containing powerful toxins. Like plants and other kinds of sedentary animals, echinoids protect themselves from predators with rigid skeletons and sharp spines, some of which also harbour poison glands. Many are conspicuously coloured, 'warning' predators away: sea urchins covered with long black spines (such as *Diadema*, Figure 1.28b) are often the most conspicuous animals in warm, shallow waters around rocky coasts.

The combination of long, sticky tube-feet and mobile spines enables these echinoderms to 'walk' over rocks and gravel. Small stones often remain attached to tube-feet, and may help to camouflage or protect some species. Other echinoids including heart urchins and sand-dollars can bury themselves in soft sand remarkably quickly. Vigorous movements of the spines and many small tube-feet liquefy the wet sand, enabling the body to sink deep enough to avoid predators and strong wave action.

Sand-dollars (Figure 1.28d and e) are flattened and rarely perfectly round in outline, so are sometimes called 'irregular' echinoids. The body is often perforated by several 'key holes' and is covered with numerous, very small spines and many thousands of tiny tube-feet. They pick up single-celled algae one by one and pass them along the rows of tube-feet to the mouth, where the food is crushed in the Aristotle's lantern. They also take in sand and digest the various kinds of microbes on the surface of the sand grains, along with the slurry of crushed protoctists. The shape of the body and the arrangement of its holes determine the animal's ability to live partly buried in shifting sands and avoid being washed away by wave action.

○ Why would sand-dollars risk being injured or washed away by remaining near the surface instead of burying themselves deeper into the sand?

● Because there are far more protoctists and microbes near the brightly lit, well-oxygenated surface than deeper in the sand, either living suspended in the water or as encrusting organisms.

Many sea urchins avoid predators and damage from wave action by wedging themselves into crevices, often surprisingly tightly, and, as divers know, the slender spines are remarkably resistant to bending or crumpling if stood on. They are able to grip very tightly to irregularly shaped holes by making their spines 'catch' in the extended position as and when required. Tissues in the body wall at the bases of these spines seems to be able to alternate flexibility and extensibility with rigidity.

Crinoids (Figure 1.29) are the other major living class of echinoderm, although they are less familiar because they occur mainly in deep water. The most abundant kinds, called sea lilies, resemble flowers in general shape and live attached to the sea floor, where they collect particles on tube-feet that protrude from long, sometimes branched arms. The tube-feet are highly mobile and the arms and stalk can also bend because their skeleton consists of numerous small plates connected by flexible ligaments. Feather stars have no stalk and even more mobile arms than bottom-living crinoids, with which they swim weakly as well as collect food.

As sessile filter feeders, crinoids face the same dilemmas as many corals and sponges: how to maximize exposure to food-bearing water currents while avoiding being broken or swept away. Such particle-feeding echinoderms, and others with similar habits, have to maintain seemingly awkward postures for long periods. Investigations into how echinoderms maintain these postures for long periods without apparently increasing energy expenditure have

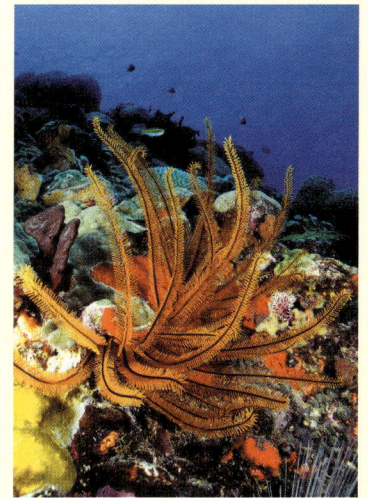

Figure 1.29 A sessile crinoid on a coral reef. These echinoderms, sometimes called sea lilies, have numerous flexible arms and are attached to the substratum. The central mouth points upwards and receives fine particles trapped on the many tube-feet. Some crinoids reach a diameter of more than half a metre.

revealed some unexpected phenomena whose physiological basis is hotly disputed (Wilkie *et al.*, 1993).

Almost all echinoderms spend at least a short period of their lives as larvae, which bear hardly any resemblance to the adults, being free-swimming, planktonic and bilaterally symmetrical without tube-feet. Bands of cilia over the body and its projections enable them to swim and collect minute particles of food (Figure 1.30). Those that feed may spend weeks in the plankton before a rudimentary skeleton begins to form and the larvae sink to the sea bed, where they develop the adult body shape and internal structure.

Figure 1.30 Echinoderm larvae: (a) The echinopluteus larva of an echinoid. The adult sea urchin forms in the central pouch of this free-swimming ciliated larva that feeds on tiny particles in the plankton. (b) The bipinnaria larva of an asteroid. Approx 1.0 mm (excluding arms).

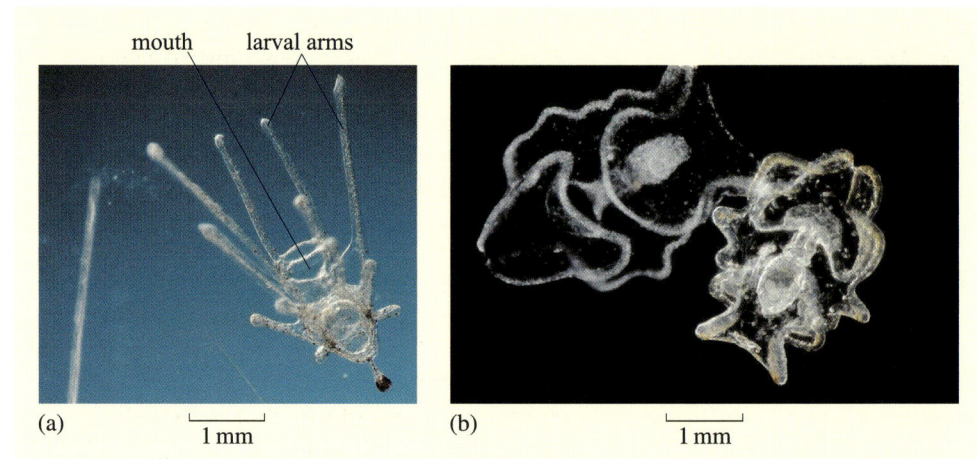

○ What would be the function of the larval stage in echinoderm life histories?

● The species is dispersed by water currents over much greater distances than the adults could move.

Actively feeding ciliated larvae (called pluteus) are the rule among intertidal sea urchins and sand-dollars in the northern hemisphere, but those of many deep-sea and arctic echinoids have no gut and assume the adult form within a few days of hatching. The structure and development of echinoid larvae have been much studied recently, and the findings suggest hypotheses about the interrelationships between the major classes of echinoderm and their relationship to other phyla, including the Chordata.

Sea-cucumbers (holothurians) have lost many of the characteristic features of echinoderms: they lack arms, and the skeleton is reduced to small spicules making the body soft and elongated, rather than rigid.

○ What could be the functional advantages of a soft body?

● The body can change shape, enabling the animals to squeeze through narrow cracks and burrow into sand, as soft-bodied worms can (see Section 1.3.2).

In the absence of segments and septa, many holothurians, including that shown in Figure 1.31, are even more extensible than annelid worms, and can wriggle through very narrow crevices.

○ Would tube-feet work within a soft, flexible body?

● No. The water vascular system must be able to apply pressure locally to small groups of tube-feet and can only work within the frame provided by the skeletal plates.

Tube-feet over most of the body are much reduced, but the mouth is surrounded by tentacles composed of large, elaborate tube-feet supported by a ring of firm ossicles around the pharynx (seen in Figure 1.31). Usually occurring in five or multiples of five, these tentacles collect particulate matter by sweeping the sea floor or by being held up in a water current. One by one, the tentacles are retracted into the mouth and any food wiped off. In most echinoderms, the tube-feet also serve as respiratory organs, but holothurians have so few of them that the larger species have special evaginations of the gut called respiratory trees that serve in gas exchange, as seen in Figure 1.32b and d (overleaf).

○ What could be the molecular basis for the respiratory trees appearing red in Figure 1.32?

● The red colour could arise from the blood pigment, haemoglobin, that binds to oxygen and so is often concentrated in respiratory organs.

In several kinds of echinoderms, especially those that burrow or live in poorly oxygenated water, the coelomocytes contain haemoglobin that is almost identical to that found in vertebrate red blood cells, and in the blood of some annelids (Section 1.3.2). As well as removing wastes and pathogens, these cells also supply oxygen to internal tissues as they circulate through the water–vascular system and other body cavities.

Being soft and flexible, the holothurian body wall lacks protective spines. Many sea-cucumbers contain toxins, but if this deterrent fails, some species expel their gut and the respiratory trees (Figure 1.32), which provide the predator with a meal while enabling the rest of the animal to escape.

○ What conditions are necessary for regeneration of large, complex organs such as the gut?

● The animal must have sufficient storage tissues, or be able to withdraw non-specialized tissues (see Pond, 2001b), to support regrowth of the extruded tissues until the gut is fully functional and feeding can resume. It must also be able to protect its wounded tissues from pathogens (see Davey and Gillman, 2001).

Deliberately injuring themselves seems a drastic way to escape from predators, but in spite of their complex internal structure and lack of a proper blood system, echinoderms, especially starfish and sea-cucumbers, can repair injuries and regenerate tissues remarkably efficiently. The capacity of invertebrates to regenerate body parts lost to predators and accidents is generally superior to that of vertebrates.

'arm'   large tube-feet
               around mouth

Figure 1.31 A holothurian from a coral reef off East Africa. The animal is being lifted out of the water, which stretches its long, tubular and highly extensible body. Five large tube-feet form the tentacles around the mouth.

(a)

(b)

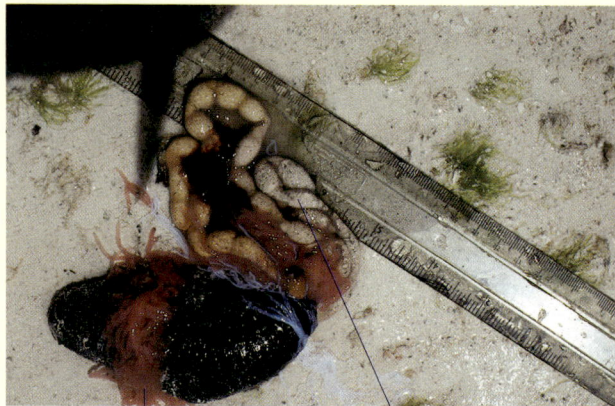
(c) respiratory tree   gut

(d)

Figure 1.32 (a)–(d) Sequence of pictures taken over about 10 minutes, showing the response to 'attack' of a holothurian on a coral reef off the coast of Kenya. The gut seen in (c) and (d) is full of sand (from which minute organisms and particles are digested away) and the respiratory trees appear dark red.

An important factor in healing is minimizing the volume of tissues that must be regenerated. We use stitches or tape to close wounds, and aligning the torn surfaces side by side greatly hastens healing. Well-vascularized mammalian organs such as the skin, liver or gut heal much more quickly than breaks and bruises to poorly perfused tissues such as tendons and joints. Biologists interested in hastening the repair of such injuries have investigated the echinoderms' ability to close and heal wounds in skeletal tissues without a blood supply.

SUMMARY OF SECTION 1.4

1   Several contrasting groups of invertebrates have hard skeletons.

2   Unique features of molluscs include the mantle, shell, radula and a muscular foot, which is usually expanded by the haemocoel acting as a hydrostatic skeleton, and is not partitioned into segments.

3   The molluscan shell is secreted by the mantle and grows from the edges. It consists of protein and crystals of calcium carbonate. Pearls are formed around foreign objects or wounds and consist mainly of protein.

4   All bivalves collect particulate food on their greatly enlarged gills. Most are sedentary, living in or on sand, rock or other substrate, protected by their

hinged, usually symmetrical shell. The main bases for diversity are the manner of attachment or burrowing, and the mechanism of food collection and selection.

5   Gastropods have a single shell, often coiled but sometimes reduced or absent; the foot is usually muscular and locomotion often involves secretion of mucus. The form of the tongue-like horny radula is diverse, reflecting the wide range of diets in gastropod species. In pulmonates, the gills are absent and the lining of the mantle acts as a lung. Many are terrestrial herbivores.

6   In most living cephalopods, the shell is reduced or absent and the mantle is very muscular. Many can swim by jet propulsion, and seize prey with long, muscular tentacles. Their behaviour is complex and parallels that of fishes in a remarkable variety of ways. The structure and properties of cephalopod eyes and blood vessels closely resemble those of vertebrates.

7   Most living cephalopods are oceanic, and live in middle or deep waters, so many species, including some of the largest have never been seen alive. Their physiological capacities must be estimated indirectly from the comparative study of isolated tissues.

8   Echinoderms are all marine and have pentamerous symmetry and an internal skeleton. Tube-feet operated by muscles and a coelomic hydrostatic skeleton serve for locomotion and food collection. The internal skeleton is much reduced in holothurians, which move more like worms.

9   Crinoids are filter-feeders that usually live attached to the substratum. Many asteroids are active predators or filter-feeders. Most holothurians are detritus feeders.

10   Echinoids rasp algae and other encrusting organisms with the Aristotle's lantern containing hard teeth. Others are detritus feeders living in or on sand. Many deter predation by sharp and/or poisonous spines, hard shells, or living in burrows.

11   Nearly all echinoderms produce numerous eggs that are fertilized externally and develop into free-swimming larvae.

1.5   CONCLUSIONS

This chapter has illustrated just a few of the enormous range of body plans and physiological capacities found among living groups of invertebrates. Although body plans are obviously important to taxonomists, they do not appear to impose insuperable constraints on the diversity of animals. Animals built to completely different body plans often live side by side and have similar diets and habitats. We should hesitate to assert that any phylum is limited to certain life forms and habits as new discoveries and more detailed study continually reveal surprises.

REFERENCES

Blackshaw, R. P. (1992) The effect of starvation on size and survival of the terrestrial planarian *Artioposthia triangulata* (Dendy) (Tricladida, Terricola), *Annals of Applied Biology*, **120**, pp. 573–578.

Bumann, D. and Puls, G. (1997) The ctenophore *Mnemiopsis leidyi* has a flow-through system for digestion with three consecutive phases of extracellular digestion, *Physiological Zoology*, **70**, pp.1–6.

Clements, M. and Saffrey, J. (2001) Communication between cells. In *The Core of Life Vol. II*, J. Saffrey (ed.), The Open University, Milton Keynes, pp. 241–304.

Davey, B. and Gillman, M. (2001) Defence. In *Generating Diversity,* M. Gillman (ed.), The Open University, Milton Keynes, pp. 151–200.

Dyson, M. (2001) Reproduction. In *Generating Diversity*, M. Gillman (ed.), The Open University, Milton Keynes, pp. 111–150.

Eldor, A., Orevi, M. and Rigbi, M. (1996) The role of the leech in medical therapeutics, *Blood Review,* **10**, pp. 201–209.

Ellers, O., Johnson, A. S. and Moberg, P. E. (1998) Structural strengthening of urchin skeletons by collagenous sutural ligament, *Biological Bulletin*, **195**, pp. 136–144.

Gray, J. and Lissmann, H.W. (1938) Studies in animal locomotion. VII. The earthworm, *Journal of Experimental Biology,* **15**, pp. 506–517.

Halliday, T. and Pond, C. (2001) Longevity. In *Generating Diversity*, M. Gillman (ed.), The Open University, Milton Keynes, pp. 201–236.

Kendall, J. M. and Badminton, M. N. (1998) *Aequorea victoria* bioluminescence moves into an exciting new era, *Trends in Biotechnology,* **16**, pp. 216–224.

Mokady, O., Lazar, B. and Loya, Y. (1996) Echinoid bioerosion as a major structuring force of Red Sea coral reefs, *Biological Bulletin*, **190**, pp. 367–372.

Pond, C. (2001a) Biological investigation. In *Introduction to Diversity,* I. Ridge and C. Pond (eds), The Open University, Milton Keynes, pp. 93–130.

Pond, C. (2001b) Dealing with food. In *Generating Diversity,* M. Gillman (ed.), The Open University, Milton Keynes, pp. 37–88.

Purcell, J. (1997) Pelagic cnidarians and ctenophores as predators: Selective predation, feeding rates, and effects on prey populations, *Annals of the Institute of Oceanography,* **73**, pp. 125–137.

Quillin, K. J. (1998) Ontogenetic scaling of hydrostatic skeletons: geometric, static stress and dynamic stress scaling of the earthworm *Lumbricus terrestris*, *Journal of Experimental Biology*, **201**, pp. 1871–1883.

Ridge, I. (2001) Diversity in protoctists. In *Introduction to Diversity,* I. Ridge and C. Pond (eds.), The Open University, Milton Keynes, pp. 55–92.

Saffrey, J. (2001) Cells and tissues. In *The Core of Life Vol. I,* J. Saffrey (ed.), The Open University, Milton Keynes, pp. 1–51.

Seibel, B. A., Thuesen, E. V. and Childress, J. J. (2000) Light-limitation on predator–prey interactions: Consequences for metabolism and locomotion of deep-sea cephalopods, *Biological Bulletin,* **198**, pp. 284–298.

Shadwick, R. E. (1994) Mechanical organization of the mantle and circulatory system of cephalopods, *Marine and Freshwater Behaviour and Physiology,* **25**, pp. 69–85.

Shadwick, R. E. (1999) Mechanical design in arteries, *Journal of Experimental Biology*, **202**, pp. 3305–3313.

Vacelet, J. and Boury-Esnault, N. (1995) Carnivorous sponges, *Nature,* **373**, pp. 333–335.

Walker, C. (2001) Microbes in the environment. In *Microbes,.* H. MacQueen (ed), The Open University, Milton Keynes, pp. 99–161.

Wilkie, I. C., Emson, R. H. and Young, C. M. (1993) Smart collagen in sea lilies, *Nature,* **366**, pp. 519–520.

FURTHER READING

Gerhart, J. and Kirschner, M. (1997) *Cells, embryos and evolution.* Blackwell Science, Oxford. [Best for molecular and developmental perspectives on invertebrates and their evolution. Quite a difficult read!]

Ruppert, E. E. and Barnes, R. D. (1991) *Invertebrate Zoology.* 6th edn, Saunders College Publishing, Fort Worth, Philadelphia, USA. [The best of several modern textbooks of invertebrate zoology.]

INSECTS AND OTHER ARTHROPODS

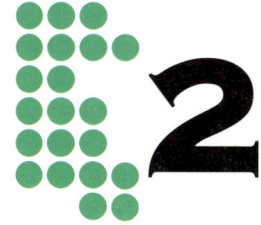

2.1   INTRODUCTION TO THE ARTHROPODS

The impact of arthropods on human activities has been the primary reason why the phylum has been studied so intensively. Widespread insect-transmitted diseases like malaria or typhus, plagues of locusts or the viral diseases of potatoes that are spread by aphids are remembered for the catastrophic effects that they have had on humans. Less prominence is given to the benefits for humans and the place of the arthropods in almost every ecosystem on the planet. The largest mammal, the blue whale, feeds almost entirely on crustaceans, while small arthropods in the soil hasten the decay of dead plant material. Sexual reproduction in many flowering plants is dependent upon arthropod pollinators while the parasitic wasps (Hymenoptera) play a major role in controlling the population density of their arthropod hosts.

The arthropods are the most morphologically diverse group of living organisms, and a substantial proportion of the known species of animal fall into this group. Of the 1.7 million animal species that have been described so far, 72% are arthropods. Of the arthropod species, over 89% are insects.

○   Which order of insects has the largest number of recorded species?

●   About 25% of all described species of animal are beetles, which makes Coleoptera the order of insects with the largest number of known species.

Charles Darwin was fascinated by the diversity of beetles, even before his voyage on the Beagle, and while at Cambridge he devoted a lot of his time to building up a collection. The sheer number of different species in a relatively small habitat continues to be a source of fascination to biologists. In a recent study of beetles in different habitats on Vancouver Island, Canada, the number of species of beetle of the staphylinid family found in a series of different habitats was counted. An example of a staphylinid or rove beetle is the devil's coach horse beetle that is found in Britain. Figure 2.1 (overleaf) shows the distribution of staphylinid beetle species at four sites in the coastal spruce forest of Vancouver Island. Both species richness and species composition differ between the sites. The number of species in the forest interior and canopy is high and some of these species are found only in one of the sites. These figures give a picture of the number of species, but not the relative abundance of each. Large though the number is, it is dwarfed by the richness of some tropical sites. In a study of a small area of tropical rain forest in southern Cameroon in 1997, researchers collected 342 species of canopy beetle, and a study of only 19 trees in Peru in 1982 produced 955 species of beetle in the canopy.

Figure 2.1 The distribution of species of staphylinid beetle at four different sites in the spruce forest of Vancouver Island. The number of species in each habitat is given in bold, with the number of species found *only* in that habitat in brackets. 'Clearcut' is where the trees have been felled, leaving only undergrowth. A line joining two habitats shows that they have species in common and the number of these species is indicated alongside the line (modified from Winchester, 1997).

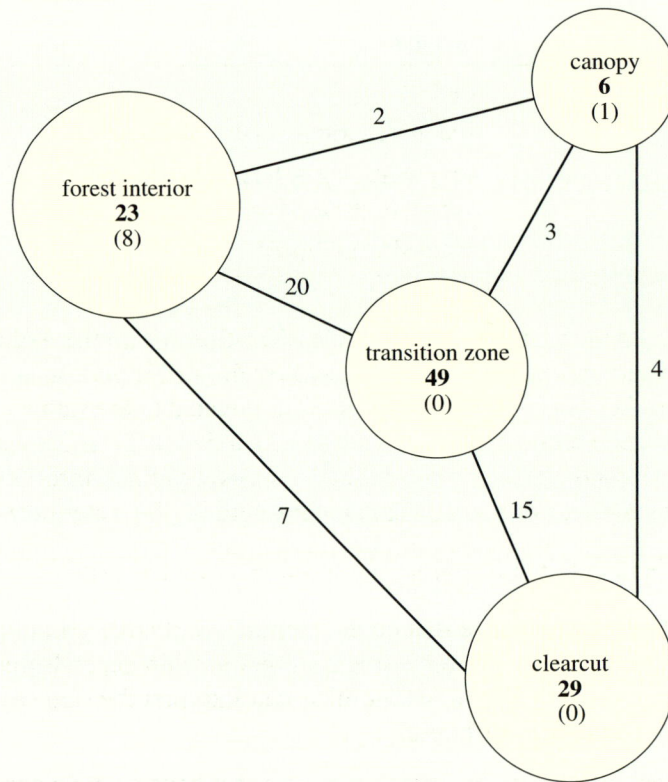

2.1.1 THE TAXONOMIC STATUS OF THE ARTHROPODA

The arthropods are a very diverse group of animals, but they share a basic body plan. That body plan is clearly remarkably flexible and adaptable since arthropods are found in a wide variety of habitats and have an impressive size range. The smallest arthropods, parasitic mites, are around 0.1 mm long, whereas fossil eurypterids (Figure 2.2) have been found that are 3 m in length.

○ What features indicate that the fossil eurypterid in Figure 2.2 can be classified as an arthropod?

● There is an exoskeleton; the body is segmented; there are jointed appendages; all these features are characteristics of the phylum Arthropoda.

We recognize four major groups of arthropods (Table 2.1). Representatives of all four groups have been found in a particularly rich fossil assemblage in the Yoho National Park in British Columbia. The rock formation is the Burgess Shale — it dates from the Cambrian and the fossils in it are exceptionally well preserved. What is so fascinating about the arthropod fossils from the shale is that all three major groups that are recognized today are already present, together with other species that do not fit into the four major groups and do not have known descendants.

Figure 2.2 A fossil eurypterid from the Ordovician. Some species grew up to 3 m.

Table 2.1 The major groups of the phylum Arthropoda.

Subphylum	Examples
Trilobita (extinct)	trilobites
Crustacea	woodlice, shrimps, crabs, barnacles
Chelicerata	eurypterids, horseshoe crabs, sea spiders, spiders, scorpions, ticks, mites
Uniramia	centipedes, millipedes, insects

The taxonomic status of the higher levels of classification within the Arthropoda is a matter of continuing debate. You may encounter the Arthropoda as a superphylum, with the subphyla listed in Table 2.1 elevated to the status of phyla. You may also find that certain groups that are not included in Table 2.1 may be included, in some taxonomic schemes, within the superphylum Arthropoda, for example Onychophora (the velvet worms) and Tardigrada (the water bears).

○ What factors lead to this uncertainty in the status of higher taxa?

● As much of the fossil evidence that could support a particular grouping is missing from the fossil record, assessment of relationship depends upon comparative anatomy. Different taxonomists can interpret the same evidence (and gaps in the evidence) differently.

The taxonomic ideal is to group together organisms that have a single common ancestor and form a distinct, closely related group. The group is then **monophyletic**. If one were to regard a phylum as a grouping with a monophyletic origin, then the Arthropoda would not fit that definition. It is almost certain that the subphyla listed in Table 2.1 have separate origins, making the Arthropoda a **polyphyletic** group. The evidence for the polyphyletic origins is complex and comes from study of the structure of the head and particularly the jaw-like appendages (strictly, the term 'jaw' should only be used for vertebrates). The jaw-like appendages of arthropods form two natural groups, those that are derived from the base of an embryonic limb and those that are derived from a complete embryonic limb. The former type is found in the Crustacea and the Chelicerata, the latter type in the Uniramia. There is no space here to delve further into this interesting and controversial area, but be aware that references to the phylum Arthropoda are references to a polyphyletic grouping, not a monophyletic one. The distinguished anatomist and taxonomist S. M. Manton famously wrote some years ago that the support of some workers for a monophyletic origin for the Arthropoda was based on:

'abundant phantasies and errors of fact'.

Taxonomists can be forthright in their views!

2.1.2 THE ARTHROPOD BODY PLAN

The arthropod body plan appeared early in the fossil record and it was well
established 570 million years ago in the Cambrian period. The essential features
of the body plan are readily visible in animals such as the centipede. The body is
divided into segments that are similar in appearance, each having a pair of jointed
appendages. The phylum takes its name from the typical jointed appendages:
arthron (joint) and *pod-* (foot, also claw or talon) are both Greek words. In many
arthropods, segments are fused together to form new functional groups called
tagmata (singular, tagma), for example the head, thorax and abdomen in insects.
The segmental origin of the tagmata often remains visible, like the segmented
abdomen visible in the locust in Figure 2.3, but the appendages may have
disappeared. The thorax of the locust is formed from three merged segments, but
the boundaries between the individual segments are barely visible. However, all
three pairs of appendages remain.

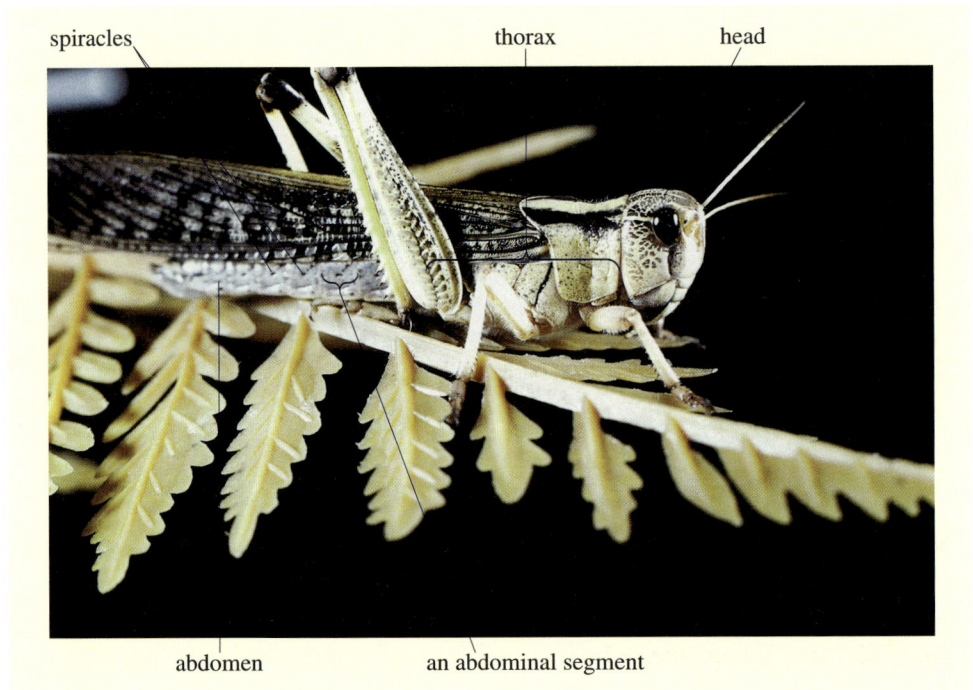

Figure 2.3   An adult desert locust, *Schistocerca gregaria,* about 4 cm long.

2.1.3 VARIATIONS ON THE ARTHROPODAN THEME

There is a huge diversity of body forms in the Burgess Shale arthropods, yet
analysis of the probable lifestyles of these marine organisms shows that the types
of habitats that they occupied were similar to modern marine environments. So
whatever selective pressure drove the proliferation of body forms, it does not
appear to be the environment alone. Nevertheless, all the forms are variants on a
basic body plan. Examples of the range of body forms are shown in Figure 2.4,
adapted from the original paper by Briggs and Whittington (1985).

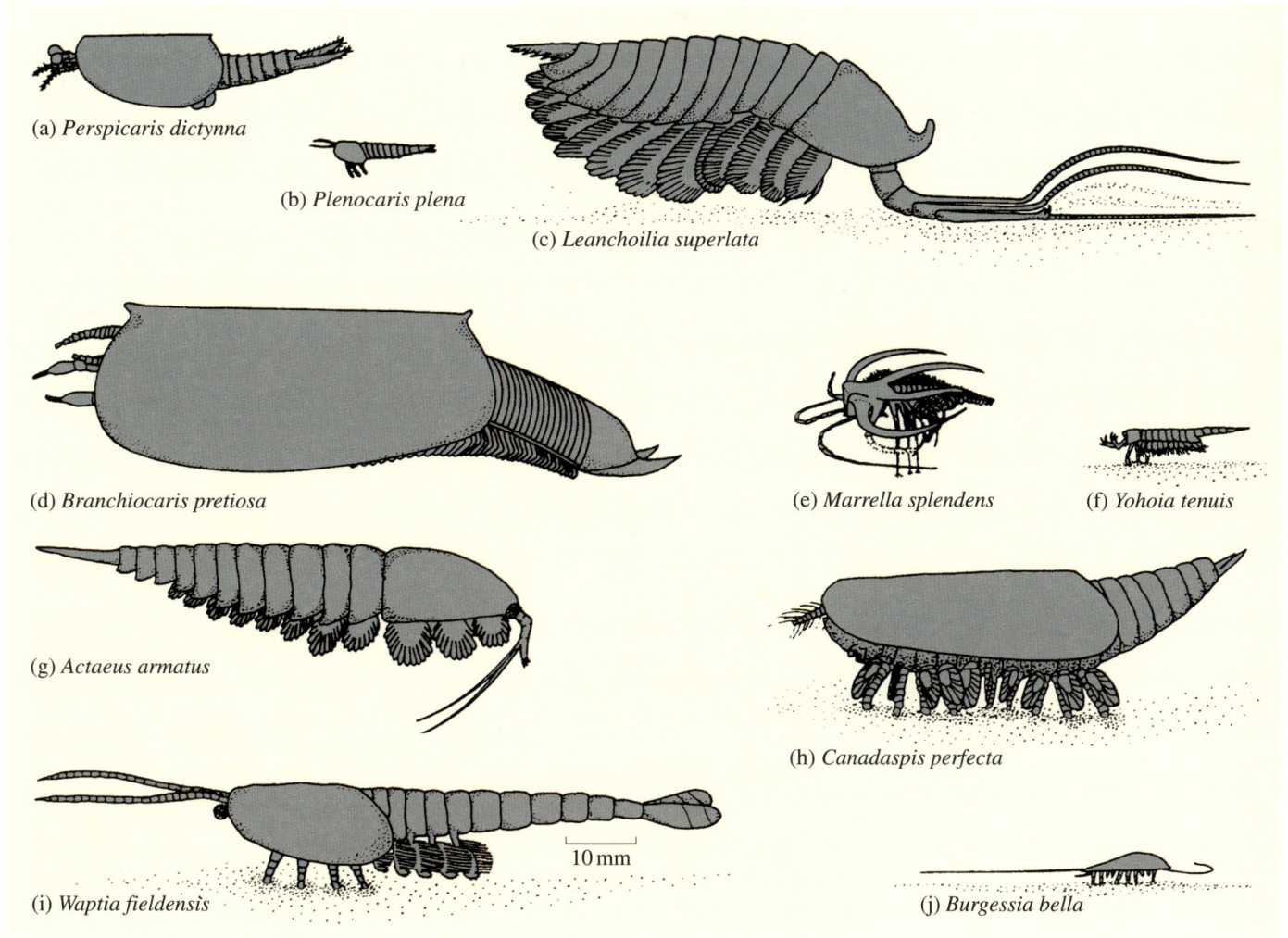

(a) *Perspicaris dictynna*

(b) *Plenocaris plena*

(c) *Leanchoilia superlata*

(d) *Branchiocaris pretiosa*

(e) *Marrella splendens*

(f) *Yohoia tenuis*

(g) *Actaeus armatus*

(h) *Canadaspis perfecta*

(i) *Waptia fieldensis*

10 mm

(j) *Burgessia bella*

Figure 2.4 Restorations of arthropods from the Burgess Shale, drawn to the same scale.

There is also huge diversity in the phylum today, with conservation of the body plan in all three living subphyla.

The crayfish *Austropotamobius* is a good example of the crustacean body plan. The European species, *Austropotamobius fluviatilis*, is generally found in calcareous streams where it lives in burrows in the banks, emerging at night to feed. Competition with introduced species has reduced the population substantially in Britain and parts of western Europe during the past thirty years.

The external features of the crayfish are shown in Figure 2.5a (overleaf). The head has six fused segments, the thorax has eight fused segments and the abdomen six articulated segments.

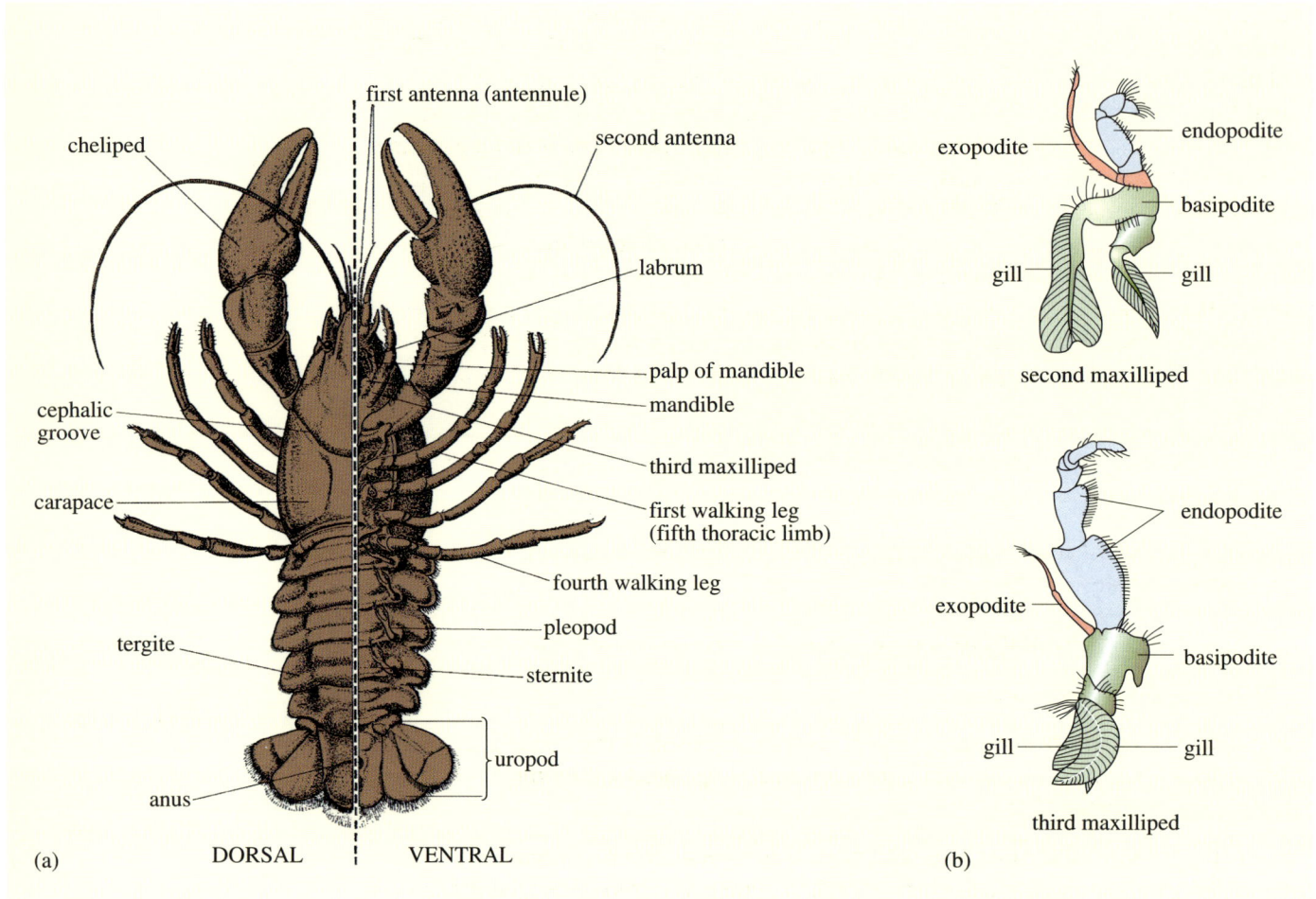

Figure 2.5 (a) The external features of both the dorsal (shown on the left) and ventral (shown on the right) surfaces of the crayfish *Austropotamobius*. (b) The second and third maxillipeds of the crayfish, showing typical features of a biramous limb. Note the expanded surfaces (lamellae) of the gills.

○ What is the technical term for these three functional regions?

● The crayfish has three functional regions, the tagmata.

The head and thorax are fused into a **cephalothorax**, but the segments that comprise the cephalothorax are still represented by the pairs of appendages. These appendages show a variety of specializations, although it is possible to show that they are all modifications of a more basic structure, a limb with two branches called a **biramous** limb. The maxillipeds are very close to the basic structure of a biramous limb, illustrated in Figure 2.5b. Attached to the base of the maxillipeds are the gills, which, in the intact animal, project into the gill chamber under the carapace. As all the appendages of the crayfish are derived from a single common type they are said to be homologous to each other. The limbs of the crayfish are, perhaps, the best example of serial homology found in animals. All the other limbs of the crayfish would originally have been biramous, but some have become secondarily **uniramous** (single branch).

○ Which limbs in Figure 2.5a are examples of uniramous appendages?

● The large chelipeds are uniramous.

The mouthparts act together in a complicated way, which demonstrates the link between the structure and function of the different appendages (see Figure 2.5). The third maxillipeds pick up and tear the food, which may be any form of plant or animal life, alive or dead. The second maxillipeds pick up pieces of the food and convey them to the mandibles, assisted by the first maxillipeds. The mandibles crush and masticate the food, which is then pushed into the mouth. Within the gut of the crayfish, food fragments are further broken down in the gastric mill, a region of the gut that contains chitinous teeth and a series of articulated plates whose movements grind the food into very fine particles. A series of fine chitinous hairs (setae) obstructs the route through into the posterior part of the gut, ensuring that particles can only enter once they have been ground finely enough by the gastric mill. The particles that are allowed to pass through are so fine that they do not cause abrasion damage.

○   What advantage does the crayfish gain from the extensive pre-treatment of food particles by grinding?

●   A wide variety of food is grist to the gastric mill and the animal can feed on calcareous algae and shelled and bony animals.

○   Why would it be advantageous for a crayfish to eat calcareous algae and shells?

●   They contain a lot of calcium.

In crayfish, like other crustaceans, the chitinous exoskeleton is thickened and stiffened by a layer of crystalline calcium carbonate deposited within the endocuticle (see Section 2.2).

The Arachnida, a class within the Chelicerata, is a diverse group of arthropods with some 70 000 species already described. Arachnids are a very old terrestrial group, having been the first arthropods to colonize the land. There are fossil scorpions from the Silurian (443–417 Ma) and spiders are known from the end of the Devonian (417–354 Ma). About half the known extant species of arachnids are spiders and the group has a worldwide distribution. Figure 2.6 (overleaf) shows the anatomy of a female spider.

The arachnid body plan comprises a fused head and thorax, the cephalothorax (or prosoma), with a mouth and four pairs of walking legs, and the abdomen (opisthosoma). The upper part of the cephalothorax is protected by a tough, chitinous carapace, the lower surface by the sternum. The abdomen shows great variation in shape, size and markings between species. It is soft and can expand as the animal feeds or as eggs develop in the female. Silk is produced in spinnerets derived from vestigial abdominal appendages. The mouthparts of spiders are relatively simple by comparison with those of insects and crustaceans.

The anterior appendages are a pair of chelicerae, each tipped with a fang (Figures 2.6 and 2.7). In most species, each fang has a small opening through which venom from the poison glands in the head region passes into the prey. There is a pair of sensory pedipalps behind the chelicerae, which also have a role in grasping and manipulating the prey. Arachnids never have true antennae. There are four pairs of walking legs; the front pair surrounds the mouth and may bear jaw-like structures on the basal segments.

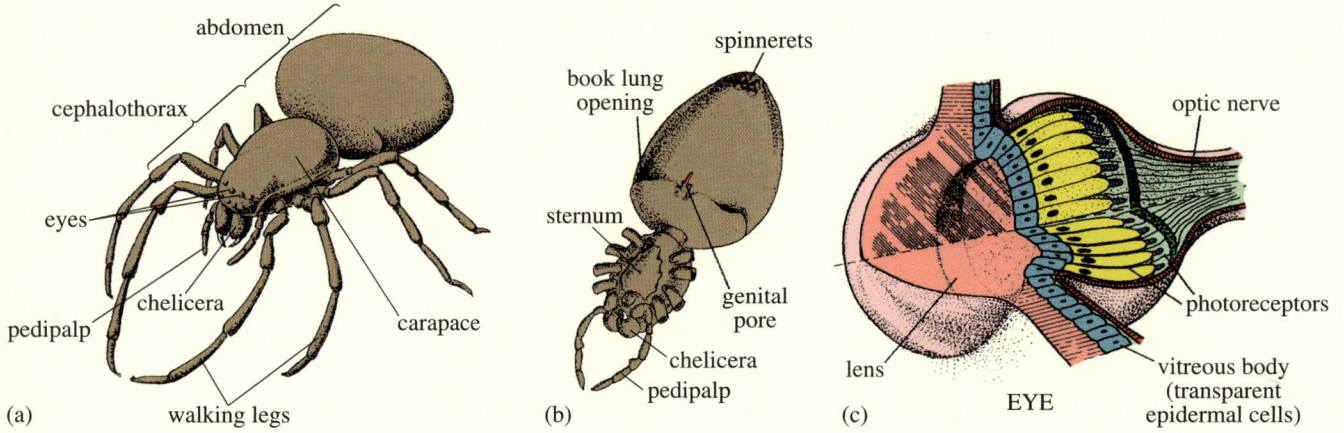

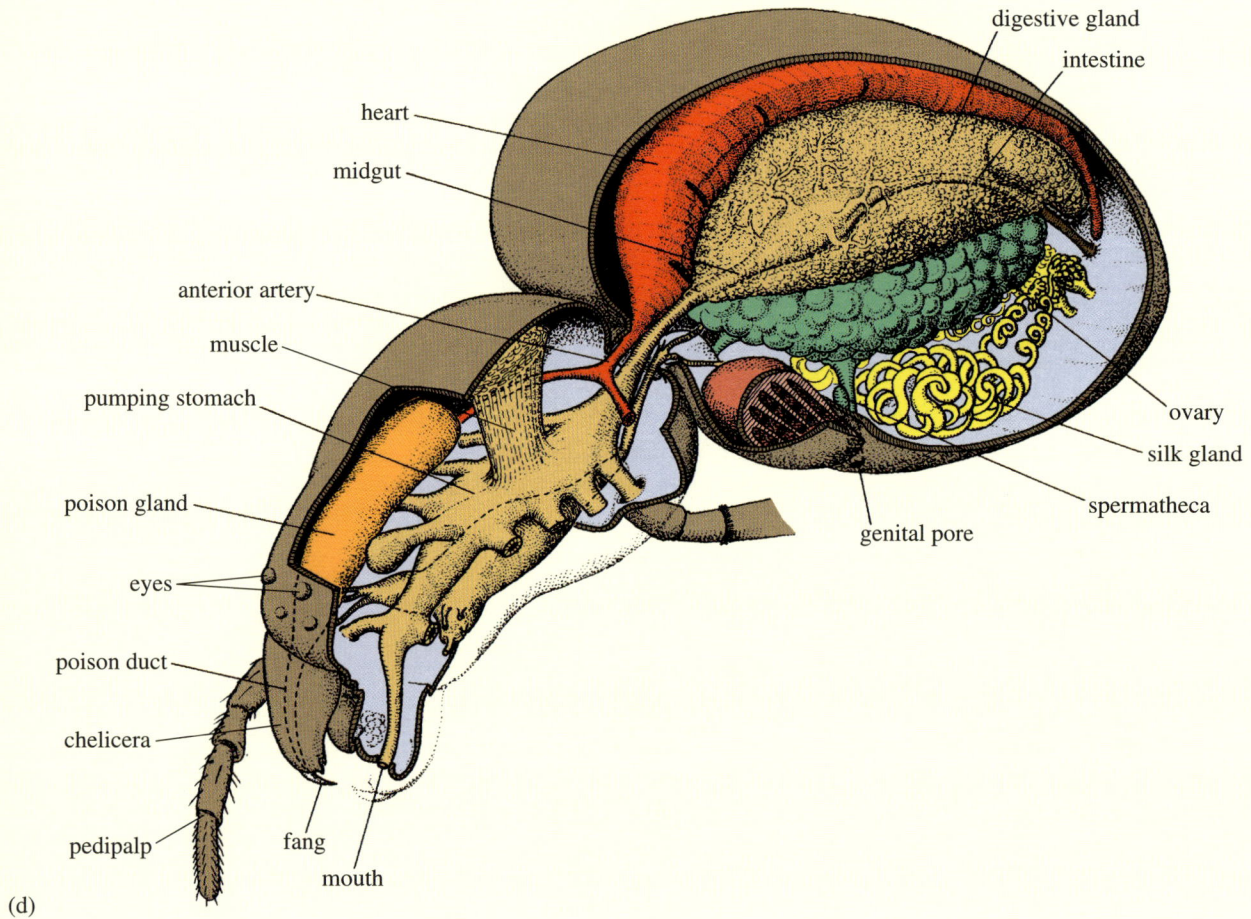

Figure 2.6 The anatomy of a female spider *Araneus*. (a) Adult female spider. (b) View of under surface of adult female. (c) Structure of an eye. (d) Internal anatomy of a female.

In the typical body plan of an insect, the head consists of six fused segments, the thorax of three fused segments and the abdomen of eight to eleven fused segments. The locust (Figure 2.8a) is a pterygote, a winged insect, and it displays this typical body plan. The head bears one pair of antennae and three pairs of mouthparts. The thorax has three pairs of legs, one pair per segment, and two pairs of wings. There are no legs on the abdomen. The huge diversity of insects derives from variations of the basic body plan and also variations in the structure of the uniramous appendages. The mouthparts of the locust, an insect that chews leaves and stalks, are relatively unspecialized (Figure 2.8b). The paired mandibles are hard and sclerotized (see Section 2.2.1) and crush and slice food by means of powerful muscles. Behind the mandibles is a pair of maxillae, which hold the food while the mandibles bite it. The labium is derived from fusion of two maxilla-like appendages and has a similar function to the maxillae. The maxillary palps are sensory and their role is to identify suitable food. The labrum overarches the mouthparts and is formed from a single plate of cuticle at the front of the head. The hypopharynx, a small tongue-like structure, is located just behind the mouth on the underside of the head, and is associated with the ducts of the salivary glands.

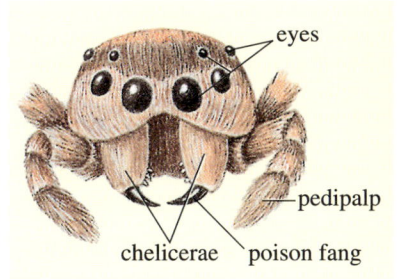

Figure 2.7 Front view of the cephalothorax of a spider, showing the chelicerae with poison fangs, the pedipalps used for manipulating prey, and the eyes.

Figure 2.8 (a) An adult locust, *Locusta migratoria* (life size). This African species is very similar to *Schistocerca gregaria* shown in Figure 2.3. (b) The head and mouthparts of a locust.

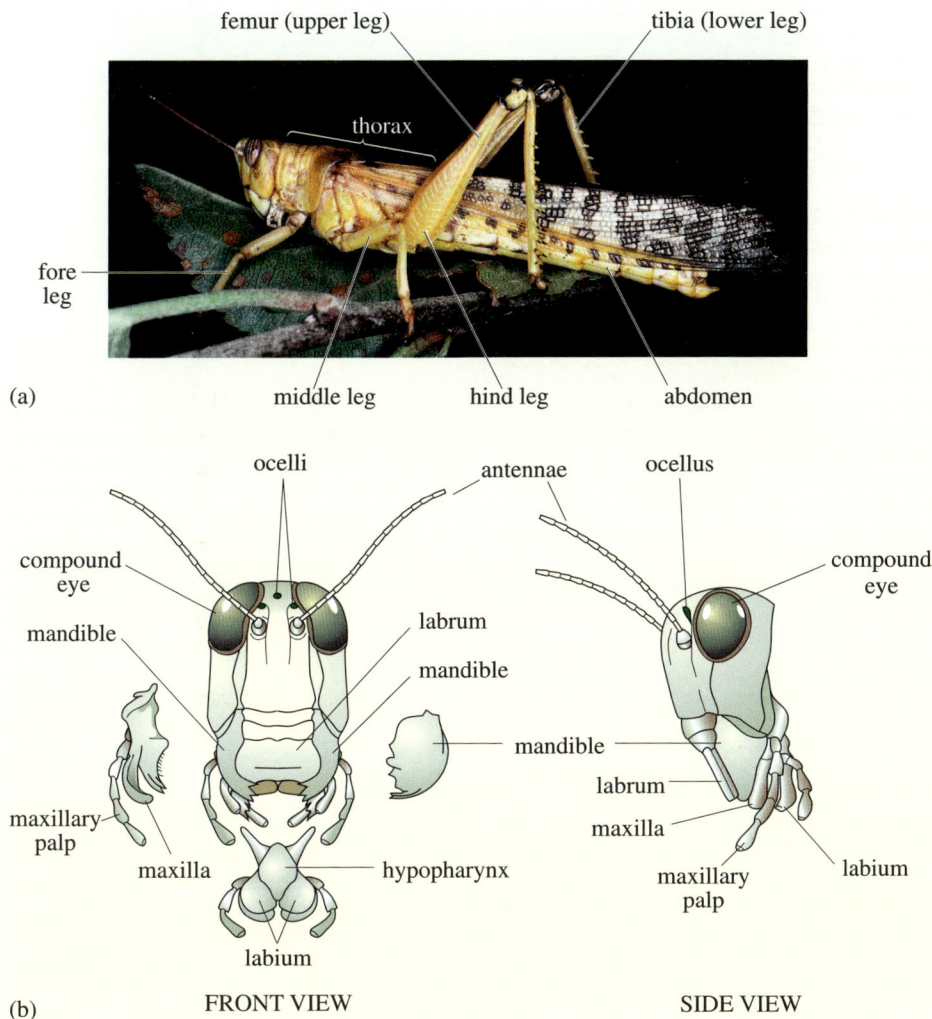

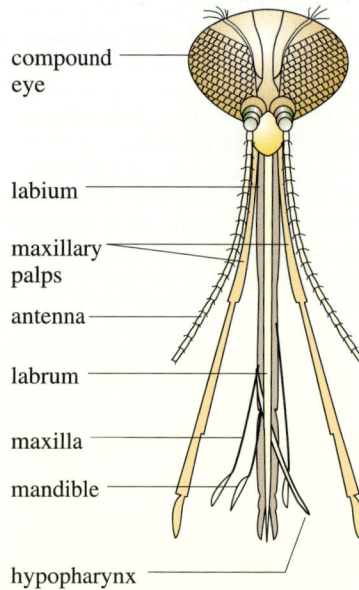

Figure 2.9 The mouthparts of a mosquito, viewed from the front and shown partially separated out for clarity.

compound eye
labium
maxillary palps
antenna
labrum
maxilla
mandible
hypopharynx

In other insect species, mouthparts can be very specialized. Mouthparts modified for sucking up liquid food are found in butterflies and moths, bugs, fleas and dipterans such as mosquitoes. The structure of mouthparts adapted for sucking shows great variety, but all are formed by development of the maxillae and labrum, for example in the mosquito (Figure 2.9). Note the prominence of the hypopharynx in the mouthparts of the mosquito. You will read more about mosquitoes in Section 2.3.3.

Providing an explanation for the diversity of arthropods is not easy. Some of the structural and physiological features of the arthropod body plan appear to confer advantages that would favour high diversity. The division of the body into segments seems to be a feature that is capable of extensive modification during evolution and you might suggest that the body plan provides an explanation for the high species diversity. However, the annelid worms too are segmented but do not show the great diversity found in arthropods. While we cannot here delve further into how this diversity may have come about, in the later sections of this chapter we shall be considering the consequences of this extensive diversity, the place of the arthropods in the diversity of life, and their economic and medical importance.

SUMMARY OF SECTION 2.1

1 The Arthropoda is the most structurally diverse of the animal phyla and many members of the phylum have a substantial impact on human life.

2 The diversity of arthropod body forms is apparent from very early in the fossil record, as exemplified by the assemblage of fossils of the Cambrian period in the Burgess Shale formation in British Columbia.

3   There is conservation of the segmental organization of the body across the whole phylum through geological time, but the body plan and form of the appendages also show a great deal of flexibility.

4   The taxonomic status of the Arthropoda is controversial, but the majority of researchers would agree that the Arthropoda is a polyphyletic group.

5   Representatives of the three living subphyla, Crustacea, Chelicerata and Uniramia, show variations of the arthropodan plan. For example, the limbs of many of the crustaceans show substantial adaptation for different mechanical or reproductive functions. Mouthparts of many different insect species show adaptation to specific methods of feeding.

2.2   THE ARTHROPODAN EXOSKELETON

A characteristic feature of arthropods is the external skeleton, formed by a layer of cuticle. Many arthropods are brilliantly coloured, particularly some insects. The colours of insects may be produced by the physical properties of the cuticle itself, or by pigments in the cuticle. The scattering of light may produce the colours. If there are surface irregularities in the cuticle that are larger than the range of wavelengths of white light then all light is reflected and the surface appears white. However, if the dimensions of the irregularities are within the range of wavelengths of white light, then the surface appears blue, as blue light has a shorter wavelength. Very fine and regular texturing of the surface, or regular laminations in the structure of the cuticle, produce reflections that interfere with each other and give an iridescent hue. There are also a number of pigments that occur in the cuticle, such as the quinones that give scale insects their red and yellow colours, or the aphins that give aphids a range of colours between yellow and blue–green.

The layer of cuticle that forms the skeleton covers the outer tissue layer, the epidermis. The cuticle is a complex of proteins, lipids and **chitin**, an insoluble acetate of a polysaccharide containing glucosamine. The structure of cuticle makes it tough, lightweight, flexible and relatively impervious to water.

2.2.1   STRUCTURE OF THE EXOSKELETON

Figure 2.10 (overleaf) shows a section of insect cuticle. The outer layer, the epicuticle, is a laminated layer containing lipids (especially waxes) and proteins that are hardened by sclerotization, a tanning process by which cross-links form between protein molecules, increasing rigidity. Sclerotized cuticle has a dark colour that is due to the hardening process, not a pigment. The surface of the epicuticle may have very fine striations. Below the epicuticle lies the procuticle, the layer that contains chitin. In insects, the procuticle is around 40% chitin and 60% protein, but in some Crustacea the proportion of chitin is much higher. The procuticle may be impregnated with mineral salts that increase the hardness but reduce the flexibility. The body 'shell' of a crab lacks flexibility as it is hardened by a substantial deposition of calcium salts in the procuticle, predominantly crystalline calcium carbonate (calcite) and calcium phosphate. The cuticle of burrowing millipedes (Diplopoda) also contains calcium. The hardening, and therefore rigidity of the cuticle is probably associated with their burrowing habit, since the non-burrowing diplopods do not have a calcified cuticle.

Figure 2.10 A section of
arthropod cuticle showing the
layered structure.

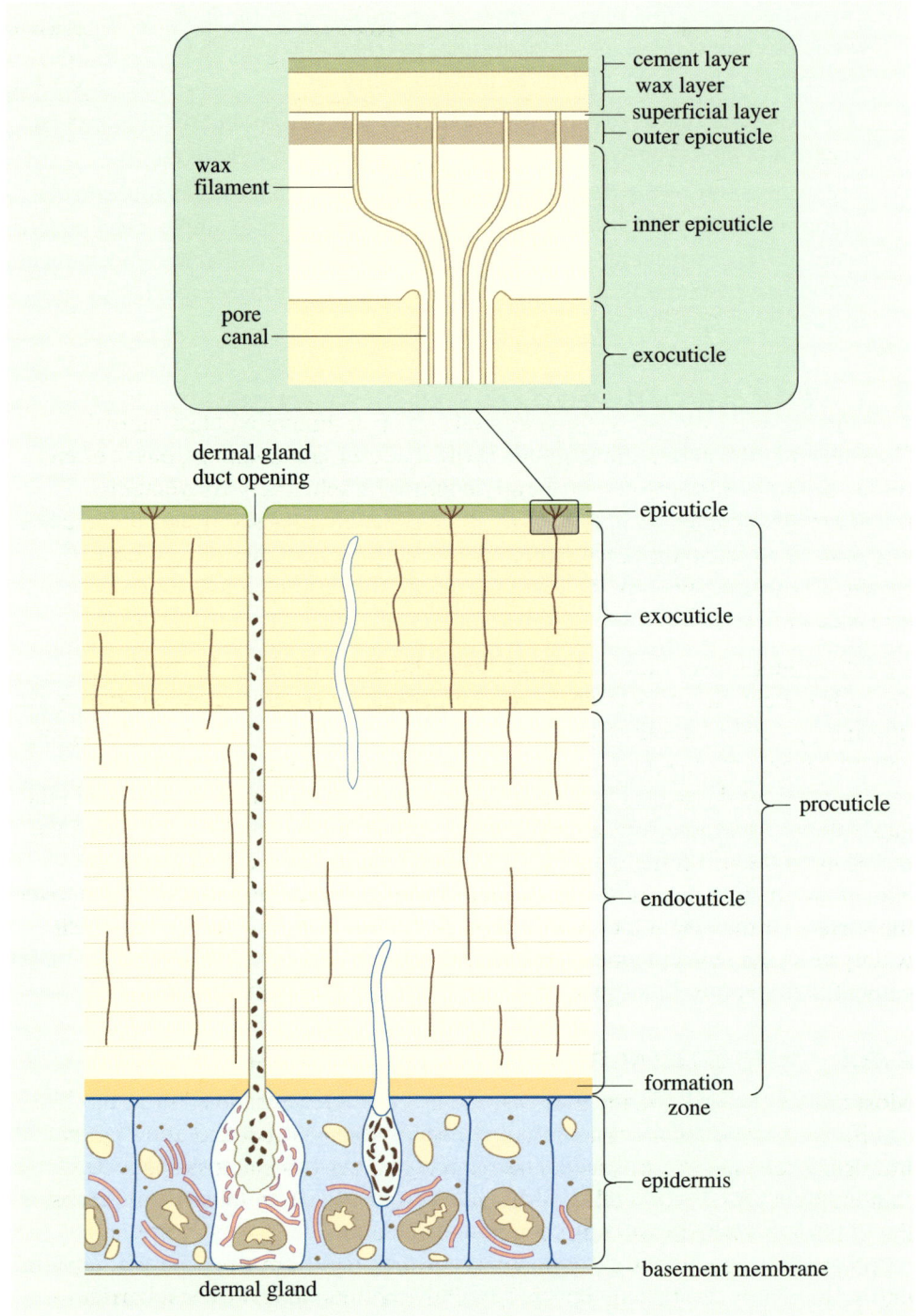

○ What three major problems arise from having an external skeleton?

● A skeleton cannot be continuously stiff, as joints are needed to allow
movement; a rigid external skeleton limits growth; and an impermeable
external skeleton prevents diffusion of respiratory gases into and out of
the body.

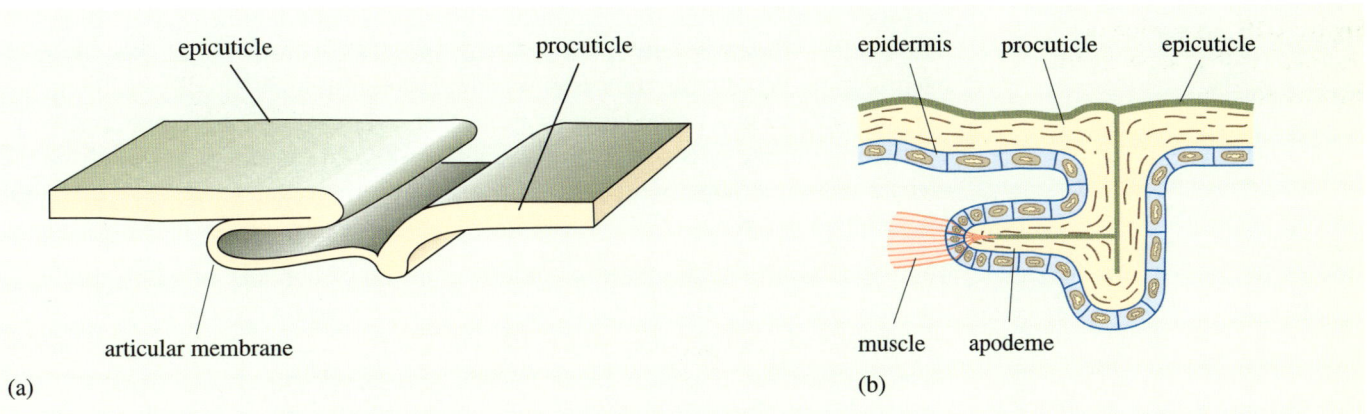

At joints between segments, the procuticle is much thinner and less tanned. Consequently the procuticle is more flexible, allowing movement, but is often folded such that the stiff plates of cuticle almost touch. The thin articular membrane is protected (Figure 2.11a) and, as it is not exposed, water loss from the joints is reduced. Similarly, joints in the appendages have a thin, flexible cuticle that permits movement but again the structure is such that little of the thin articular membrane is exposed.

Cuticle cannot grow in the same way that cellular tissue does. An arthropod discards the cuticle and secretes a new, larger one on several occasions during its lifetime, generally moulting between four and seven times. Before shedding the cuticle, an enzyme is secreted that starts to break down the old cuticle and separate it from the underlying tissue. A new layer of epicuticle is secreted in the gap, followed by the upper layer of the procuticle, the exocuticle. Following the moulting of the old cuticle, a new layer of endocuticle is secreted below the exocuticle. Just before moulting, the cells produce a secretion that flows out over the surface of the epicuticle. A wax layer is formed from this secretion, which waterproofs the new epicuticle. The dermal glands (Figure 2.10) secrete a cement layer that provides protection for the underlying waterproof layer.

2.2.2 GAS EXCHANGE IN CRUSTACEANS AND INSECTS

Most crustaceans are aquatic animals, taking up oxygen dissolved in water. The calcified exoskeleton of crustaceans is impermeable to water and allows very little gas exchange to take place. Gas exchange takes place across specialized flattened surfaces, the lamellae of the gills (Figure 2.5b). In lobsters and crabs, the gills are enclosed within gill chambers, and a paddle-like structure, the scaphognathite, oscillates to pump water through the chambers. The thin layer of uncalcified chitin that coats the gill lamellae does not appear to be a barrier to gas diffusion.

Like all arthropods, crustaceans have an open circulatory system. In many larger species, the fluid within the circulatory system, the haemolymph, contains the respiratory pigment haemocyanin, which is colourless when deoxygenated but blue when oxygenated (see Section 1.4.1). So crabs and lobsters have blue blood! The blue colour derives from copper within the molecular structure of haemocyanin. Note that the respiratory pigment is not contained within blue blood cells — haemocyanin is in colloidal suspension in the blood. Deoxygenated haemolymph flows from the systemic circulation into blood spaces inside the gill lamellae and returns to the heart as arterial (oxygenated) haemolymph. Actively ventilating marine crustaceans can extract 30–40% of dissolved oxygen from the water stream in contact with their gills. Freshwater crayfish can extract up to 80% of dissolved oxygen.

Terrestrial insects obtain their oxygen from the air. A network of tubes, the tracheal system, permeates the tissues, enabling respiratory gases to be transported to and from respiring cells. The tracheal system opens to the exterior through **spiracles**, generally a pair in each segment (Figure 2.12). In some insects, the **tracheae** in each segment remain separate, but in most insects the tracheae from all the segments are fused to form an interconnected system (Figure 2.12a). The finer branches are called **tracheoles** (not shown in the figure) and are very numerous. For example, in the larva of the silkworm there are estimated to be around 1.5 million. Each muscle fibre can have several attached tracheoles. It is the supply of oxygen direct to the tissues, together with the rapid removal of carbon dioxide, that permits a high metabolic rate and hence a high level of activity in some insects.

Figure 2.12 (a) The tracheal system of the flea *Xenopsylla*. (b) A typical spiracle attached to a trachea.

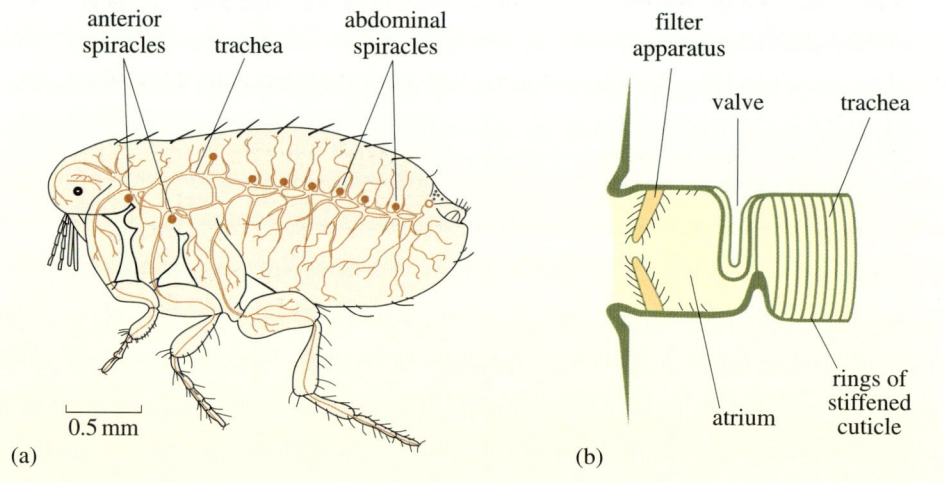

○ What role does insect haemolymph play in transporting oxygen to respiring tissues?

● A minor role for most insects. The tracheal system carries air directly to the respiring tissues. Gas exchange occurs at the ends of the tracheoles within the tissues.

However, in some insects, aerated haemolymph plays an important part in supplying oxygen to the ovaries.

Tracheal systems can be either open or closed. A closed system, such as that found in some endoparasitic or aquatic larvae, has no opening to the exterior and relies on gas diffusion through the cuticle into the tracheae. However, most insects exposed to air have open tracheal systems in which gas exchange occurs through openings in the cuticle, the spiracles, that connect directly to the tracheae (Figure 2.12b). A spiracular valve closes off the entrance to the tracheal system to minimize water loss when the respiratory rate is low, and there may be a filter on the exterior side of the valve. The tracheae are formed by invaginations of the epidermis and so the inner lining of the cuticle is continuous with the body surface. Thus tracheal linings are shed along with the rest of the cuticle at moult. The tracheae are thickened at regular intervals by rings of stiff cuticle, giving them both flexibility and resistance to sideways compression.

○ Can you think of a household object that is a good model for an insect trachea?

● The tube of a vacuum cleaner has to be flexible but resistant to sideways compression — it has thin walls strengthened at intervals by stiff rings.

SUMMARY OF SECTION 2.2

1 Arthropods have an exoskeleton, a layer of chitinous cuticle covering the epidermis, of variable stiffness and permeability.

2 The cuticle is made up of several layers and is a complex of proteins, lipids and chitin, secreted by the epidermal cells. In crustaceans, the cuticle is hardened by deposition of calcium salts; in insects, the cuticle is hardened by sclerotization.

3 The cuticle is thin and soft at joints, facilitating movement of the body and limbs.

4 Moulting and replacement of old cuticle enables growth.

5 The arthropodan exoskeleton is largely impermeable to water and gases. Gas exchange for respiration is usually by means of specialized structures. Crustaceans, aquatic arthropods, have gills. Insects, mostly air-breathers as adults, have tracheae and tracheoles that carry air directly to the respiring tissues.

2.3 LIFE CYCLES OF ARTHROPODS

A defining feature of arthropods is that they have a life cycle that involves many stages, often including larval forms. The crustaceans are a very large group of arthropods that mostly live in water and have great ecological significance in marine and freshwater food chains. Of course they also have gastronomic significance to humans. The larval stages are generally very small and are carried around by water movements — larvae are a dispersal stage for many species, for example barnacles. Few crustacean species have exploited the terrestrial environment successfully and those that have are restricted to damp areas where their essentially aquatic physiology can continue to function.

Spiders do not have a distinctive larval stage in their life cycle; the eggs hatch into tiny immature spiders, called nymphs. The newly hatched spiders soon disperse by means of ballooning (see Section 2.3.2). In some insect species the larval stages move comparatively little and the winged adult is the predominant dispersal stage.

Insects are the arthropod group that has the greatest impact on humans. Study of life cycles of insect pests such as locusts has provided strategies for their control. Understanding of the timing of the stages of the life cycle of blowflies has applications in forensic entomology — a case study is included in this section. Insects are also of significance as **vectors** that transmit diseases of both humans and domestic animals. The number of species that transmit disease is not large but their impact is great, and the study of medical and veterinary entomology is an important branch of the biological sciences. Section 2.3.3 considers just one example: the link between the life cycle of mosquitoes and malaria, and the adaptations of the mouthparts of the mosquito that facilitate feeding on blood.

2.3.1 LIFE CYCLES AND LARVAL LIFE OF CRUSTACEANS

In the genus *Austropotamobius*, mating takes place in the autumn, with sperm conveyed in a capsule, the **spermatophore**. The eggs are deposited by the female underneath her abdomen in the space formed by the curved abdominal tergites (Figure 2.5a). Once fertilized, the eggs attach to specialized hairs on the pleopods and are cemented in place by fluid released from cement glands. The cement forms a protective coat around the eggs while development takes place. In the spring the eggs hatch. The young are almost identical to the adults although, of course, much smaller.

This life cycle contrasts markedly with that of most Crustacea, where there are a number of larval stages after the egg hatches. The lifestyle of the larvae may be very different from that of the adults. The first larval stage of the prawn *Palaemon,* for example, is the zoea (Figure 2.13). The zoea larvae live in the marine plankton and go through a number of moults. The thoracic legs are very different from those of the adult and are used for swimming. By contrast, the adults swim using the pleopods. The zoea larva moults into the post-larva, still within the plankton. Subsequently, the post-larva settles out from the plankton and continues to grow on the seabed, eventually becoming an adult.

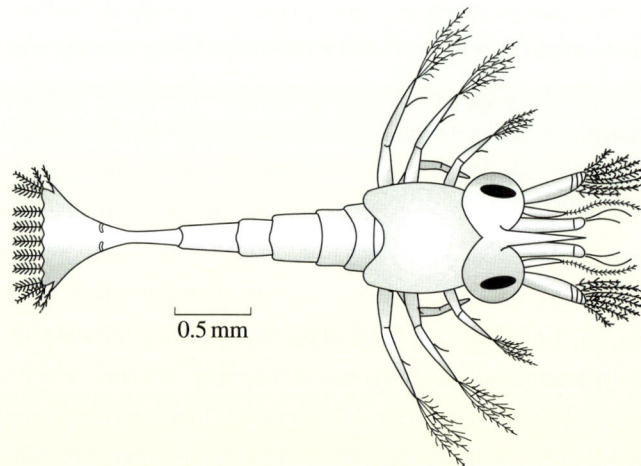

0.5 mm

Figure 2.13   The zoea larva of *Palaemon elegans*.

For the prawn the larval stage in the plankton is both a phase of dispersal and a stage in the life cycle when the animal exploits a very different food source. However, for some other crustaceans the planktonic larva represents the motile stage of an otherwise sedentary existence. Barnacles (Figure 2.14a and b) are common on rocky shores in the intertidal zone and are the only non-parasitic crustacea to have a sedentary adult. The animal lives inside a protective array of calcareous plates formed from the carapace and cemented to the rock surface. The top plates open when the tide is high and the feeding appendages of the animal sweep through the water, removing small particles of food, mostly small planktonic organisms. The feeding appendages are modified thoracic limbs, they are long, with many joints, and covered with fine branches.

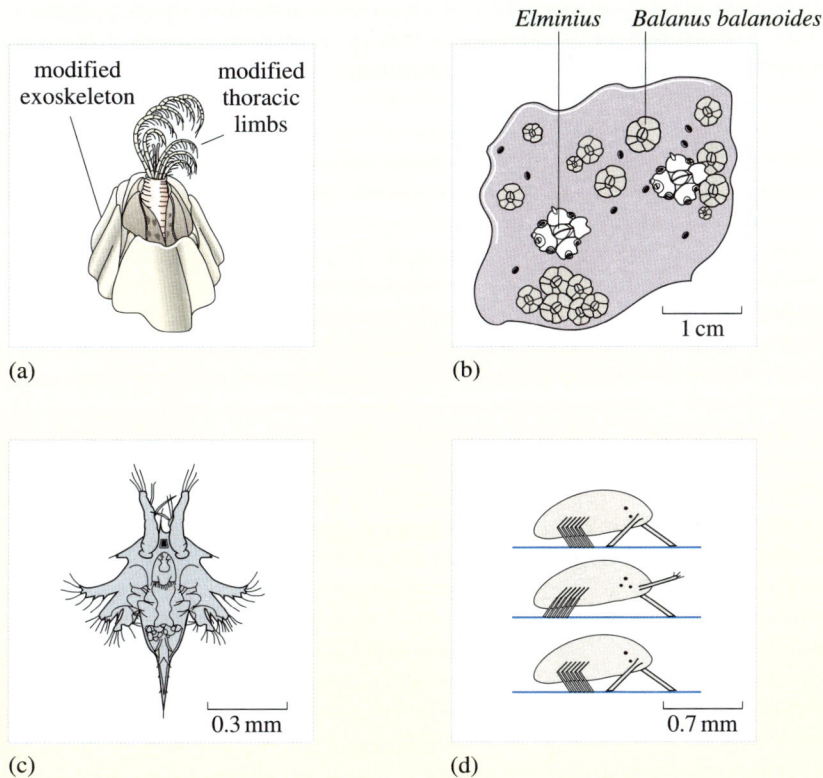

(a)

(b)

(c)

(d)

Figure 2.14   (a) Adult barnacle with thoracic feeding appendages sweeping the water. (b) Top view of adult barnacles and settled cypris larvae. Adults with four side plates are *Elminius modestus*, those with six plates are *Balanus balanoides*. (c) Nauplius larva of *B. balanoides*. (d) Successive movements of the cypris larva of *B. balanoides* when 'selecting' a site for attachment.

The planktonic nauplius larvae of the barnacle (Figure 2.14c) are triangular-shaped animals with three pairs of legs. They are active swimmers and positively phototactic, that is, they swim towards the light and so remain close to the surface of the water. The nauplius larvae grow by means of a series of six moults. The final moult results in an oval, laterally compressed larva having a bivalved, hinged shell and six pairs of walking legs. This cypris stage is intermediate between the free and fixed stages of the life cycle. The cypris larvae are carried to the shore by the tide and sink to the bottom, where they walk about the surface by means of their six pairs of legs (Figure 2.14d), searching for a suitable spot for settlement. They attach themselves to the substratum by means of a sticky secretion from glands in the appendages that were the antennae in the nauplius larva. Once settled, the cypris larva metamorphoses into a sedentary adult, but is able to delay metamorphosis for up to two weeks if there is no suitable place to settle. The larvae do not settle randomly. Experiments have shown that the larvae make choices based on the type of substrate and, more importantly, the presence of other barnacles of the same species. Figure 2.15 shows the results of a classic experiment on the settling of barnacle larvae on limestone rocks near Swansea, Wales. The distance between the centre of each newly settled cypris larva and the edge of the nearest settled neighbour was measured. A frequency distribution of the results was then plotted. The cypris larvae were all of the same species, *Balanus balanoides*, but the adults were of the three species, *B. balanoides*, *Chthamalus stellatus* and *Elminius modestus*.

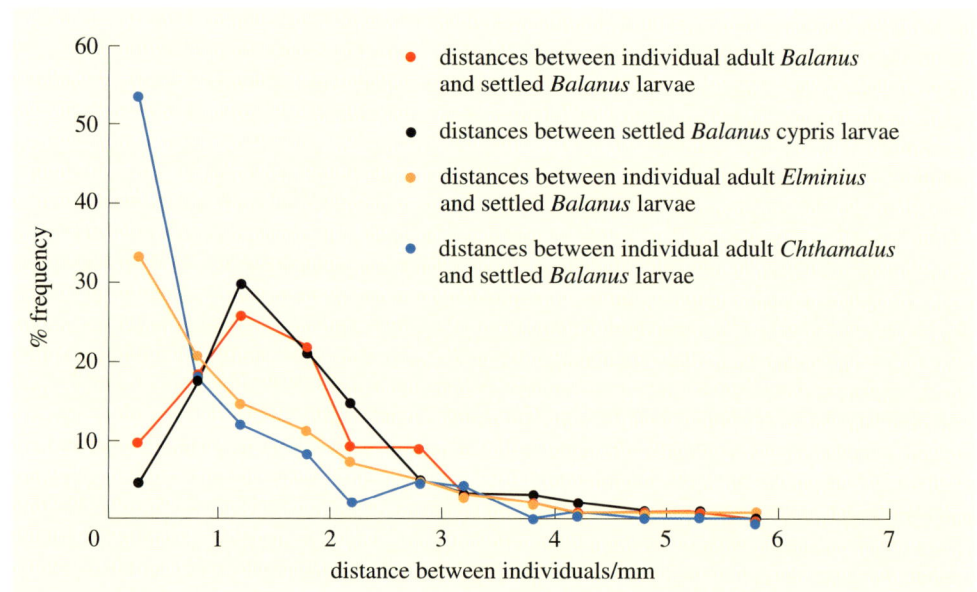

Figure 2.15 Settling behaviour of the cypris larvae of the barnacle *Balanus balanoides*. The graphs show the frequency distributions of the distances between the centre of newly settled cypris larvae and the edge of the nearest previously settled adult barnacle. Data from Knight-Jones and Moyse (1961).

○ What type of curve would you expect if settling behaviour was completely random? Would you expect the curve to be similar to that shown for *Elminius*, for *Balanus*, or some other curve?

● If the barnacles settled at random, then you would expect there to be an equal chance of the larvae settling at any point, so the larvae would be uniformly distributed and the curve would be a straight line parallel to the *x*-axis.

○ What can you deduce about the distribution of the cypris larvae when they settle, from the data in Figure 2.15?

● The larvae are not settling randomly and must be making a choice of settling site.

○ From the graph, how do the larvae of *Balanus* appear to be behaving towards the adults of *Elminius*?

● The large number that are settling at less than 1 mm from the edge of an adult, compared with the relatively small numbers settling at a greater distance, shows that the larvae are not settling randomly but appear to be selecting a site on or close to adult *Elminius*.

○ What does the fact that the greatest frequency of recently settled larvae is found at a distance of 0.2 mm from the edge of adult *Elminius* suggest?

● The larvae appear to be actively seeking out adults on which to settle. A distance of 0.2 mm from the centre of a settled cypris and the edge of an adult indicates that cypris larvae were settling selectively on the edge of *Elminius* adults.

○ Are the larvae actively seeking out *Balanus* adults to settle on?

● The peak in the frequency curve for both *Balanus* larvae and adults is around 1.2 mm. The position of this peak suggests that the larvae are actively avoiding settling on adults of their own species, or too near to settled larvae of their own species. The position of the peak also suggests that the larvae are choosing a place to settle that is close to an adult, but not too close.

The results of this and other experiments show that the cypris larvae of barnacles are making a sophisticated choice. They are able to distinguish between their own species and other species and it is probable that this discrimination is chemical. *Balanus* larvae can distinguish between settling sites soaked in extract of adult *Balanus* and those soaked with extract of adult *Elminius*. The results also show two different processes going on in the barnacle colonies. Competition is occurring between differing species, whilst conspecific individuals are aggregating, but maintaining a certain distance between individuals. It has been suggested that aggregation is advantageous to barnacles, as the roughening of the surface of the rock by the presence of many conical protuberances increases turbulence in the seawater and improves conditions for filter feeding. In addition, the large numbers of individuals all sweeping their feeding limbs through the water at the same time creates enhanced feeding currents that benefit the whole colony, although there is no suggestion that the feeding movements of individuals are in any way synchronized. The spacing of the newly settled larvae reduces overcrowding as the larvae develop.

○ What is the advantage of aggregation for reproduction, especially since *Balanus* has internal fertilization?

● As the adults are sedentary, cross-fertilization is enhanced if the animals are aggregated.

The aggregation of barnacles is an interesting example of members of the same species cooperating in the exploitation of their environment.

2.3.2 LIFE CYCLES AND DISPERSAL OF SPIDERS

Reproduction in some species of spider involves displays and ritualized dancing, but for males approaching a female across her web there is also an element of risk. Communication by appropriate vibrations through the silk of the web is certainly one way in which mate recognition is achieved, but females do eat male spiders. The lifespan of a male is shorter than that of conspecific females and after several matings he gets weaker and may be eaten by a female of his own or another species. Female *Agriope* wrap the much smaller male in silk during copulation and may start eating him before mating is complete.

Before mating, the male transfers sperm from the genital opening to a small, silk net of three or four strands. He then picks up the blob of sperm that is clinging to the silk with his palps, which have a sperm reservoir, and inseminates the female. The female stores sperm in a spermatheca, for subsequent use in fertilizing eggs. The males of some species, following insemination, ensure that their sperm is used by sealing the female opening with a secretion that forms a hard plug. In other species, where no plug is formed and the females are capable of multiple matings, the male stays with the female and fends off other suitors.

The eggs of all species are fertilized as they are ejected from the female. They are usually then wrapped in silk and either guarded by the female or carried around by her. The egg sac may contain only two or three eggs or may have thousands. The embryos develop into pre-larvae, which continue to develop within the egg sac, moulting to become larvae, which at this stage are still nourished by the yolk. Finally the larva moults again, still within the egg sac, and becomes a nymph — a fully developed spider, except for its size and the lack of functional sexual organs.

Once out of the egg sac, the nymphs may stay with the mother, as in some wolf spiders where the mother carries them on her back. In the genus *Araneus,* the newly hatched spiders form a ball held together by silk, but eventually all newly hatched spiders disperse, often by ballooning. Young spiders move onto a suitable high point on a warm day when air currents are rising. Pointing the abdomen upwards, the spider releases silken threads, which, if conditions are right, are carried upwards by the air currents. When there is sufficient pull from the threads the spider releases its grip and is pulled off the ground. Once airborne, the spider is a passive traveller in air currents, which can take it up to great height, certainly a few thousand metres. Point of descent is entirely dependent on the air currents, so there must be a very high mortality due to landing in inhospitable environments, such as water. Ballooning is used in many species and would appear to be an effective mechanism for dispersal.

2.3.3 LIFE CYCLES AND LARVAL LIFE OF INSECTS

Insects are primarily terrestrial, and the larval stages seen in the present orders bear little trace of the ancestral aquatic forms of larvae that must have existed in the past. So the insect larval stages are not comparable with the crustacean larvae described in Section 2.3.1. However, some insects do share with crustaceans a structure and mode of life distinctly different between larva and adult.

The butterfly and housefly life cycles are representative of one of the two distinct types of life cycle found in the winged insects (Pterygota). The larvae, which undergo several moults, lack most of the features of the adult, such as wings, compound eyes and reproductive organs, which develop after a larva has pupated and are visible only in the final adult instar (Figure 2.16). The wings are derived from internal wing-buds, the feature that gives the name to the type of life cycle, **endopterygote**, meaning internal wings. The earliest endopterygote insect known is a beetle from Upper Permian rocks (Permian, 290–248 Ma). The adaptive value of the endopterygote type of life cycle is the opportunity it provides for the adult larva to exploit other very different ecological niches, coupled with a period of substantial growth in the larval stage, without the need for a series of moults of a complicated external skeleton.

In insects such as cockroaches (Dictyoptera), bugs (Hemiptera, Figure 2.17, overleaf) and locusts (Orthoptera), the larval stages are often called nymphs and resemble the adults, although they lack reproductive organs and functional wings. The wings develop externally as wing-buds, increasing in size at each moult. Not surprisingly, this type of development is called **exopterygote**. The exopterygote life cycle is probably the more ancient, as cockroaches similar to present day genera have been found in rocks from the Carboniferous period (354–290 Ma).

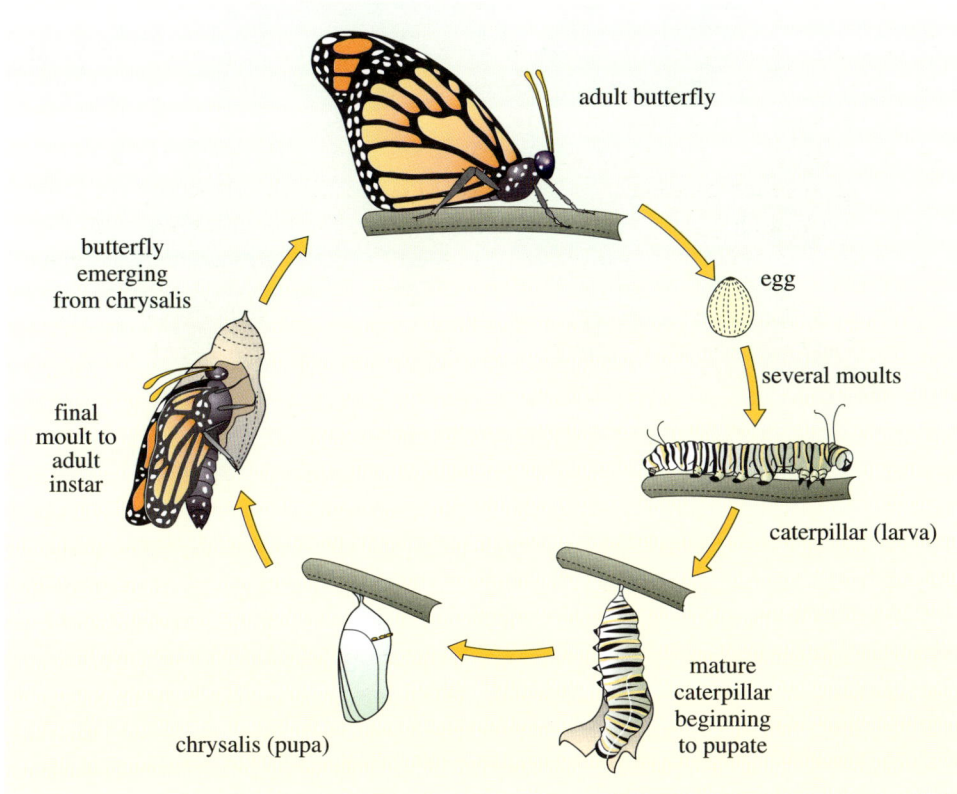

Figure 2.16 The life cycle of a typical endopterygote insect, *Danaus plexippus* (monarch butterfly) order Lepidoptera. Each stage is called an instar.

Figure 2.17 The life cycle of a typical exopterygote insect, the bug *Nezara viridula* on a tomato plant, not to scale.

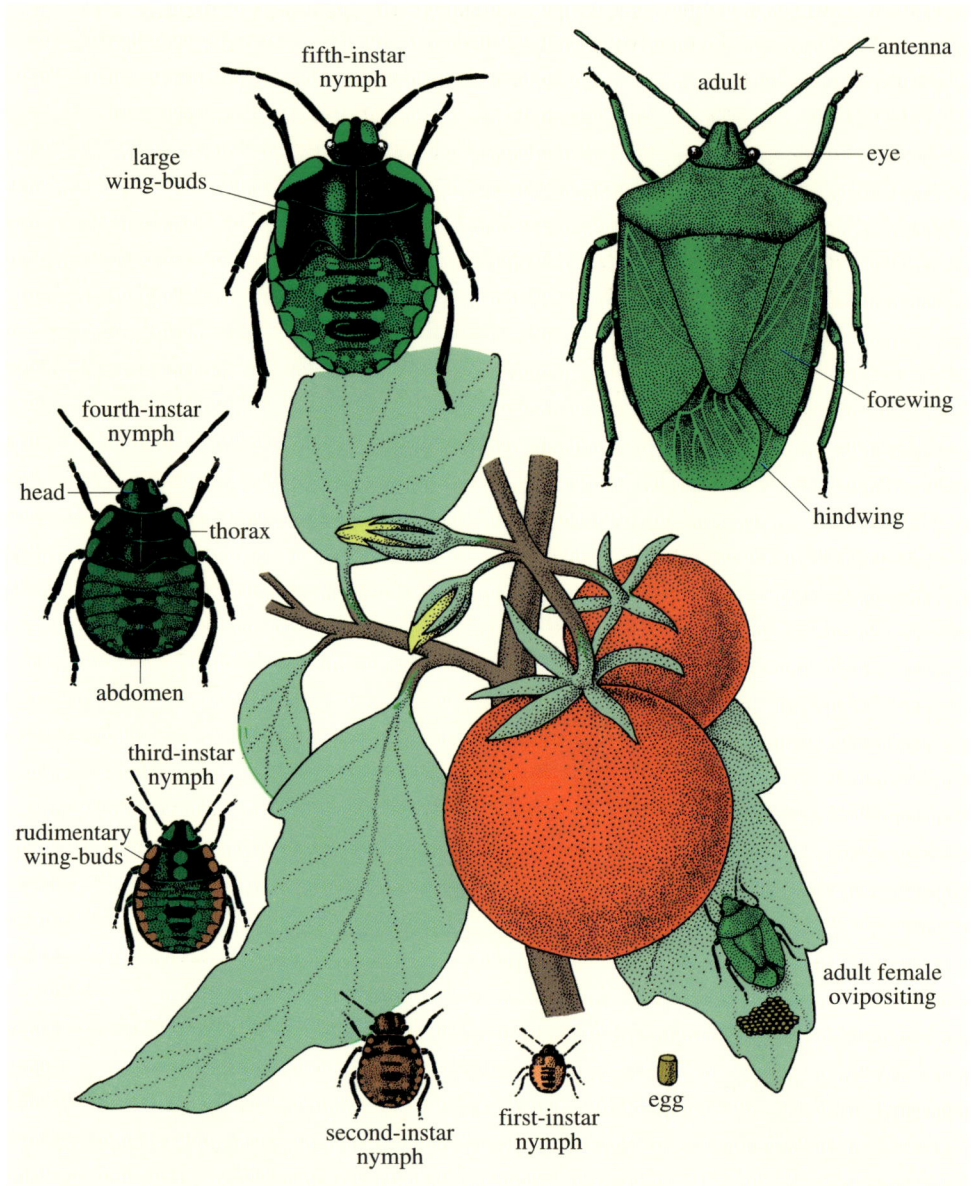

LIFE CYCLE OF THE BLOWFLY AND APPLICATION TO FORENSIC SCIENCE

The study of insect larvae, particularly dipteran larvae, has a very practical application in the field of forensic science. The speed of development of fly larvae (maggots) is primarily determined by temperature. The maggot can increase in size without moulting as the cuticle is thin and flexible, so its age can be determined from the overall length when fully extended. Published graphs of temperature against age in days for maggots of different lengths enable the age of a maggot to be determined quite accurately. For example, a maggot of the blowfly (*Calliphora vicina*) that is 6 mm long is 0.8 days old if it develops at an air temperature of 30 °C but 4.4 days old if the temperature is 10 °C. Where the air temperature is variable, an age range can be estimated from the maximum and minimum temperatures recorded.

The developmental stages of *Calliphora vicina* are shown in Table 2.2. The female lays her first batch of up to 300 eggs four to five days after emerging from the pupa. As the adults do not fly at night eggs are laid during the day, and can be laid on corpses within a few hours of death. The three larval instars are distinguishable from each other on the basis of the spiracles (Section 2.1.2). The first instar has only one slit present in each spiracle, whereas the second instar has two slits in each and the third instar three. When the maggots are mature they leave the corpse and migrate up to 100 m to find a site to pupate. Prior to pupation, the maggot becomes less active, contracts, and forms a prepupa. The cuticle then goes through a tanning process during which it stiffens and darkens in colour. The pupa forms within this puparium. The adult fly emerges from the puparium by inflating a sac on the head with body fluid, rupturing the tip of the puparium and making an exit hole.

Table 2.2 Development stages of *Calliphora vicina* at 22 °C. The figures were collected over five generations of fly (see also Figure 2.16).

Stage	Mean duration	Range
egg	24 hours	20–28 hours
first instar	24 hours	18–34 hours
second instar	20 hours	16–28 hours
third instar	48 hours	30–68 hours
prepupa (non-feeding larval stage)	128 hours	72–290 hours
pupa	11 days	9–15 days
total time immature	18 days	14–25 days

BOX 2.1   FLY MAGGOTS AND THE TIME OF DEATH

Detailed study of the structure and physiology of flies that are associated with corpses can enable a forensic entomologist to deduce much about the history of the corpse after death, as illustrated by an example from the casebook of one of the pioneers in this area, Professor Pekka Nuorteva. He was sent a sample of full-grown maggots of *Lucilia sericata* and *Calliphora vicina*, obtained from a human body that had just been discovered by the police. The maggots were alive so he reared them to adults. On the basis of the entomological evidence he was able to say that (Nuorteva *et al.,* 1967):

1   the person had been dead for more than seven to eight days, as flies first deposit eggs on or around the second day after death and development of the maggots to the migration stage takes five to six days at the air temperature measured at the site of discovery;

2   the body had been in the south of the country in a city, or on a small island, the only places where *Lucilia sericata* occurs in Finland; and

3   the body had been in sunshine, since *Lucilia sericata* does not usually lay eggs onto surfaces at a temperature less than 30 °C.

This type of information is very valuable for the investigation of crime, but it is worth noting the limitations as well. For example, the first conclusion above does not give the real date of death (it was actually 30 days before discovery) as it was not possible to determine, just from the samples provided, that the larvae were from the first eggs laid. Accurate sampling of the insect fauna is necessary and the identification of the samples may be very difficult, particularly if live samples cannot be obtained. Determining the rate of development of maggots on corpses can also be difficult since most crimes are not committed close to weather recording stations and so estimates of temperature need to be made. Forensic entomology, although a very specialized application, is a research area that is growing in importance.

LIFE CYCLE OF MOSQUITOES AND THE TRANSMISSION OF MALARIA

With 120 million new cases a year, malaria is a devastating disease. The causative organism is a protoctist, a parasitic member of the phylum Apicomplexa, from the genus *Plasmodium*. Four species of *Plasmodium* (Table 2.3), all of which are transmitted by mosquitoes, cause malaria. The vector mosquitoes come from a single genus of dipteran fly (two-winged or true fly), *Anopheles*.

Table 2.3   The species of *Plasmodium* that cause malaria.

Species	Occurrence of disease	Specific features of disease
P. falciparum	common in warmer areas of the world	fever recurs at 48-hour intervals; fatal if untreated
P. vivax	more widespread than *P. falciparum*, with a wider temperature tolerance	fever recurs at 48-hour intervals; rarely fatal; relapses can occur up to 8 years after infection
P. malariae	widespread but rare	fever recurs at 72-hour intervals; fatal if untreated; relapses can occur up to 50 years after infection
P. ovale	rare	long incubation period and limited effects

Some 30 of the 400 known species of *Anopheles* spread the four species of *Plasmodium* that infect humans. However, some of these species have little significance as they are geographically highly localized, and some can only transmit the disease when infected experimentally. A species may not be a vector across its whole geographical range, for example *Anopheles gambiae*, which is an African species with a wide distribution across the continent but a more limited vector range. The probable explanation is that the species is a complex of several sibling species, not all of them vectors of malaria. These sibling species are indistinguishable from each other on morphological grounds but are revealed by examining chromosomes in the salivary gland or the nurse cells of the ovary, and also by protein differences detectable in studies using gel electrophoresis for protein separation.

The mosquito life cycle is typical of endopterygote insects, such as other flies, butterflies and beetles, although the aquatic larva is unusual. The egg hatches, producing an air-breathing larva that hangs from the surface film in water, filtering particulate food such as bacteria and protoctists from the surface water (Figure 2.18).

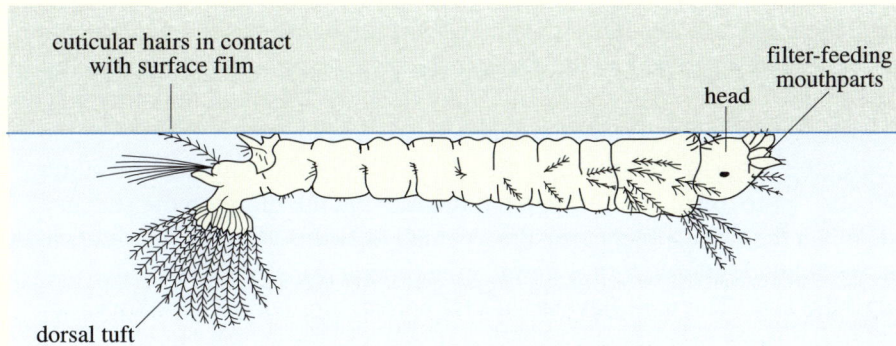

Figure 2.18 The attitude adopted by a mosquito larva in water, suspended on the underside of the surface film.

The larva moults several times as it grows, and at the end of the larval stages a pupal case forms and a massive reorganization of cells takes place within the pupa before the adult emerges. As the pupal stage of the life cycle takes place in water, the adult mosquito has to emerge without getting caught by the surface tension of the water.

The development of mosquitoes can be very fast in the tropics. A female can lay eggs in batches of around 100 every two to three days and the resulting adults emerge 10 days later. Eggs can be laid and develop in any small pool of water containing detritus or microbes, so in seasons of frequent rainfall, the mosquito population can increase very rapidly over a period as short as two weeks. In addition to increasing the number of egg-laying sites, rainfall increases the humidity of the atmosphere, enabling adults to survive for longer. Mosquitoes are not strongly resistant to desiccation, one of the reasons why they are generally active at night.

BOX 2.2 CONTROLLING MALARIA BY DESTROYING MOSQUITO LARVAE

The destruction of mosquito larvae is seen as one of the best ways of controlling malaria, but it is a very difficult task. There are two possible methods, either chemical or biological control. Chemical control has been used in the past, including such methods as spraying watercourses and ponds with the insecticide DDT. Using chemical sprays to change the surface tension of the water is effective as it prevents the larvae hanging from the surface. The drawback of chemical control is that it contaminates the water, and water has many other uses. The preferred method for the future would appear to be some form of biological control. There are many predators that feed on the larvae, particularly fish, but they are unlikely to make a great impact because they cannot colonise small transient bodies of water where mosquitoes often breed. The toxin derived from *Bacillus thuringiensis* could possibly be used, but it is very expensive and, as with other insecticides, not species-specific.

Only the female mosquito feeds on vertebrate blood and she must take a blood meal before ovulating. The structure of the mouthparts (Figure 2.9) enables the female to penetrate the skin of the host and to suck out blood. The mouthparts are shown separated out in Figure 2.9, but in the live insect, the labrum and the labium form a tube surrounding the two mandibles and the hypopharynx. The salivary canal runs down the hypopharynx. The tips of the mandibles are sharp and cut through the skin. If the source of blood is an animal infected with malaria, the mosquito becomes infected only if the blood has the gamont stage of *Plasmodium* present. The gamont stage is the only one that is resistant to the digestive enzymes in the gut of the mosquito (Figure 2.19); all the other stages are digested. Following sexual reproduction of *Plasmodium* within the gut of the mosquito, the zygotes bore through the gut wall and eventually a large number of small cells, sporozoites, are formed that migrate via the blood to the salivary gland. When the female feeds, she ejects small amounts of saliva into the wound made by the mandibles and thus an infected female transfers sporozoites into the blood of the host. The well-developed Malpighian tubules are necessary for excreting the large quantities of excess water derived from the blood.

Figure 2.19   The gut of the mosquito.

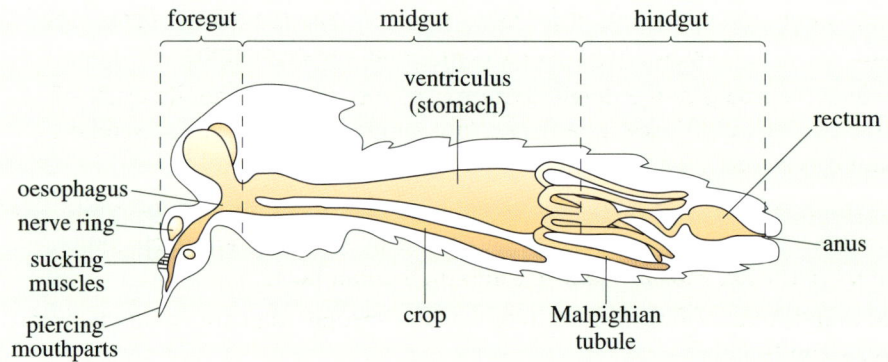

BOX 2.3   CONTROLLING MALARIA BY DESTROYING ADULT MOSQUITOES

The control of adult mosquitoes may be possible through biological means in the future, but at present the most effective way of combating malaria is to prevent mosquitoes biting people. An example of a successful method is one used in China. In 1995, the Bulletin of the World Health Organization (Cheng *et al.*, 1995) reported that 2.42 million bed nets in Sichuan province in China were being treated with deltamethrin each year. Deltamethrin is a synthetic pyrethroid and is a contact insecticide that can produce 100% mortality of mosquitoes. Treating bed nets with deltamethrin can reduce the incidence of malaria by 50%, so keeping mosquitoes away from their human hosts is effective.

Given the problem of the developing resistance of *Plasmodium* to antimalarial drugs, research into the biological control of mosquito populations is likely to continue to be important and may well expand.

SUMMARY OF SECTION 2.3

1   The life cycles of most crustaceans, such as barnacles and prawns, include a planktonic larval stage, with the larva being very different in structure from the adult.

2   Barnacles have a planktonic nauplius larva that provides a dispersal stage in the life cycle, but the behaviour of the cypris larvae when settling out of the plankton to metamorphose into adults is complex and appears to ensure that aggregations of barnacles of the same species are formed.

3   The life cycle of spiders does not include a free-living larval stage — eggs hatch into nymphs, sexually immature miniature spiders. A single female may lay thousands of eggs and the young spiders are quite vulnerable during the dispersal phase.

4   The newly hatched spiders disperse, often by ballooning (being carried aloft on a few silken threads).

5   There are two principal types of larval development in insects. Exopterygote insects have larval stages that are similar in structure to the adult, although lacking functional wings and reproductive organs. In the endopterygote insects, the larvae are very different organisms from the adults. The winged adults are the dispersal stage rather than the larvae or nymphs.

6   An understanding of the detailed timings of the stages of the life cycle of blowflies, e.g. *Calliphora*, has important applications in forensic science, enabling estimation of the time of death of murder victims.

7   Malaria is a devastating disease spread by an insect vector, the mosquito. The female mosquito feeds on human blood and while doing so injects the malaria parasite, *Plasmodium* sp., into the human host.

8   The mouthparts of female mosquitoes show special adaptations that facilitate feeding on blood.

9   Understanding of the life cycle and habits of mosquitoes has helped in development of methods of control. Destruction of mosquito larvae has proved effective although there are problems with this method. To date the most effective method of controlling malaria has been by preventing contact between mosquitoes and humans.

2.4 SENSES

Arthropods have a variety of sensory systems, some of which are highly developed and also ancient, in evolutionary terms. The typical **compound eye** of insects is a very complex organ, yet the eyes of some fossil trilobites appear to have the same structure of multiple facets (Figure 2.20, overleaf) and presumably functioned in the same way. Evidence of this similarity highlights the contrast between the huge diversity of the phylum and the conservation of the body plan, with some features being conserved over enormous stretches of geological time. Compare the trilobite eye with the compound eye of a living insect and the similarities are apparent.

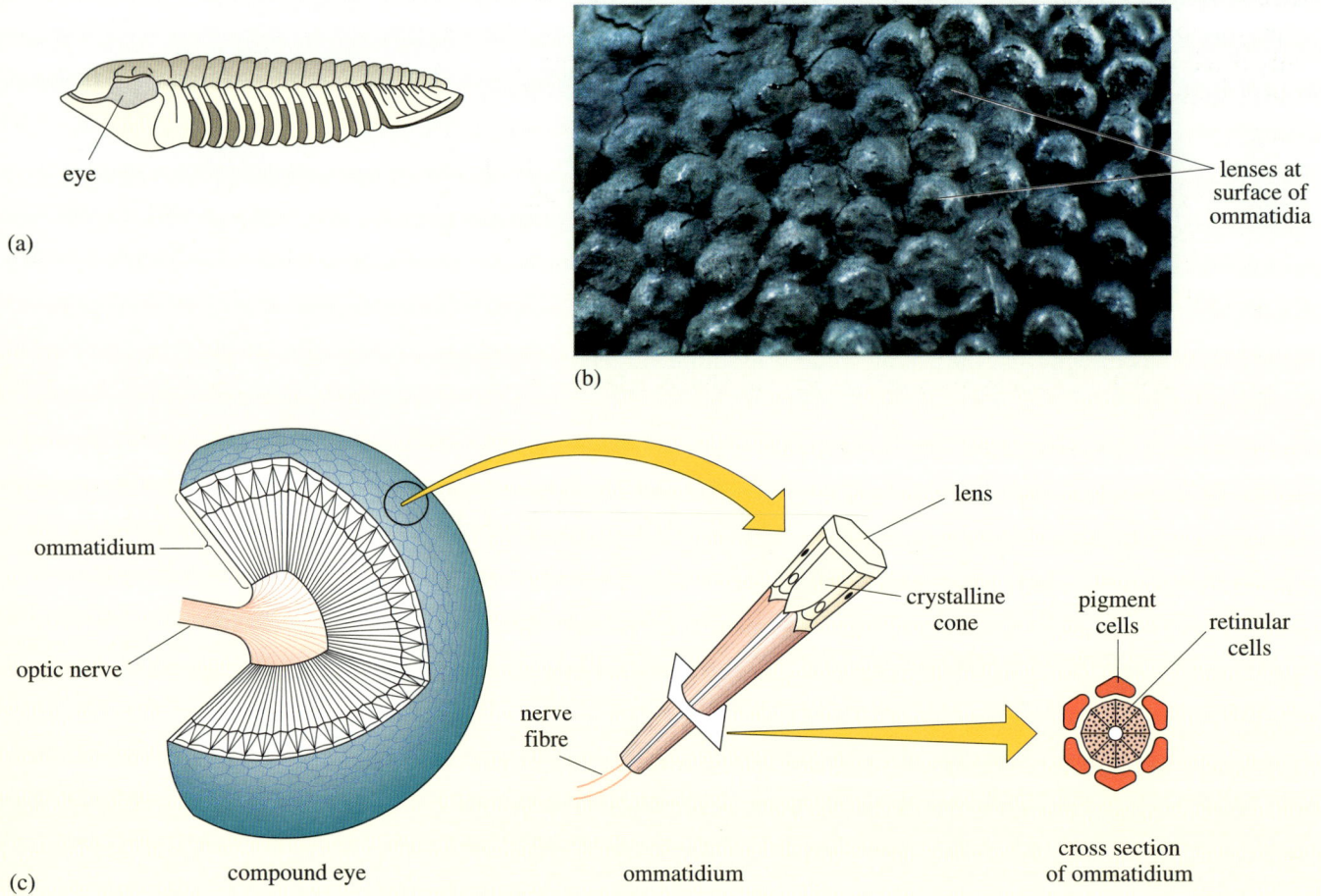

Figure 2.20 (a) and (b) The eye of a Silurian (443–417 Ma ago) trilobite. (a) Lateral view showing position of eye. (b) Close-up showing individual lenses. (c) The compound eye and an individual unit (ommatidium) of a modern insect, sectioned to show the optic nerve and the arrangement of the light-absorbing retinular cells and the surrounding pigment cells that screen each ommatidium from its neighbours.

2.4.1 ARACHNIDS

Spiders have simple eyes with a single lens (Figure 2.6c), in contrast to the compound eyes of insects and trilobites. Typically there are two rows of eyes on the cephalothorax (Figure 2.7). In most species vision is poor, providing information about movements within the immediate environment. However, the hunting spiders and jumping spiders (Salticidae) have much more sophisticated eyes that provide binocular vision, probably detect colour and polarized light and have the greatest visual acuity of any arthropod eye. A typical salticid is the zebra spider (Figure 2.21), which is found throughout northern Europe. It has four large eyes on the front of the carapace and two much smaller eyes on each side. The smaller eyes do not have the sophistication of the eyes on the front and primarily detect movement. If movement is detected, the spider turns rapidly to bring the large eyes at the front to bear on the source of the movement.

For most spiders, sensory hairs all over the body provide much more information than do the eyes, particularly about the small vibrations of the web. There are ultra-fine hairs, each set in a socket, that are very sensitive to air currents and vibration. They may be arranged in rows of hairs of different lengths. The length of the hair governs the way in which it vibrates, so each length is tuned to a particular range of frequencies. This sensitivity to vibration is exploited by some species for communication. Tapping on a surface produces vibrations that are readily detectable, even if that surface is water. Web-spinning spiders are very sensitive to movements in the silken threads of the web that indicate that a passing insect has become trapped, but signals between individuals can also be transmitted along the threads of the web, the individuals plucking at the thread to produce vibrations. Some males are able to produce sound by **stridulation** (the sound produced by rubbing two rough, hard surfaces together, see Box 2.4), but when spiders are close together, touching and stroking play a role in communication.

Some communication between spiders is by pheromones, chemical stimuli produced by an individual and released into the air. Unless released in vast quantities such molecules disperse to give very low concentrations, but the sensitivity to pheromones is so great in certain moths that even a single molecule reacting with a receptor may produce a response. It is not known if spiders are as sensitive as moths, but female wandering spiders, *Cupiennius salei* from Costa Rica, are known to impregnate silk drag-lines with pheromones, which may result in a more controlled spread of the molecules (Coghlan, 2001).

2.4.2 SENSES IN INSECTS

Almost all adult insects have compound eyes, which are made up of many individual units called ommatidia (Figure 2.20d). The eye has good all-round vision but is rather shortsighted. Each ommatidium can form an image from the field of view, suggesting that the insect's compound eye perceives an overall 'mosaic' image. The six to ten nerve cells in each ommatidium are insufficient to form a detailed image. Each ommatidium records light intensity and so the eye is sensitive to pattern, shape and contrast rather than to fine details. The more ommatidia an insect has, the better its perception of movement and the more detail it can see. Adult dragonflies, predators that detect their insect prey by sight, have up to 30 000 ommatidia in each eye. *Aeshna* adults detect the movement of a paper pellet thrown up in the air and turn towards it, but turn away when they see it is not a flying insect.

Insects use sensory information in many ways. Some of the most interesting and sophisticated uses of the senses are associated with the finding of a mate or the avoidance of predators. True sound reception organs, tympanic organs, have been found in species of seven orders of insect, for example grasshoppers, crickets and bush crickets (Orthoptera), cicadas (Hemiptera) and most moths (Lepidoptera). A group of receptors in the antenna of insects, Johnston's organ, acts primarily as a proprioceptor, providing information about the movement of the antenna, but in mosquitoes and midges sound is detected as well. The tympanic organs are very different from the ears of vertebrates and generally consist of a pocket of air that is acoustically connected to the exterior by a thin tympanic membrane. The sense cells that detect the sound are attached directly to the membrane or to some structure associated with it, and the number of cells varies from two in some moths to around 1500 in some cicadas.

Figure 2.21 The zebra spider (*Salticus scenicus*).

The location of tympanic organs on the body is not the same in all groups of insect that possess them. In the crickets and bush crickets, the tympanic organs are located on the tibia of the forelegs, whereas in grasshoppers they are on the first abdominal segment. Some moths have tympanic organs on the abdomen whereas others have them on the thorax.

The positioning of the tympanic organs in moths allows for some sophisticated analysis of sound, even though there are only two receptors (Figure 2.22). The moth can detect the ultrasonic pulses produced by predatory bats when hunting. As the tympanic organs are very sensitive, the moths can detect the bats before the bats are close enough to have picked up the weak echo of their returned signal. The nervous system of the moth compares the signals from the two tympanic organs to determine the direction of the bat. The distance of the bat from the moth is indicated by the rate of firing of the sensory receptors. The louder the sound detected by the tympanic organ, the greater the number of impulses passed down the tympanic nerve.

The two receptors have different thresholds. The A1 receptor fires at very low sound levels while the A2 receptor only responds to loud sounds. Activity in the A2 receptor triggers an escape response in the moth. It plunges down towards the ground and the echo that the bat receives from it is then masked by a myriad of echoes from the ground. The tympanic organs are masked by the wings during each wing beat, which modulates the signal from the receptors in the tympanic organs. Thus the moth can distinguish between a bat that is below it, when the signals from the receptors are not altered by the wing beat, and a bat that is off to one side, where the signal from the receptor increases and decreases in synchrony with the wing beat.

Orthopteran insects make sound for communication between the sexes and for bringing the sexes together for mating. The tympanic organs in the forelegs of crickets and bush crickets are much more complicated than those of moths. The sense cells provide sufficient information to the nervous system for the individual to identify a sound as being made by a member of the same species, and to find their location. Figure 2.23 shows the tympanic organ of the bush cricket *Ruspolia differens*, a common species in Africa. Although the tympanic organ itself is on the foreleg, sound can reach the tympanic membranes by two routes. Each tympanic membrane forms the wall of a trachea, with the outer surface in a cavity open to the air through a slit in the leg, and the inner surface in contact with air in the trachea. The trachea opens to the exterior via the large spiracle on the prothorax, through which sound can enter. The precise mechanics of the system are unclear, but experiments have shown that the tympanic membrane is affected by sound reaching it from both inside and out. Bush crickets have very good directional hearing and good discrimination that enables them to pick out the song of their own species in environments that are full of noise.

Figure 2.22 The tympanic organ of a noctuid moth showing the four air sacs, derived from enlarged tracheae. Each tympanic organ is situated in a cavity between the thorax and abdomen below the hindwing.

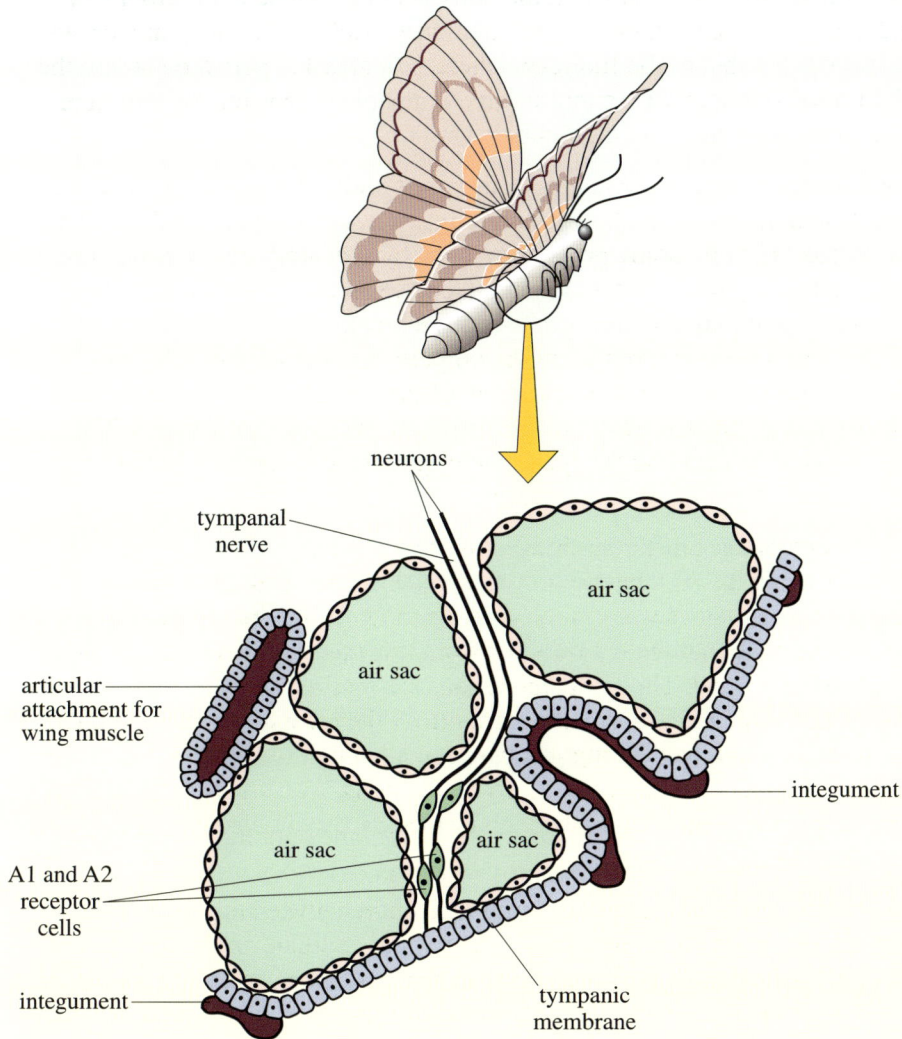

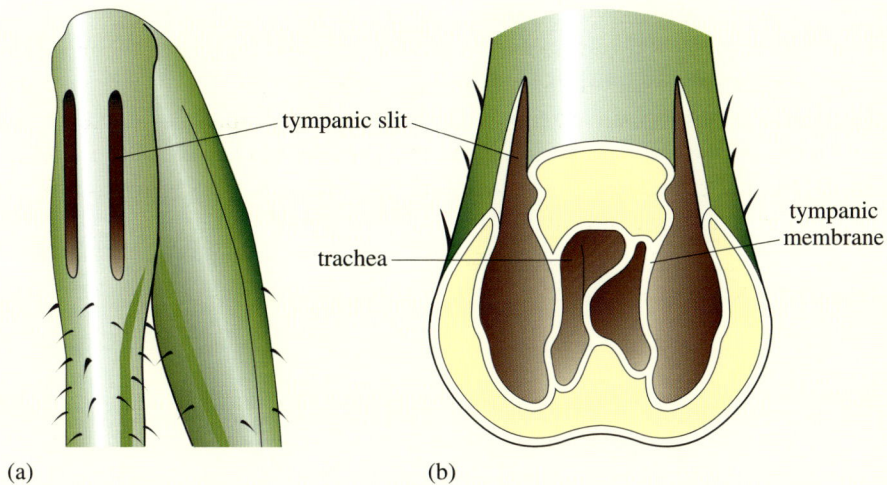

neurons

tympanal nerve

air sac

air sac

articular attachment for wing muscle

integument

air sac

air sac

A1 and A2 receptor cells

integument

tympanic membrane

tympanic slit

trachea

tympanic membrane

(a)

(b)

Figure 2.23 The tympanic organ of *Ruspolia differens*. (a) Surface view. (b) Cross-section.

BOX 2.4   SOUND PRODUCTION

Sound production in insects has evolved many times. Some sounds are produced merely as side-effects of an activity such as feeding or moving, and have no known behavioural significance. Other sounds do have a purpose but are not produced by specialized organs. The warning hiss produced by the cockroach *Gromphadorhina portentosa* is a deterrent to predators, and is produced by the sudden expulsion of air from the spiracles. Among the specialized organs that produce sound, frictional ones are the most common. Two surfaces of cuticle are rubbed together producing sound, a process commonly called stridulation. In male bush crickets, a scraper on one wing is dragged over a line of teeth on the other wing as the wings move past each other, producing the characteristic song. The wing movements for stridulation are different from the flight movements since during flight the wings do not contact each other. The sound produced by stridulation can be complex as it is influenced by the resonances from different areas of the wings. In a few species of bush cricket the females also produce sound, but the sound-producing apparatus has evolved separately from that of the male. As the interchange of songs between males and females is part of the process that brings them together prior to mating, some elements of sound production and reception in the sexes must have coevolved (see Section 2.7).

SUMMARY OF SECTION 2.4

1   Spiders have simple eyes typically arranged in two rows on the cephalothorax.

2   While most species of spider have poor vision, the hunting and jumping spiders have sophisticated eyes providing great visual acuity.

3   Sensory hairs on the body of spiders detect small vibrations, both in the air and the web.

4   A few spiders are known to communicate by pheromones.

5   Insects have compound eyes made up of multiple units, ommatidia. As each ommatidium points in a slightly different direction, each looks at a slightly different part of the environment, and it is likely that an insect sees a mosaic image.

6   Sound reception organs, tympanic organs, are found in seven insect orders. Their structure is very different from the vertebrate ear.

7   Location of the tympanic organs is not the same in all groups of insects that have them.

8   The tympanic organs of moths are used to detect their predators, bats.

9   Orthopteran insects use sound for communication as a means of bringing males and females together for mating.

2.5   TERRESTRIAL PREDATORS

There are a large number of predatory arthropods. All spiders are carnivores, preying mostly on other arthropods, mainly insects. Predatory behaviour has evolved independently in several insect groups, such as the dragonflies

(Odonata), some families of beetle (Coleoptera) and praying mantids (Dictyoptera). Predatory insects often show great agility and speed, for example the robber flies and dragonflies that catch their prey in flight.

2.5.1   PREDATION IN SPIDERS

Spiders are well known for their ability to build webs, sticky nets that trap flying insects. Predation by web-spinning spiders is described by Pond (2001). Unlike web spinners, hunting and jumping spiders (Salticidae) do not sit and wait for their prey — they actively stalk an insect and jump on it.

The zebra spider (Figure 2.21) is very active in warm weather and is often seen hunting for food on walls. When the spider jumps, either to catch its prey or to escape from danger, it uses the third and fourth pairs of legs. The legs are first drawn into the sides of the body, while the spider lays down one end of a silk thread that acts as a dragline when it takes off. The flexor muscles in the leg are relaxed at this stage. The legs then extend very rapidly as haemolymph is pumped into them, and the spider is projected forward. Some small species of spider can jump as much as 20 body lengths, which is a large distance for spiders but is less than the distances that some insects can jump.

There is a species of snake, the spitting cobra, which can eject venom from the fangs, disabling at a distance of over a metre. Similarly, there is a spitting spider, *Scydotes*, which can eject a liquid from the fangs. The liquid is produced in a set of glands in the head and the fluid is glue-like. It can be squirted to a distance of over 1 cm and it sticks the prey down, enabling the spider to capture it.

2.5.2   PREDATORY INSECTS

Praying mantids and dragonfly larvae are stealthy predators that remain motionless until a meal is within reach and then strike with astonishing speed. There is impressive neural coordination in the mantis. The forelegs are modified for prey capture; the contraction of the leg is directed using information from the compound eyes about distance and feedback from the sensory spines in the neck region about the orientation of the head in relation to the longitudinal axis of the body.

For fast strikes, the absolute distance to the prey must be known. The dragonfly nymph extends the labium under hydrostatic pressure and the extension distance is fixed. As a consequence the dragonfly has to be in exactly the right position relative to the prey for the strike to be successful. The mantis has a little more flexibility, as the forelegs do not strike to an absolutely fixed distance, though clearly they have a maximum range. The mantis can jump towards the target before striking, and the mechanism used to estimate distance has been studied. The position of the eyes on the head suggests that mantids have stereoscopic vision. The first demonstration that they had stereoscopic depth perception came from work in which tiny prisms were attached to the eyes of a praying mantis. The mantis then failed to estimate the distance to its prey correctly, and missed by precisely the amount calculated from the vertical disparity caused by the prisms. More recent work, however, has shown that mantids can distinguish between targets on the basis of their distance. A mantis nymph was placed on a small,

Figure 2.24 The absolute number of jumps at a target for a total of 14 second-instar mantids of the genus *Polyspilota*. Data from Poteser and Kral (1995).

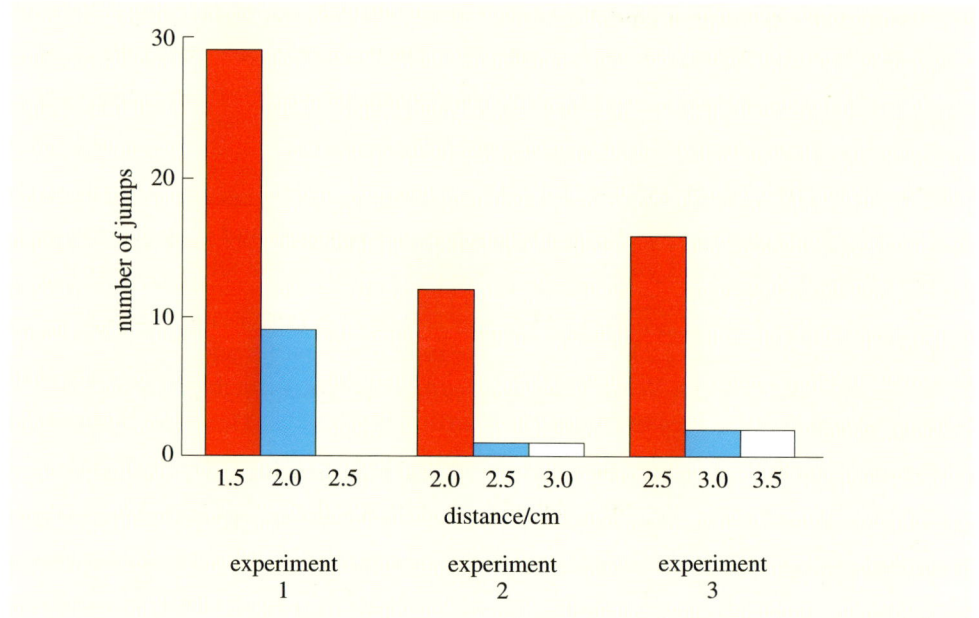

circular platform with three black, oblong targets at the same height above the ground but at different distances from the mantis. The target that the mantis jumped to was recorded. The results of three experiments are shown in Figure 2.24.

○ What conclusion can you draw from the results shown in Figure 2.24?

● The results show a strong preference for the closest of the three targets irrespective of absolute distance.

○ How do you know that the jump is not based only on absolute distance?

● The same absolute distances (2.0, 2.5 and 3.0 cm) appear in at least two of the experiments, but the results are very different. For example, in Experiment 2 the target at 2.5 cm was largely ignored, whereas in Experiment 3, that distance produced the greatest number of jumps.

It is clear that, offered a choice between stationary targets at different distances, the mantids choose the nearest. How do they estimate distance?

The two researchers conducted a series of experiments that showed that the peering movements that the mantids made prior to jumping were probably part of the distance measurement process. The rate at which the image of a distant object moved across the retina of the eye would be less than the rate for a nearer object. Interestingly, the research also showed that the mantids retained a memory of distance for long enough to assess each target in turn and decide which was the closest.

Certain insects trap their prey. The ant-lion larva digs a shallow, cone-shaped pit in the ground where the soil is fine or sandy. The ant-lion buries itself at the bottom of the cone-shaped pit and waits until a small insect either enters the pit or stands on the rim, making its presence known by disturbing sand grains. The ant-lion responds by creating a landslide into the centre of the pit, bringing the prey within reach of its mouthparts. The aquatic larvae of some caddis flies build silk nets through which the water can flow. Small organisms get swept into the net, trapped and eaten.

SUMMARY OF SECTION 2.5

1   Most of the terrestrial arachnids are predators. Predation has evolved independently in several orders of insect.

2   Hunting and jumping spiders use their eyes to detect prey — they stalk insects and jump on them.

3   Praying mantids and dragonfly nymphs remain motionless until a victim is within reach and then strike rapidly.

4   Praying mantids have stereoscopic vision and can measure distance very accurately. They also appear to be able to compare distances, which implies the existence of memory.

5   Certain species of predatory insect trap their prey, e.g. the ant-lion larva.

2.6   MASTERY OF THE AIR

At some time during the Carboniferous period, around 300 million years ago, a series of evolutionary steps took place that produced the first flying insects. It is arguably this single development that has led to the enormous diversity of species that is found today. It is probably quite appropriate to call this development unique, as the structure of the wings of all orders of flying insects appears to be derived from a common ancestral form.

Fully developed wings in insects are only found in adults (or, in the exceptional case of mayflies, final instar nymphs). A typical wing is a thin sheet of cuticle supported and strengthened by a number of tubular veins. The major veins are conduits for tracheae and nerves (Figure 2.25). The space between the epidermal layers of the wing, where they expand to form a vein, is continuous with the haemocoel and carries haemolymph.

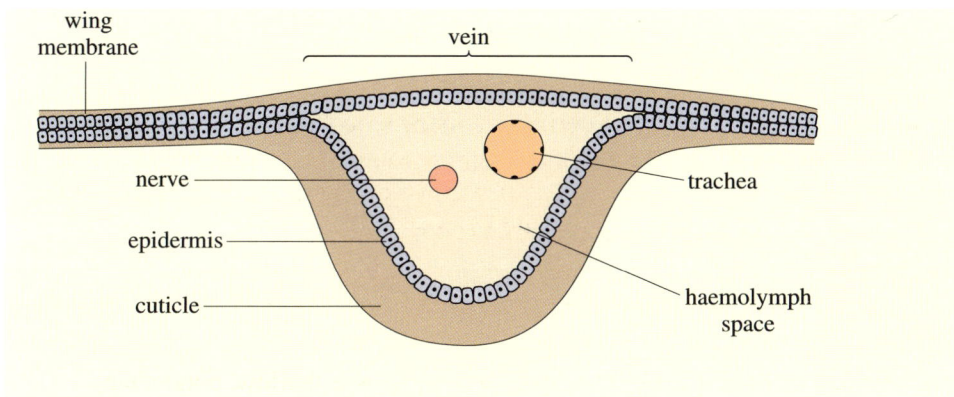

Figure 2.25   Cross-section through a wing vein.

Some examples of wing shapes are shown in Figure 2.26. Usually, insects have two pairs of wings, located on the second and third thoracic segments. However, there are many modifications to this basic structure. The true flies (the Diptera) have only the anterior pair of wings and the posterior pair is reduced to small balance organs, the halteres. The veins on the anterior part of the dipteran wing (Figure 2.26b) are thicker and closer together and radiate away from the point of attachment to the thorax, where the flight muscles act. This pattern of venation strengthens the wing at the point where the greatest forces are exerted. In other orders of insect the wings have also been modified during evolution. Grasshoppers and cockroaches have thick, leathery forewings (tegmina), while the hindwings are larger and much thinner (Figure 2.26a). The forewings of beetles are hardened and form protective covers (the elytra; singular, elytrum) over the more delicate hindwings when the insect is not flying (Figure 2.26c).

Figure 2.26 Examples of insect wings. (a) The left forewing and hindwing of a locust. (b) The left forewing of a dipteran fly. (c) The left elytrum of a beetle.

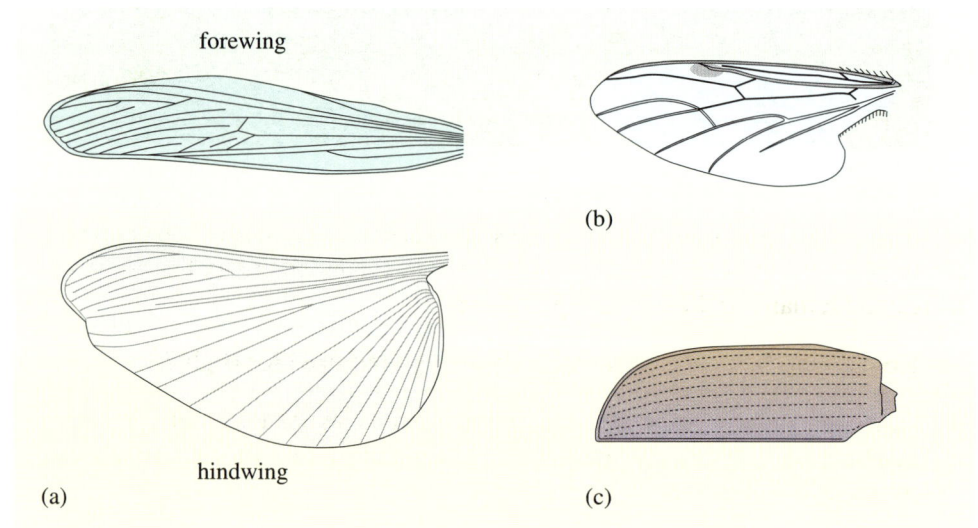

In termites and ants, which are social insects, there are wingless forms and winged reproductive forms. The winged forms, the alates, have large wings that are shed immediately after the nuptial flight. Some insects have become secondarily flightless and have wings that are very reduced (Figure 2.27). Two orders of insect that diverged from the rest before the evolution of flight have never had wings, but all the fleas and lice have secondarily lost their wings. Those insects that have never had wings form a group called the Apterygota, which includes several small orders of silverfish, bristletails and others. All the other insect orders are grouped as Pterygota.

2.6.1 FLIGHT AND TEMPERATURE

Wings are sometimes preserved in the fossil record and there are particularly fine specimens of dragonfly from the Carboniferous coal deposits (354–290 Ma). The fossil dragonflies appear very similar to present-day forms, except that some species were giants. The largest species of the fossil genus *Meganeura* had a wingspan of about one metre, ten times larger than the span of the largest dragonfly that exists now. All the evidence from the fossils suggests that *Meganeura* was a fast-flying predator like modern dragonflies, which raises some interesting questions about the energy developed in the flight muscles.

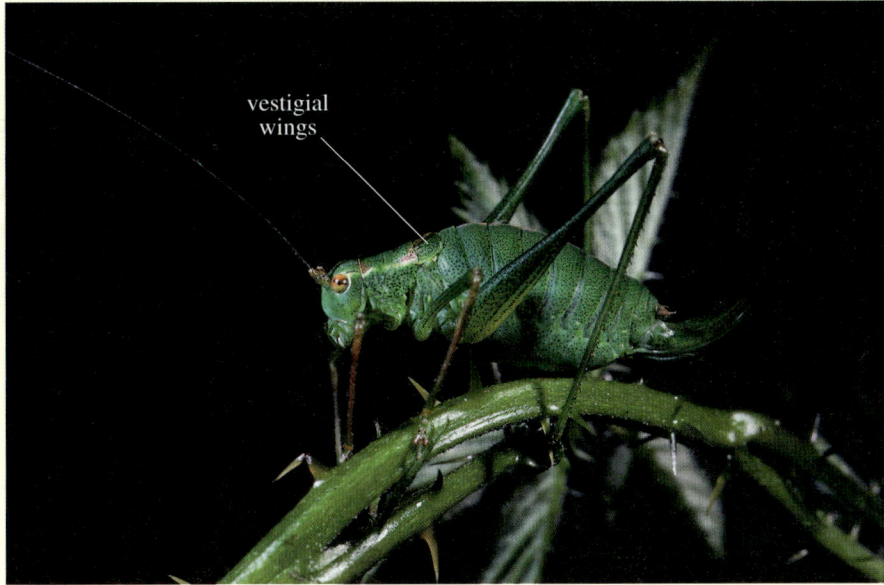

Figure 2.27 Adult of the bush cricket (*Eupholidoptera chabrieri*), an insect that has become secondarily flightless.

A general study of insect flight muscle suggests that about 6% of the energy developed by flight muscles is converted into mechanical energy that moves the wings; the remaining 94% is dissipated as heat.

○ How do the flight muscles obtain the energy required for flight?

● Glucose and/or lipid is broken down in the muscle tissue. Breakdown of fuel generates ATP in muscle mitochondria, which provides the energy required for contraction of the muscles.

Insect flight requires a large amount of energy, and with so much heat generated within the thorax there is a risk of overheating.

○ Which insect, a large dragonfly or a small fruit-fly, would you expect to have a lower thoracic temperature during flight?

● The fruit-fly. Intuitively, you might have expected the larger insect with the larger flight muscles to produce more heat and therefore have a higher thoracic temperature. The smaller insect has a larger surface area relative to volume, and so heat is lost to the air by convection much faster than in the larger insect.

Measurements of bees in flight have shown that only the large species heat up substantially in flight, with a measured thoracic temperature of 15 °C in excess of air temperature. For comparison, the temperature in the thorax of a flying mosquito was less than 1 °C above ambient. This effect is not related to the amount of energy produced per unit mass of flight muscle. For a mosquito, the power output of the muscles is large and a lot of heat is produced. It is the rate at which heat is lost to the air that keeps the thorax close to air temperature.

○ What temperature would you expect the thorax of a large bee to be at on a warm day in a temperate climate?

● On a warm day, the air temperature might be as high as 25 °C. The temperature in the thorax would be 15 °C above that, at 40 °C.

A temperature of 40 °C is not excessive, but a further rise would get close to the point where some proteins would start to denature. In even larger insects the thoracic temperature would certainly rise above that at which proteins would be damaged, if there were not ways of counteracting the temperature rise. Michael May studied the flight behaviour of the large green darner dragonfly, *Anax junius*, which weighs about a gram. When the temperature in the thorax is high, the insect switches to a form of gliding flight, which reduces the energy required for wing movement and reduces heat production by about 30%. In addition, heat is shunted into the abdomen, which is long and narrow and can function as a radiator.

The fossil dragonfly *Meganeura* would have had much more of a problem with heat load, as a consequence of its size. May has extrapolated his data from *Anax* and attempted to work out the thermal relationships of *Meganeura*. He estimated that in flight, *Meganeura* would have had a thoracic temperature that was 62 °C in excess of air temperature. On a hot day, flight would have raised the thoracic temperature close to the boiling point of water, clearly an impossible temperature. But even in an air temperature of 0 °C, the thorax would still have been too hot. We do not know how the dragonfly managed to reduce the thoracic temperature to a safe level, but given the size of the wings it might have spent much of its time gliding rather than in flapping flight. In addition, the size of the wings, and in particular the diameter of the wing veins, would have allowed much larger flows of blood through the vessels inside the veins than is possible in present-day dragonflies. So it is possible that the wings also acted as radiators.

2.6.2 WING-BEAT FREQUENCY

In gliding flight, the airflow over the outstretched wings is sufficient to provide lift that counteracts the forces of gravity. The leading edge of each wing must meet the airflow with a positive angle, called the angle of attack. This angle can be as much as 50°, which is much more than the maximum of 20° that an aircraft wing can sustain without stalling. Most insects are too small to make use of gliding flight and fly by beating their wings. The rate at which the wings beat shows large variations between species. For example, large butterflies have a wing-beat frequency of around 5 Hz (5 times each second) whereas some midges beat their wings at almost 1000 Hz. Table 2.4 shows some typical wing-beat frequencies and Table 2.5 some typical flight speeds.

A single wing beat has several components. In the locust, the cycle starts with a downward and forward movement and the tip of the wing starts to trace out a path in the form of a figure-of-eight (Figure 2.28). The wing then moves upward and backward, rotating around the point of attachment to the thorax, which adjusts the angle of the wing, producing power and lift on both the down stroke and upstroke. The wing may also flex during the cycle, in response to localized pressure changes, thus altering the aerodynamic characteristics.

Table 2.4 Typical wing-beat frequencies of various species of insect (collected data).

Order	Genus	Wing-beat frequency/Hz
Odonata	*Aeshna* (dragonfly)	22–28
Coleoptera	*Coccinella* (ladybird)	75–91
Diptera	*Aedes* male (mosquito)	587
	Culex (mosquito)	278–307
	Musca (housefly)	190–330
Hymenoptera	*Apis* (honeybee)	190–250
	Vespa (wasp)	110
Orthoptera	*Locusta* (locust)	18–24

Table 2.5 Typical flight speeds of some insects (collected data).

Order	Insect	Flight speed/km h^{-1}
Odonata	large dragonfly	25.2
Diptera	mosquito	3.2
	housefly	6.4
	blowfly	11.0
Lepidoptera	hummingbird hawk moth	18.0
Hymenoptera	bumblebee	3.0
	honeybee	22.4
Orthoptera	small grasshopper	1.8
	desert locust	16.0

○ Do the data in Tables 2.4 and 2.5 indicate a direct relationship between frequency of wing beat and speed of flight?

● No. *Aeshna*, a dragonfly, and the locust appear to show a similar relationship between wing-beat frequency and flight speed. Wing-beat frequency of *Aeshna* is 22–28 Hz; flight speed is 25.2 km h^{-1}. The locust has a wing-beat frequency of 18–24 Hz and a flight speed of 16 km h^{-1}. However, the honeybee, with a flight speed of 22.4 km h^{-1}, in the same range as the dragonfly and the locust, has a wing-beat frequency of 190–250 Hz. The mosquito, *Aedes*, has a wing-beat frequency of 587 Hz and a flight speed of only 3.2 km h^{-1}. Therefore there is not a clear relationship between frequency of wing beat and speed of flight.

Figure 2.28 The path traced out by the tip of the left forewing of a locust during flight. The wing itself has been omitted, for clarity. The numbered points on the line indicate successive positions of the wing tip during the wing-beat cycle.

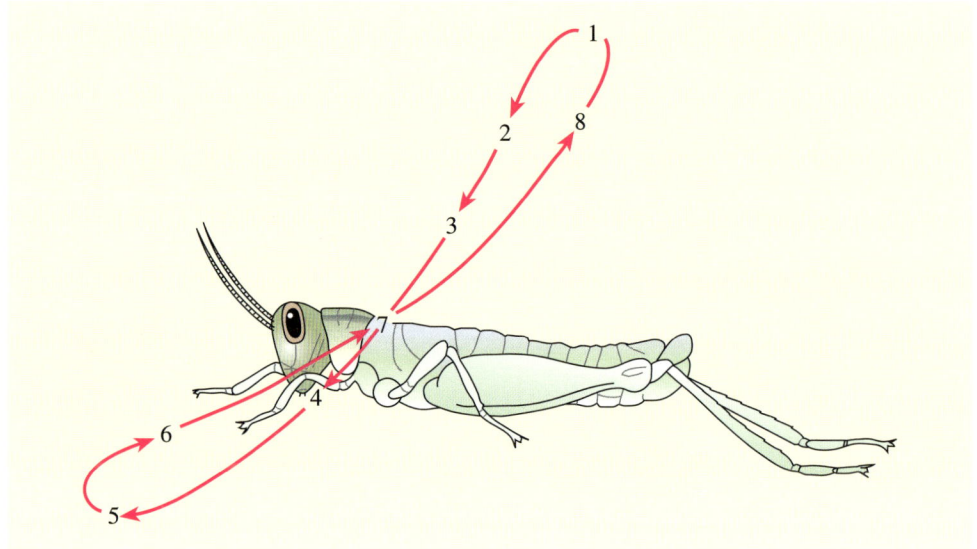

2.6.3 THE FLIGHT MUSCLES

There are two types of arrangement of the muscles that provide the power for flight. In dragonflies a pair of flight muscles is attached *directly* to each of the wings. The elevator muscle raises the wing when it contracts, which stretches the depressor muscle. Then the depressor contracts, lowering the wing and stretching the elevator muscle. This cycle is illustrated in Figure 2.29a and b.

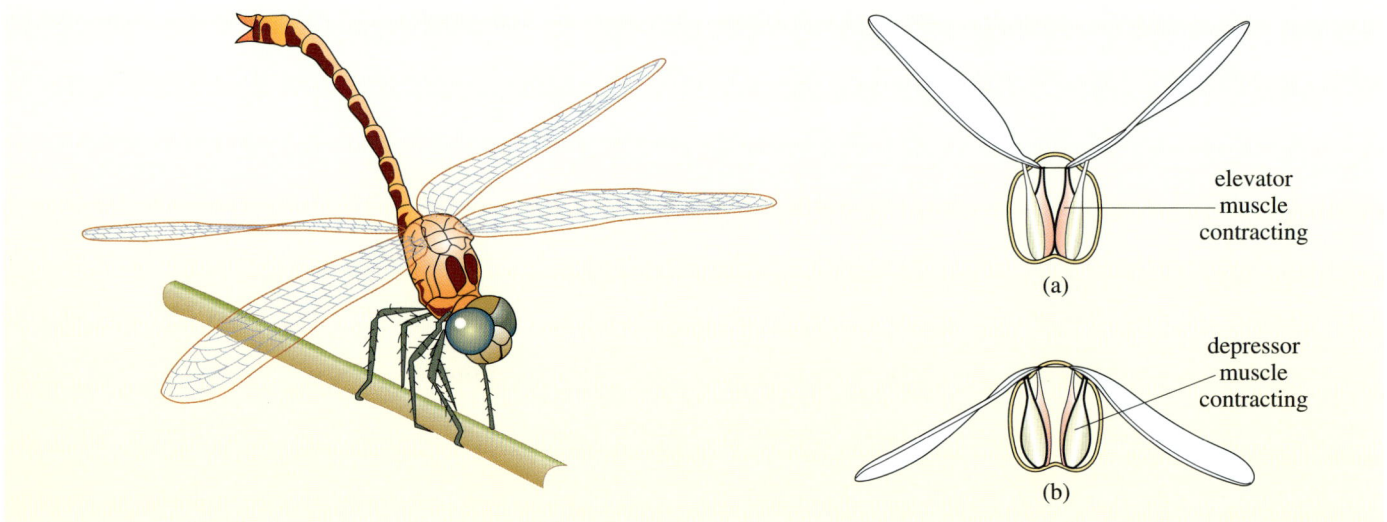

Figure 2.29 The direct flight muscles of the dragonfly at (a) the peak of the upstroke of one pair of wings and (b) the end of the downstroke.

In insects such as the dipteran flies, the flight muscles are indirect. The thorax acts like a distortable box, with one pair of muscles squeezing the top and bottom of the box together and the other pair squeezing the ends together. The movement of the thorax and the wings is illustrated in Figure 2.30. The structure of the wing joint is complex and although the basic wing beat is produced by the indirect muscles, direct muscles are present and also contribute to wing movements. The direct muscles attached to the basalar and the subalar sclerites (Figure 2.31)

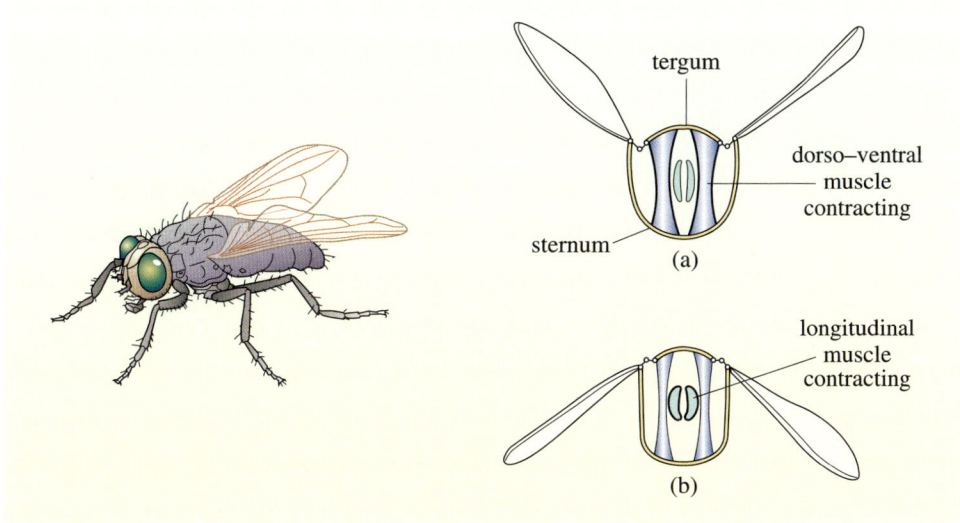

tergum

dorso–ventral muscle contracting

sternum

(a)

longitudinal muscle contracting

(b)

Figure 2.30   The action of the indirect flight muscles in a housefly. In (a) the dorso–ventral muscles have contracted, pulling the upper cuticle, the tergum, closer to the lower cuticle, the sternum, and flipping the wings up. In (b) the longitudinal muscles have contracted, bowing the tergum upwards and flipping the wings downward.

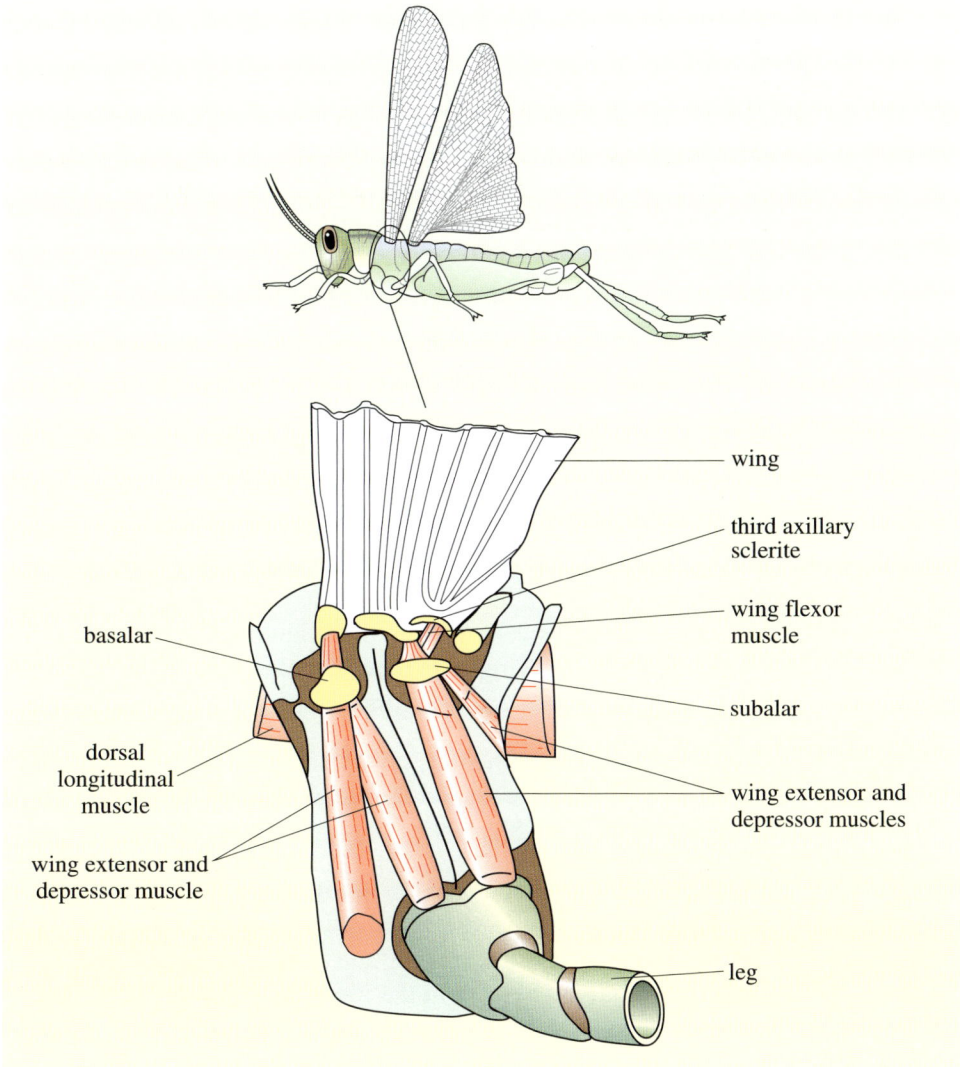

wing

third axillary sclerite

wing flexor muscle

subalar

basalar

dorsal longitudinal muscle

wing extensor and depressor muscles

wing extensor and depressor muscle

leg

Figure 2.31   A lateral view of the thorax of a typical ptergyote insect dissected to show the direct wing muscles.

extend and depress the wing. Contraction of the muscle attached to the third axillary sclerite produces flexion of the wing, to allow it to fold over the abdomen. Most pterygote insects can fold the wings back over the abdomen, the exceptions being dragonflies, damselflies and mayflies.

So far the account has descibed events associated with one pair of wings. Most insects have two pairs of wings but there is a clear trend in the evolution of insects for the two pairs to be reduced to a single functional unit. A. K. Brodsky reviewed the evolution of insect flight in 1994 and he concluded that the pressure for a transition from flight using four wings to flight using wings that were functionally two was due to the inefficiency of the hindwings. The hindwings lie in the path of the turbulent air that spills off the trailing edge of the forewings and so have reduced aerodynamic efficiency. When the forewings move before the hindwings, as in locusts, the hindwings are moving through turbulent air. However, in dragonflies the wing beats are reversed and the hindwing moves before the forewing and is less affected by turbulence.

Dipteran flies have only one pair of wings. However, other insect groups with two pairs of wings have adaptations that enable the two pairs to function as a single unit. The fore- and hindwings of the honeybee are physically linked. A line of hooks (the hamuli) on the leading edge of the hindwing engages with the trailing edge of the forewing (Figure 2.32). The hindwing is smaller than the forewing and this reduction in size is also seen in other groups of insects that have wing-coupling mechanisms, for example the moths. Although beetles have two pairs of wings and they are not coupled together, the forewings are so modified as elytra (see Figure 2.26c) that they generally play no part in propulsive flight, although they do act as an aerofoil, generating lift.

In some insects having two pairs of wings, there are adaptations to the control systems, which enable synchronous movements of the two pairs of wings. This method of control is not as effective as physically coupling the wings together and is likely to have developed earlier in the evolution of flight.

2.6.4   THE CONTROL OF FLIGHT

The patterns of wing movements in insects have been studied using high-speed cameras and stroboscopic light. Large insects like locusts can be tethered in the beam of a stoboscope and electrodes inserted into the flight muscles (Figure 2.33a and b). The movements of the wings can then be linked to the electrical signals recorded in the muscles, as shown in Figure 2.33c.

○ What is the delay between the electrical activity in the two depressor muscles (numbers 128 and 98) in the hindwing and the forewing in Figure 2.33c? Does this delay bear any relationship to the wing movements?

● The delay is about 7 ms. This matches the phase difference between the hindwings and the forewings, that is, the time between two identical points in the cycles of the two wings. The peak upward movement in the hindwing occurs at 11 ms and the equivalent peak in the forewing occurs at 18 ms.

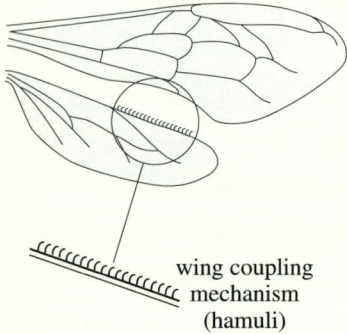

Figure 2.32   The wings of a honeybee, separated to show the position of the wing-coupling mechanism.

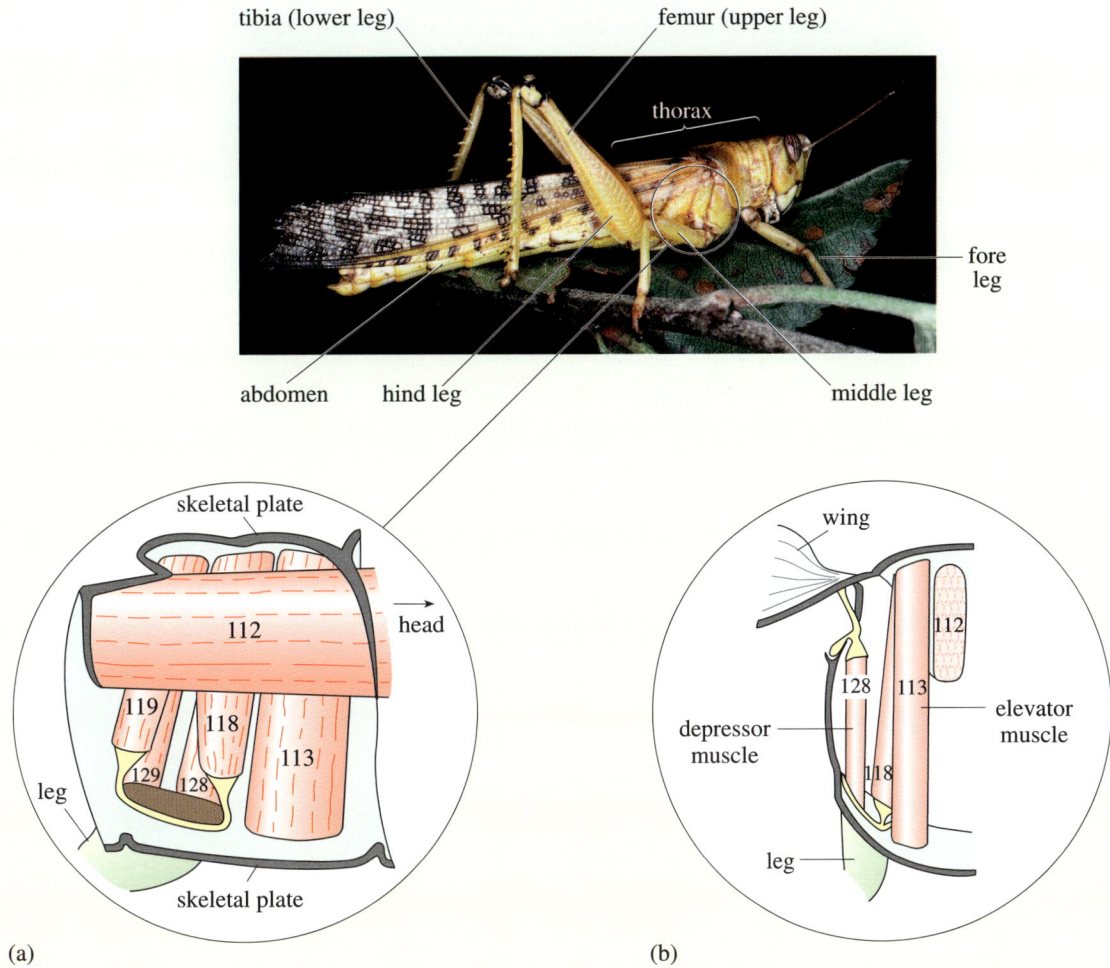

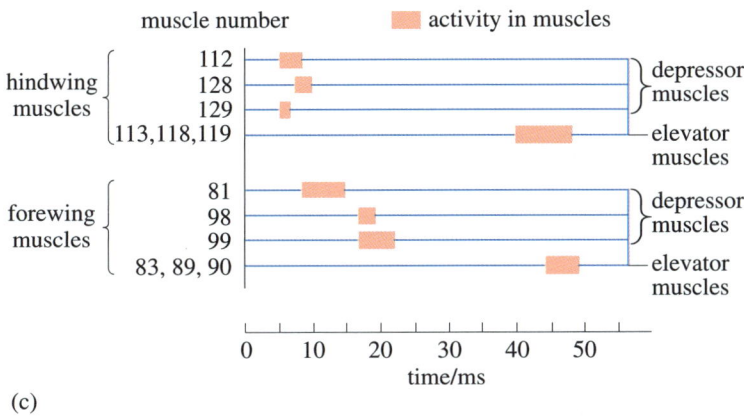

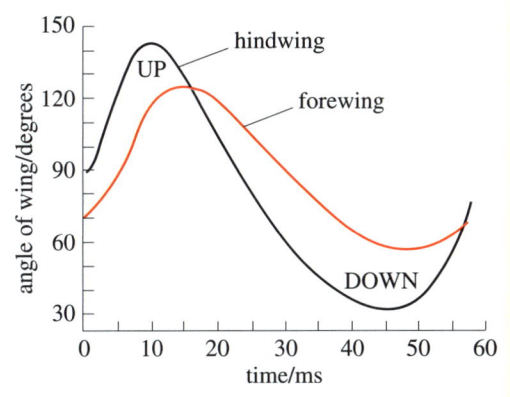

Figure 2.33   (a) A longitudinal section through the metathorax of a locust showing the major flight muscles. (b) A transverse section through the same region. The muscles are numbered according to a standard anatomical system. (c) The pattern of electrical activity in major muscles of a locust during flight and the wing positions during one complete cycle of the wing beat.

Maintaining the timing of contractions in the muscles requires sophisticated control by the nervous system. To study the structure of the key **neurons** in the flight system, a single neuron is penetrated using a glass microelectrode full of stain, for example a cobalt salt. The stain spreads only through that neuron, so it is possible to stain individual neurons and study their three-dimensional structure and the connections to other neurons. Such a stained neuron in the central nervous system of the locust is shown in Figure 2.34a.

The major features of the locust nervous system are shown in Figure 2.34b. The ventral nerve cord is expanded in each segment to form a ganglion. Most neuronal cell bodies are located in the ganglia and sensory information is routed to the ganglia, while motor neurons branch out of the ganglia to muscles. The **interneurons** are found only in the ganglia and nerve cord and the connections that they make enhance integration within the nervous system.

Figure 2.34   (a) A drawing of a stained preparation of a single neuron in a thoracic ganglion of a locust. Each neuron has a single axon, a cell body and dendritic tree of many dendrites. Note that this image is not a section through the ganglion, but is a whole ganglion, so you see the complete three-dimensional structure of the neuron. (b) The gross structure of the anterior central nervous system of the locust.

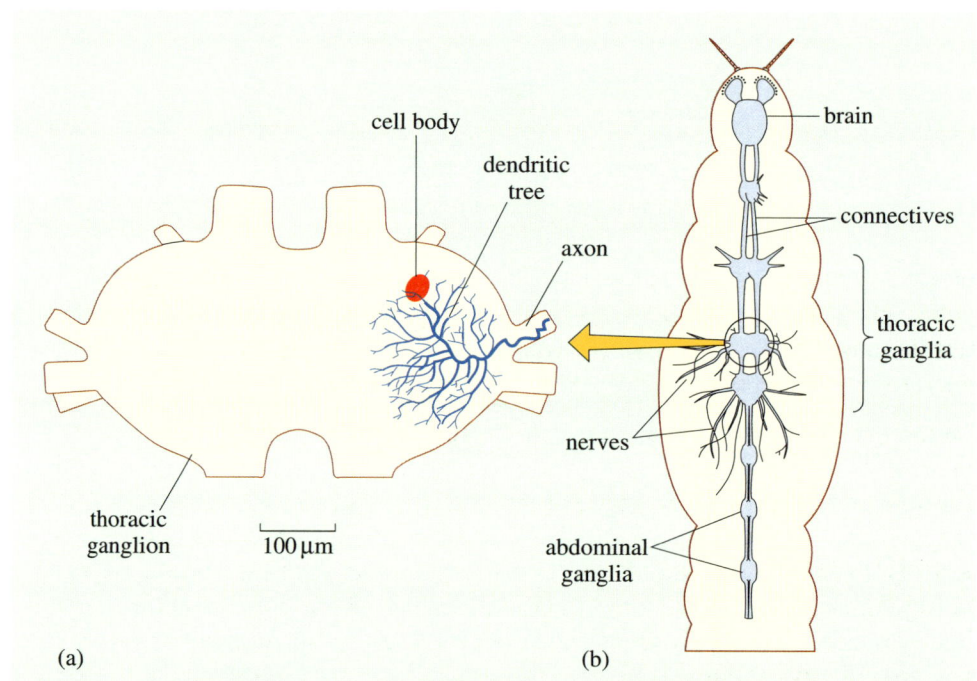

The control of flight has been studied extensively in the locust, partly because of its size but also because of its economic importance. With advances in miniaturization of electronic components it is now possible to mount a radio transmitter on a free-flying locust and obtain true in-flight data. However, much of the detailed study of the neuronal elements of the flight control system still has to be carried out on tethered animals. There is evidence that a central pattern generator composed of interneurons in the second and third thoracic ganglia controls the basic flight pattern (Figure 2.35) by providing a framework of signals for the generation of the wing-beat cycle. The fine-tuning of the pattern and the matching of it to external conditions is carried out by a group of flight interneurons. For example, information about the state of the wings is fed back from sense organs within the wings and the flight interneurons modify their output as a consequence. The air currents produced by the wing beats are also

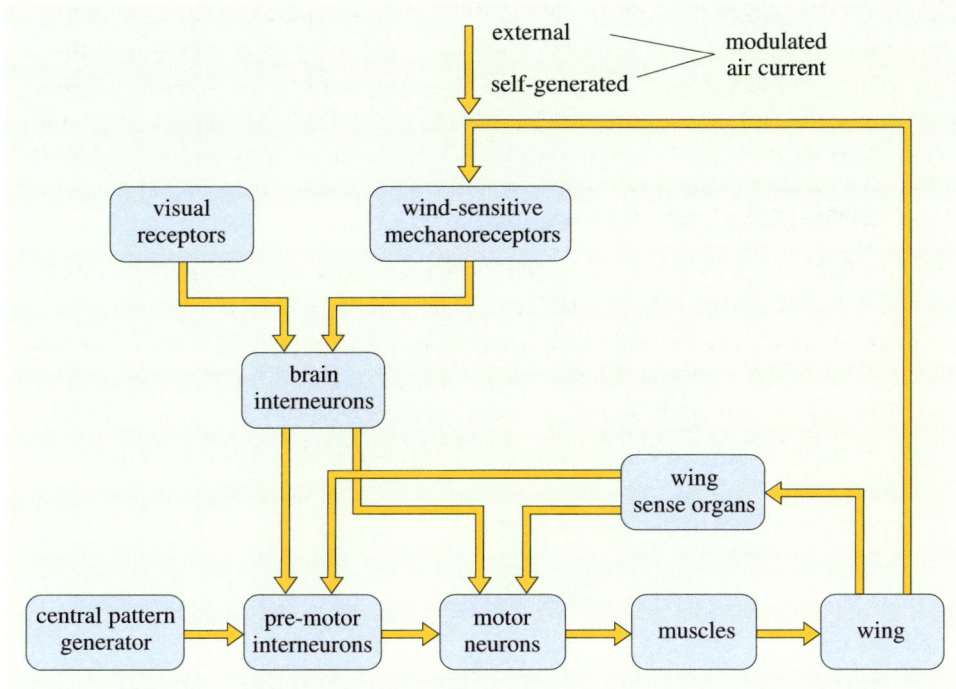

Figure 2.35   A highly simplified diagram of the functional relationships between the neural elements involved in the control of the flight pattern generator in locusts.

detected by hair receptors on the head and, as the pattern of air movement is cyclic, information about the wing-beat frequency is fed back to the flight interneurons. Their output is an integration of the input from the central pattern generator and information derived from sensory systems.

The movements of the wings in locusts are synchronized with the stimulation of the muscles by the nervous system. The maximum wing-beat frequency in locusts is around 24 Hz, and similar frequencies are found in dragonflies (see Table 2.4 for typical wing-beat frequencies). At much higher wing-beat frequencies, such as the 190 Hz of houseflies, the frequency of the stimulation of the flight muscles by the flight neurons is much lower than the frequency of wing beats, and not directly related. This type of muscle is called asynchronous. In dipteran flies, there are only two stable positions of the wings, fully up or fully down. Through the elasticity of the thorax, the wings flip from one state to another, with any intermediate position being unstable. Thus the flight muscles only have to contract a small amount to flip the wings from one state to another. This mechanism is called the 'click' mechanism and it amplifies the changes in length of the flight muscle up to 600 times in the wing movement. The fibrillar flight muscles respond to nervous stimulation by maintaining a prolonged tension and respond to stretching by contracting. Thus the nervous system maintains the tension in the muscles by stimulation at a much lower frequency than that of the wing beat, while the click mechanism, by stretching each of the antagonistic set of flight muscles in turn, stimulates their subsequent contraction. The muscles oscillate, contracting at a much higher rate than the stimulation from the flight neurons.

There is a fascination in the study of insect flight and the senses associated with it. The interest comes partly from the wide range of subject areas that contribute to the analysis of flight, from biochemistry through to physics. However, perhaps the main reason is the one mentioned at the start of this section. Flight is probably the one evolutionary development that has led to the unique species diversity of insects and their central role in most terrestrial ecosystems. Insects and their ecosystems are the subject of the final section.

SUMMARY OF SECTION 2.6

1   The wings of all orders of flying insect appear to be derived from a common ancestral form. It is possible that development of wings led to the enormous diversity of insect species that are found today.

2   Insect flight requires a large amount of energy; ATP is produced in the mitochondria within the flight muscles, but most of the energy released by breakdown of fuel is dissipated as heat.

3   Contractions of the flight muscles generate heat within the thorax; large flying insects such as dragonflies and bees may be at risk of overheating during flight.

4   Wing-beat frequencies vary in different species, as do flying speeds.

5   There are two types of flight muscle, direct and indirect, which produce wing movements by different means. The direct muscles are older in evolutionary terms than the indirect muscles.

6   The neural control of flight has been studied in the locust. A central pattern generator of interneurons controls the basic flight pattern, which is modified by flight interneurons in response to external conditions detected by sense organs.

2.7   ECOSYSTEMS AND COEVOLUTION

Many insects have associations with other organisms, at the individual level, the population level and over an entire community. So although evolution is essentially about the adaptation of an individual to the environment, where that individual has another association there is a link between the evolution of one and the evolution of the other. This reciprocal relationship was recognized by Charles Darwin and he used the example of the honeybee and the flowering plant to illustrate it.

> …a flower and a bee might slowly become, either simultaneously or one after the other, modified and adapted in the most perfect manner to each other …

Darwin was describing, though he did not give it this name, the process of coevolution. Coevolution is a reciprocal relationship. If two populations A and B are associated and a change occurs in A that affects the relationship with B, and *if* that change produces an evolutionary response from population B, then the two populations are coevolving.

2.7.1   POLLINATION AND COEVOLUTION

Insects pollinate many angiosperms. The flowers provide a reward of nectar to the visiting insect and the insect involuntarily transfers pollen between plants. The pollinating insect does not often feed on the pollen. The relationship between one plant species and one insect species is rarely an exclusive one. Usually, plants are visited by more than one species of pollinator and pollinators visit more than one species of plant. In a study of bees and flowering trees in the tropics, pollen from over 200 species of plant collected by colonies of bees was found at just one site. It is difficult to see this result as evidence for coevolution. However, there are some examples of tight evolutionary coupling that do represent coevolution. There is a star orchid in Madagascar whose floral spurs are around 30 cm in length. The nectar is at the tip of the spur. The flowers are pollinated by a large sphingid moth (hawkmoth), *Xanthopan morgani*. The moth has a long proboscis, almost 25 cm long. By pushing its head into the flower the moth can reach the nectar and, while doing so, it pollinates the orchid. *Xanthopan morgani* is the only moth that can pollinate this orchid, which suggests very strongly that the plant and the moth have evolved together.

Perhaps the most widely quoted examples of coevolution in pollinating insects are the *Heliconius* butterflies and the fig wasps. Heliconid butterflies pollinate two vines of the Cucurbitaceae family (*Anguria* and *Gurania*). The vines are relatively rare and do not occur at high densities. In these species, the plants are of a single sex. A significant feature of the flowering pattern of the vines is that the female plants flower at certain times of the year whereas the males flower continually.

Heliconius is long-lived for a butterfly, with a lifespan as an adult of about 6 months. The butterflies assemble at night at traditional roosting sites and travel along established tracks during the day, foraging for food. It is thought that the young butterflies learn the routes from older ones.

The male flowers of the vines produce large amounts of pollen and nectar, although the flowers only last a few days. Measurements were made of pollen and nectar production of a single plant of *Anguria umbrosa* under greenhouse conditions. The pollen grains are about 80 µm in diameter, which is unusually large. With 10 000 flowers produced in one year, the plant provided 20 g of pollen and 145 g of sucrose, measured as the dry mass. The pollen is used as food by the butterflies, a habit that is known only in heliconids. Pollen yields many amino acids when digested and these contribute to egg production. *Heliconius* feeding on a normal diet of pollen and nectar can have a five-fold increase in egg production over those taking nectar only.

○   Why is feeding on pollen and nectar advantageous to both the plants and the long-lived insects?

●   Butterflies that live longer have a longer period learning the foraging routes and also longer in which to transfer this knowledge to younger adults. The plant benefits, as there is a greater probability of successful pollination.

The relationship between *Heliconius* and the vines is a strongly stabilizing one. The butterflies improve the chance of seed being set and become established as regular visitors. The food provided by the plants improves both the nutrition and the constancy of the egg production of the butterflies. In neither case does the relationship increase the carrying capacity of the environment for either species. It is a stabilizing and self-centring relationship that maintains the population sizes in equilibrium.

The relationship between the fig tree (*Ficus*) and a family of wasps (the Agaonidae) is perhaps the most remarkable of the examples of coevolution. The five stages in the development of the fig inflorescence appear to have evolved such that they favour the life cycle of the wasp. In fact, most of the life cycle of the wasp takes place inside the fig.

The fig is a false fruit. The inflorescence has an enlarged receptacle that contains the small flowers. At the top is a small opening, the ostiole (Figure 2.36). Once the ostiole in the inflorescence of a fig is open, a female wasp carrying pollen can enter. Generally, only wasps of the family Agaonidae can pass through and each species of fig has only one species of wasp that pollinates it. As the female squeezes through the ostiole she loses her wings and parts of her antennae. She has a long ovipositor that enters the style of a flower and penetrates to the ovary, where she lays her eggs. There are two types of flower inside the fig, those with short styles and those with long ones. It is generally the short-styled ones in which the female lays her eggs, while the pollen she is carrying fertilizes the long-styled flowers (Figure 2.37). At this point, the female flowers have developed and can be pollinated, but the male flowers have yet to develop. When the wasp eggs hatch within the ovaries, galls form and the larvae develop within them. The males emerge first and then find an ovary with a female inside. The male has a long abdomen and a long, flattened head, adapted to rasping. He cuts his way into the ovary and mates with the female. She leaves the gall and collects pollen from within it, for by now the male flowers have matured. The female then leaves the fig through an exit hole bored by the male and flies off to find another fig with the ostiole newly opened and the flowers ready to be fertilized. Lacking wings, the males never leave the fig where they hatched.

The intricate relationship that has evolved between the fig and the fig wasp shows a remarkable degree of coevolution. It is significant that the relationship is developed between one species of fig and one species of wasp, because the nature of the relationship provides reproductive isolation.

○ What is the principal consequence of reproductive isolation in this example?

● Reproductive isolation is likely to enhance speciation, which probably explains the diversity of species in both groups, *Ficus* and the family Agaonidae.

Tight relationships between two species are also found in a number of parasitic arthropods and here, too, one might expect coevolution to have taken place.

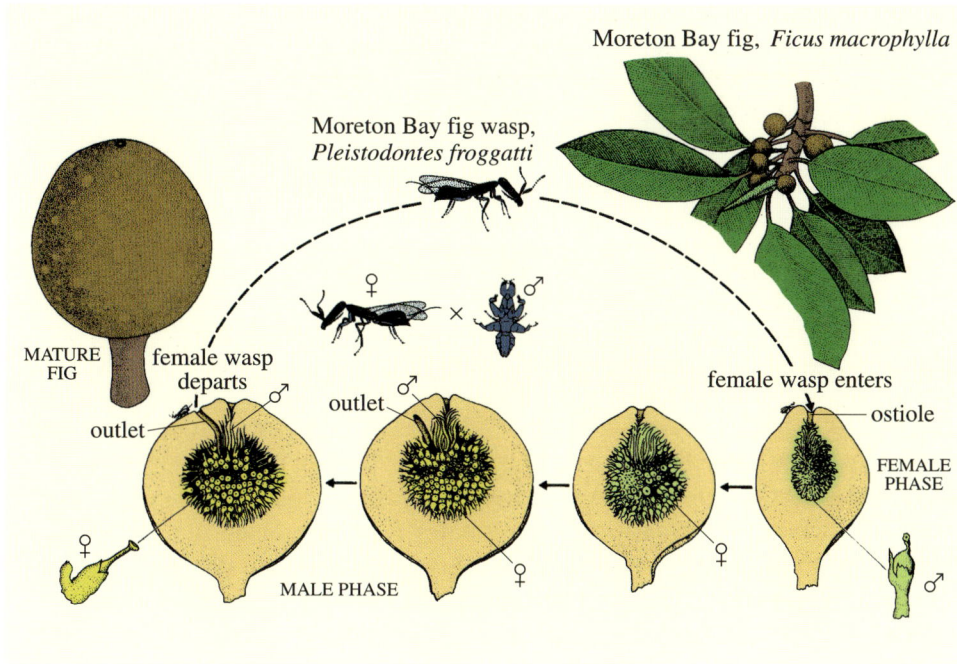

Figure 2.36 The development of the fig and the links with the life cycle of the fig wasp.

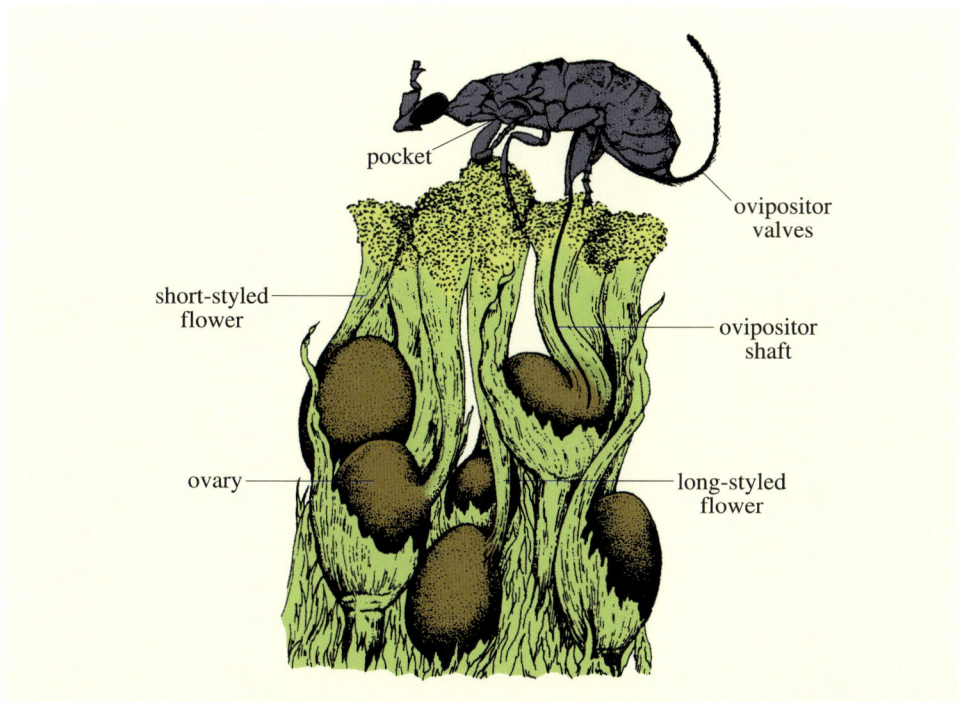

Figure 2.37 A female fig wasp laying eggs in the ovary of a short-styled flower.

2.7.2 PARASITES AND COEVOLUTION

The Phthiraptera are ectoparasitic lice (Figure 2.38), whose life is spent entirely on their host, and they are mostly **monoxenous**, that is they have only one host species. In an analysis of the ectoparasites of 26 families of bird, with a total of 127 species, there was an average of 1.1 species of louse for each bird species. These results would lead you to expect substantial coevolution between parasite and host, and it has been suggested that so close is the linkage that if you plot out the evolutionary trees for both species, you get the same pattern.

Wherever a speciation event has affected the host, the parasite has undergone a synchronous speciation. This suggestion has been tested in a family of lice that are parasitic upon mammals, the Trichodectidae. On a sample of 224 host species there were 337 louse species, with 34% of hosts having more than one parasitic species. So some independent speciation must have taken place. One way in which this process could happen is if there was ecological isolation on the host. For example, there are two species of louse that feed on humans and they are ecologically separated. The head louse, *Pediculus capitis*, and the body louse, *Pediculus humanus* (Figure 2.39), are sibling species that are similar morphologically but isolated ecologically and so isolated reproductively as well.

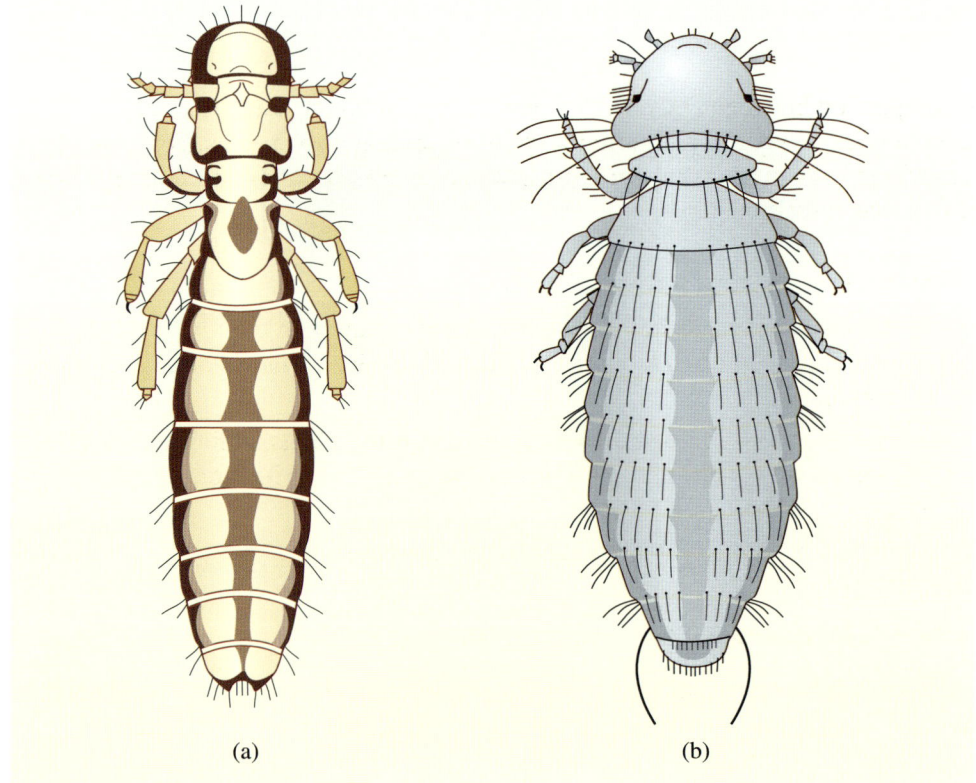

(a)   (b)

Figure 2.38 Examples of parasitic lice. (a) *Lipeurus caponis,* about 2.5 mm in length. (b) *Menopon pallidum*, to the same scale.

Figure 2.39   The body louse, *Pediculus humanus.* This species is one of two sibling species of louse that are ectoparasitic on humans, the other being the head louse, *Pediculus capitis,* which grows to about 2 mm in length.

A comparison between lice and fleas shows that, although some fleas are monoxenous, many are **polyxenous** and have several host species. For example, some fleas have over 30 known hosts and one species of rat has more than 20 species of flea that parasitize it.

SUMMARY OF SECTION 2.7

1   Where two organisms have some form of association in their life cycles evolution can act upon them in a reciprocal fashion. This phenomenon is known as coevolution.

2   Coevolution is well illustrated by some flowering plants, which appear to have coevolved with their insect pollinators, particularly the heliconids and the fig wasps.

3   Coevolution is often a feature of parasites and hosts.

2.8 CONCLUSION

Arthropods have colonized almost every ecological niche imaginable, and yet this diversity is based on a body plan that, though highly variable, nevertheless is recognizably the same in all groups. This feature should not imply that the arthropods are monophyletic in origin; they are not, but the external skeleton with jointed legs is clearly a very adaptable plan. The plan is based on a series of identical segments, and the variety of body styles seen in arthropods arises from the different ways in which fusion of segments has taken place, together with functional adaptation of the limbs. The crayfish is a good example of how the limbs of each segment have become adapted for feeding, swimming, walking and several other uses. The life cycles of arthropods are often complex, with a series of larval stages. In general, the larval stages exploit different niches from the adults, and may represent the dispersal stages of the species. However, in some insects, like the dipteran flies and the moths, the adult is the dispersion stage and the larva is primarily a growth stage.

The importance of arthropods to humans is very great and not always fully appreciated. The spread of disease by vectors, the role of insects in pollination and the use of arthropods in forensic science, are all interactions between humans and arthropods. The arthropods are the only invertebrates that have colonized the land, sea and air, so for that reason alone they merit our attention. However, it is the incredible diversity of form and life style that exerts the greatest fascination. That fascination is obvious when one looks at the great butterfly collections of the Victorian period or the mushrooming of present-day amateur societies specializing in keeping live cockroaches, spiders, mantids and many other exotics. It is also obvious when, as biologists, we try to account for such diversity, and recognize that the process of natural selection has produced arthropods in almost all habitats on the planet.

REFERENCES

Briggs, D. E. G. and Whittington, H. B. (1985) Modes of life of arthropods from the Burgess Shale, British Columbia, *Proceedings of the Royal Society of Edinburgh,* **76**, pp. 149–160.

Brodsky, A. K. (1994) *The Evolution of Insect Flight,* Oxford University Press, Oxford.

Cheng, H., Yang, W., Kang, W. and Liu, C. (1995) Large-scale spraying of bed-nets to control mosquito vectors and Malaria in Sichuan, China, *Bulletin of WHO*, **73(3)**, pp. 321–328.

Coghlan, A. (2001) Sex on eight legs, *New Scientist*, **169**, p. 6.

Knight-Jones, E. W. and Moyse, J. (1961) Intraspecific competition in sedentary marine animals, *Symposium of the Society for Experimental Biology,* **15**, pp. 72–95.

May, M. (1995) Dependence of flight behaviour and heat production on air temperature in the green darner dragonfly *Anax junius* (Odonata: Aeschnidae), *Journal of Experimental Biology,* **198**, pp. 2385–2392.

Nuorteva, P., Isokoski, M. and Laiho, K. (1967) Studies on the possibilities of using blowflies (Diptera) as medico-legal indicators in Finland, *Annales Entomologici Fennici,* **33**, pp. 217–225.

Pond, C. (2001) Dealing with food. In *Generating Diversity,* M. Gillman (ed.), The Open University, Milton Keynes, pp. 37–88.

Poteser, M. and Kral, K. (1995) Visual distance discrimination in praying mantis. *Journal of Experimental Biology,* **198**, pp. 2127–2137.

Winchester, N. N. (1997). Arthropods of coastal old-growth *Sitka* spruce forests. In *Forests and Insects,* A. D Watt, N. E Stork. and M. D Hunter (eds), p. 368.

ACKNOWLEDGEMENTS FOR CHAPTER 2

The author would like to thank Dr Patricia Ash for her contributions to this chapter and Professor Pekka Nuorteva for his help with the literature on forensic entomology.

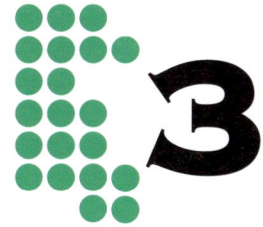

PARASITES

3

3.1   INTRODUCTION

All organisms exist in some kind of association with others. Sometimes the association is very intimate and long-lasting whilst at other times it is merely transitory.

○ Briefly list the kind of associations an animal makes with members of its own species.

● You could have included herding together or living in groups for protection culminating in colonial life, pairing and bonding for reproduction and parental care.

Such associations between the same species are known as intra-specific associations and are characterized by the fact that all their members share the same gene pool. Associations between members of different species (inter-specific associations) can be very loose or very intimate and are termed **symbiosis** (living together).

Upon what basis are symbiotic associations maintained? Associations have either a positive or negative aspect. **Mutualism** is where both organisms benefit. In others, one organism benefits at the expense of the other. At one extreme is the predator–prey relationship, which can vary from predators taking many kinds of prey and prey having many predators to one where the predator is totally dependent on one kind of prey. At another, **parasitism** involves a tight and intimate symbiotic association between host and parasite, where all the benefits are to the latter. **Helotism** is a relationship where the invading organism is enslaved and all the benefits go to the host, e.g. the mitochondrion is an enslaved 'bacterium' which is essential for aerobic respiration in all eukaryotic cells.

The word 'parasite' is commonly used to describe an acquaintance in a rather derogatory manner. In biology, it means an organism that lives in or on another organism and derives food and other biological necessities from it without rendering it any benefit in return.

A parasite infection certainly can reduce the evolutionary fitness (survival and/or reproductive success) of the host, the extent of which is referred to as the parasite's **virulence**. Thus, parasites can cause morbidity (general weakening), mortality (death) or reduced reproductive success. However, many hosts can withstand the worst effects of invasion or may even eliminate the parasite altogether, which is a measure of the host's **resistance** (Davey and Gillman, 2001). These effects and counter-effects are the hallmark of parasite–host interactions, with the parasite evolving features that promote its own infectivity and survival, and the host evolving more sophisticated protective measures. If, as a result of this coadaptation, both mechanisms are working efficiently, then a parasite is tolerated by its host, even though there may be some reduction in the host's fitness as a result of nutrient depletion or tissue damage.

Parasites themselves may be subjected to invasion by other organisms and although the ditty

> 'big fleas have little fleas upon their backs to bite them,
> little fleas have lesser fleas and so on *ad infinitum*'

Jonathan Swift

is not literally true, there are many examples of parasites living in, or on, other parasites. For example, certain fluke and tapeworm parasites are known to harbour protoctistan microsporidian microbes of the genus *Nosema*. It is possible that such microbes themselves are infected with bacteria which may be invaded by viruses, so giving four levels of infection, a situation referred to as **hyperparasitism**.

Parasites may attach themselves to the outside of their host (**ectoparasites**) or be found within the tissues of the host (**endoparasites**), usually within cavities such as the lungs, gut or blood system or occasionally between muscle cells.

○ What are some advantages and disadvantages of the parasitic way of life?

● Advantages include a relatively safe and stable habitat and a plentiful supply of food. Disadvantages include difficulties in completing a life cycle because of the need to transfer stages between one host individual and another, the need for attachment, a total dependency of the parasite on a host organism, and attacks from the host's immune system.

Certainly the feature that we have already emphasized as defining parasitism, that of obtaining nourishment, means that the parasite has an obligatory metabolic dependency on its host. We explore this intimate biochemical association in Section 3.3.

In Section 3.2 we deal with the difficulties of transmission, i.e. how parasites make the hazardous journey from host to host. For the majority of parasitic species, infection passes between unrelated hosts (horizontal transmission) but some parasites transfer between a parent host and its offspring (vertical transmission). Some parasites have *direct* life cycles: they occur only in a single **definitive** (= final) host where sexual reproduction, if any, occurs. However, many parasitic species have *indirect* life cycles: they require additionally one or more **intermediate** hosts which may belong to different classes or phyla from the definitive host. Some parasites are able to move via free-living forms directly between different hosts to complete different stages of their life cycle, but others depend on being transported by a **vector**. Others transfer from intermediate to definitive host through one or more **paratenic** hosts in which no anatomical development occurs. Occasionally, it is to the advantage of the parasite to suspend its development through dormancy; for example, certain larval nematodes delay maturation in hosts inhabiting harsh environments or in those with marked seasonality in their own life cycle, a situation which favours maturation of the parasites and release of their eggs during good conditions and when potential new hosts are plentiful.

First of all, we examine the extent of parasitism amongst animal phyla and review some of the features that make this way of life so special and so fascinating, especially those related to the mechanisms of transfer between hosts.

3.2   PARASITISM — A NORMAL WAY OF LIFE

There are more parasitic species in the world than free-living species. Parasites are found in every major phylum of the animal kingdom, in many of the phyla of the kingdom Protoctista and also among fungi, bacteria and a few plants. Thus, parasitism cannot be regarded as an unusual way of life and has undoubtedly evolved many times. Because of the weight of evidence accumulated, it can be said, almost certainly, that you would find parasites in any animal you examined in detail. Although the same thing cannot be said for plants, there are nevertheless many noteworthy cases of parasitic associations between plants, especially those of economic importance, such as mistletoe (*Viscum album*) on apple trees and between a plant and animal parasites, e.g. eelworms (nematodes) which occur in root cells of both celery and potato.

The study of parasites (parasitology) is a huge area of science incorporating the tools of taxonomy, molecular biology, physiology, ecology and evolutionary biology. It should not surprise you that most effort and resource is directed to research into those parasites of economic importance, particularly those whose presence causes disease in humans and their domestic animals.

Let us now look at examples of the range of parasites throughout the kingdom Animalia and their life-cycle strategies before looking in more detail at the nutritional nature of the relationship (Section 3.3), and the evolution of the parasitic habit and intimacy of the relationship in terms of the behaviour of host and parasite (Section 3.4).

3.2.1   HELMINTHS

The word 'helminth' is a general term for worms but by convention refers to parasitic worms, mostly members of the three phyla that include the majority of animal parasites found living on or within vertebrates. The three major phyla are the Platyhelminthes (flukes and tapeworms), the Nematoda (roundworms) and the Acanthocephala (spiny-headed worms). Helminths represent a phylogenetically diverse range of animals that show different structural organization and life-cycle strategies and, with the exception of the Acanthocephala, whose members are exclusively parasitic, contain both free-living and parasitic members. Although the Platyhelminthes and Nematoda contain species which are amongst the most destructive in terms of the debility, deformity and death that they cause to humans and domestic animals, it should be remembered that most parasites exist in a kind of uneasy truce with their hosts. However, where infective stages are disseminated from a host, the longer the host survives the more infective 'units' can be dispersed.

Chapter 1 described the structural characteristics of free-living platyhelminths (Section 1.3.1) and nematodes (Section 1.3.3).

○ From the list below choose those characters which are found (i) in free-living platyhelminths and (ii) in nematodes.

(a)   there is cellular mesoderm;

(b)   they possess a body cavity called a pseudocoel;

(c)   the body covering is a cuticle periodically replaced by ecdysis;

(d)   the body surface is covered with cilia;

(e)   the gut has one opening, a mouth;

(f)   the gut has two openings, a mouth and an anus;

(g)   they possess both circular and longitudinal muscle layers enabling the body to twist, shorten and lengthen;

(h)   they possess only a longitudinal muscle layer allowing whip-like motion.

● (i) Free-living platyhelminths have characters a, d, e and g; nematodes have characters b, c, f and h.

Platyhelminths, then, have a ciliated ectoderm, which is often spiny as well in parasitic members, whereas the external surface of nematodes is cuticle. Some aspects of the arthropod cuticle were described in Section 2.2. A tanning process may occur in both arthropods and nematodes but whereas the arthropod cuticle is a complex of protein and chitin, the predominant structure of the nematode cuticle is an inner layer of collagen fibres, with some collagen also extending into the outer layer, which may be keratinized. These fibre layers are vital to the unique whip-like motion of the nematode and act antagonistically with the longitudinal muscle by bracing against the fluid-filled pseudocoel, which is thus always under pressure. Furthermore, in arthropods, the hardened exoskeleton has to be continuously moulted in order for adult growth to occur, whereas in nematodes by comparison, increase in adult size involves continuous growth of the cuticle, although the larvae moult as they grow, often forming distinct stages with different structures and habits.

We shall discover in later sections how the leaf-like body of platyhelminths and the cylindrical, tapered body of nematodes are both well adapted to living as internal parasites. In particular, we will see how the body plans of cestodes (tapeworms) have evolved to a state where they lack a gut and maximize reproductive capacity by having a long segmented body.

Indeed, reduction in structural organization is a general feature of parasites. For example, in addition to the loss of a gut in tapeworms, many helminths lose the eyespots and locomotory cilia characteristic of their free-living relatives. Thus they are sometimes referred to as being 'degenerate'. However, as we shall see in this chapter, parasites are highly specialized to their way of life and have evolved appropriate physiological and morphological features as well as losing other, inappropriate ones. In particular, the feeding mechanisms of parasites allow them specifically to cope with a wide variety of animal and plant tissues, from which nutrients are absorbed either through a gut or, in the case of many parasites found in the intestines of animals, directly across the body wall. Thus, the biochemical specializations of parasites can be quite complicated. Furthermore, parasites often live in situations where they are challenged by the host's defence mechanisms or by its digestive enzymes and many parasites have evolved sophisticated mechanisms to deal with these hazards. Because they are in constant danger of being dislodged from their specific site in a host, parasites have also evolved elaborate attachment organs. In the following sections, we will see how complicated parasite life cycles can be, often involving up to six different stages in the life cycle as they transfer between a number of hosts. All of these features should convince you, therefore, that the use of the word 'degenerate' for parasites should be treated with caution!

Acanthocephalans (Figure 3.1) possess a pseudocoel but, unlike nematodes, have no gut. They therefore represent a highly specialized group, all species of which are parasitic in the small intestine of various vertebrates (mainly birds and fishes, but sometimes in reptiles and small mammals) where the hooked proboscis embeds within the gut lining. After fertilization, eggs released in faeces from the host gut are ingested by insect or crustacean intermediate hosts, where hooked larvae hatch and penetrate into the arthropod's haemocoelic cavity. There the larvae encyst and are released into the small intestine of the vertebrate host when it eats the infected insect or crustacean.

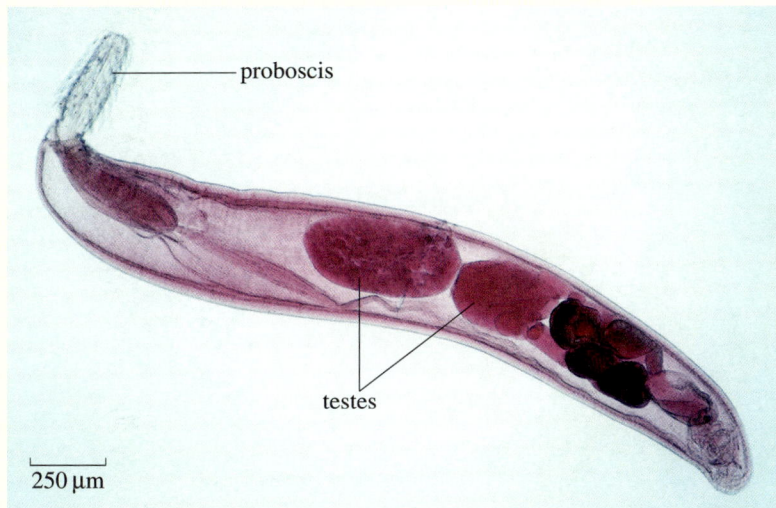

Figure 3.1 Morphology of an adult male *Acanthocephalus* sp.

Acanthocephalans are mostly species-specific, but a very few, mostly accidental, infections of humans occur. On the other hand, as mentioned earlier, both platyhelminths and nematodes include free-living members but some of the parasitic species cause extremely important human diseases. We will examine a few of the particularly important ones which illustrate some complex life-cycle strategies.

Figure 3.2 Parasitic platyhelminths (not to scale). (a) The anatomy of a fluke *Clonorchis* sp. (Digenea). (b)–(d) Tapeworm *Taenia solium* (Cestoda): (b) parts of whole worm; (c) one ripe proglottid; (d) the scolex (head).

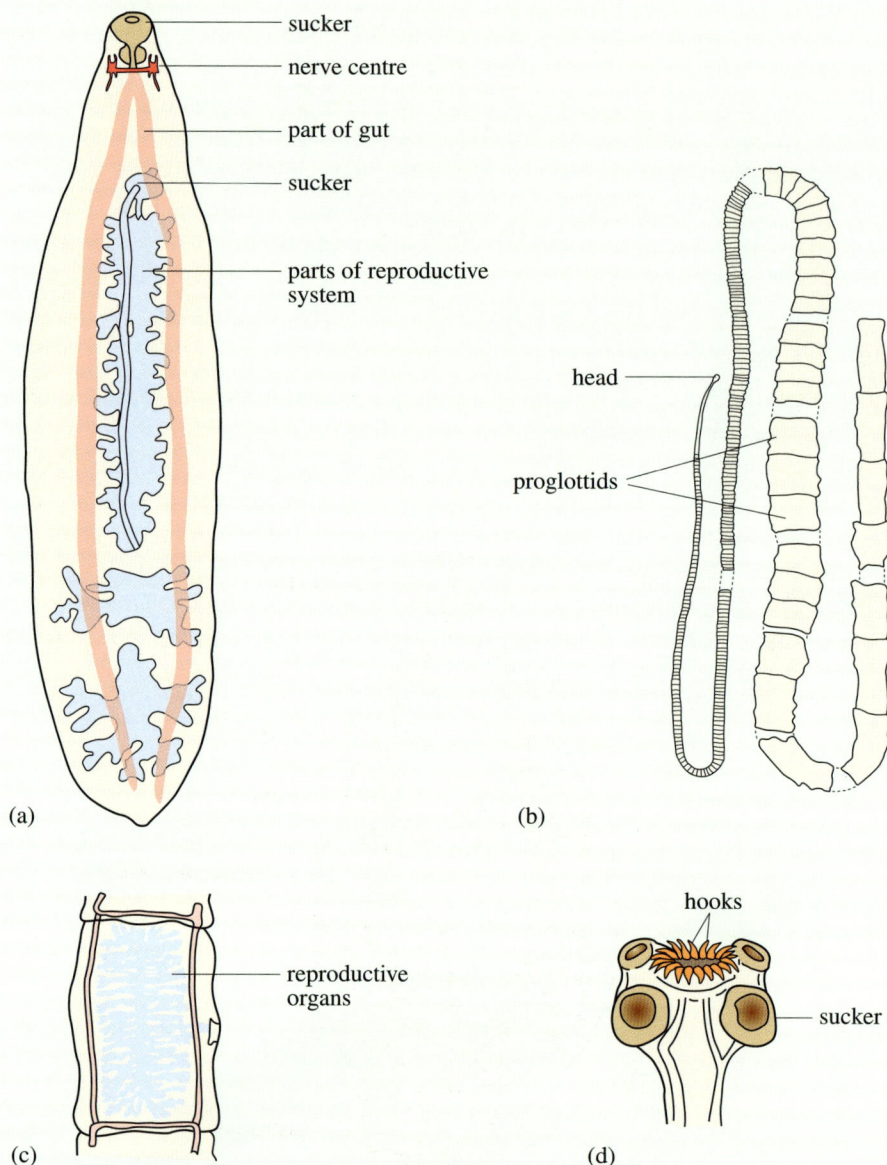

(a)

(b)

(c)

(d)

3.2.2 FLUKES AND TAPEWORMS

The parasitic platyhelminths are compared in Figure 3.2. Some platyhelminths (Figure 3.2a) have a flattened leaf-like body (Monogenea and Digenea) with one or more suckers for attachment. Others (cestodes; Figure 3.2b–d) have a body divided into a head end called the **scolex** (plural, **scoleces**) with hooks and/or suckers and a series of specialized segments called **proglottids**.

Monogeneans and digeneans are called flukes as adults (e.g. Figure 3.2a). They are almost always endoparasites of vertebrates (i.e. live internally), inhabiting a wide range of organs including the gut, liver, lungs and circulatory system. Monogeneans (Figure 3.3) are typically parasites on the external surface of fish,

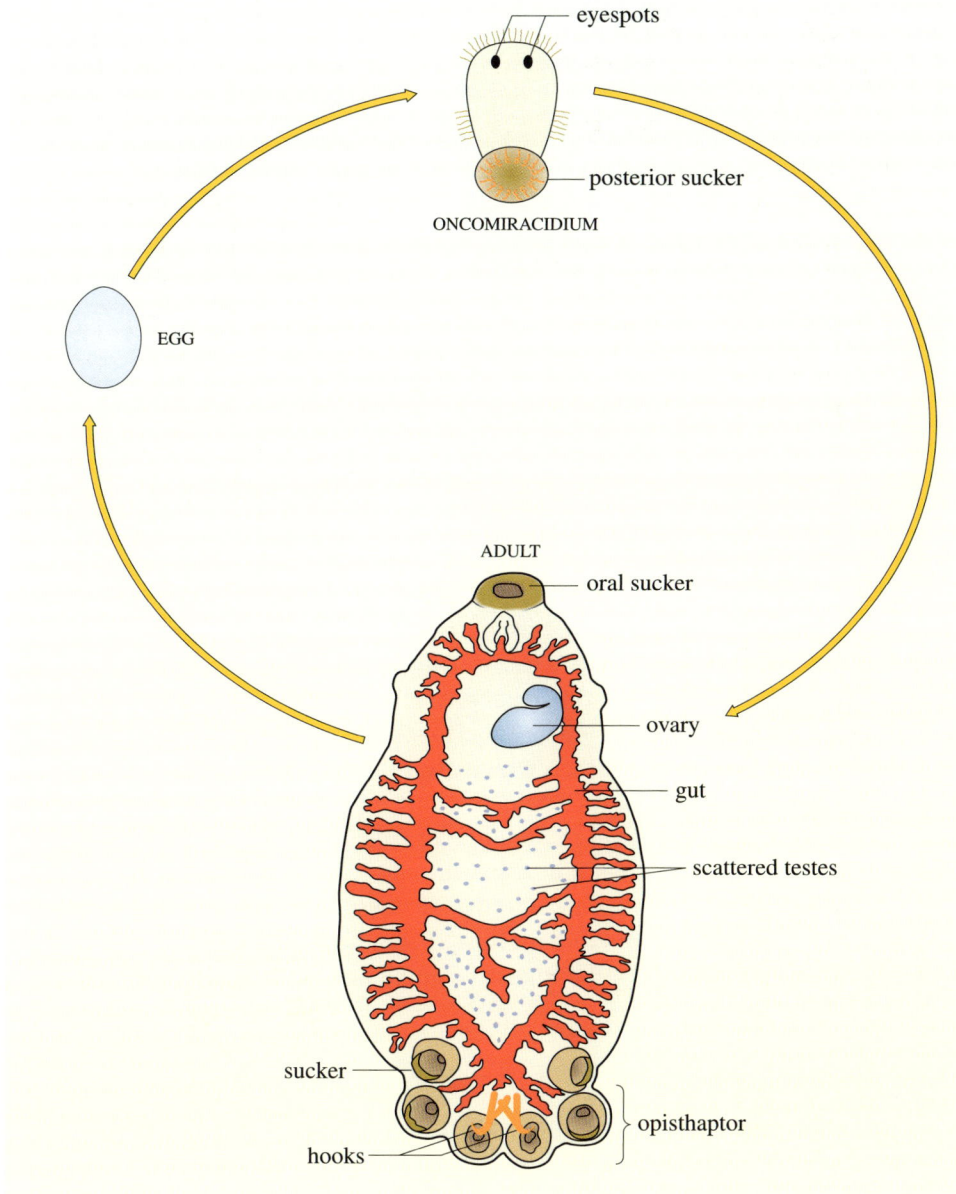

Figure 3.3   Anatomy of a monogenean fluke (*Polystoma intergerrimum*) and stages in the life cycle (not to scale).

living either on the skin or on the gills, i.e. they are ectoparasites. However, a few species, e.g. *Polystoma intergerrimum* have invaded the cloaca and bladder of amphibians.

○ What is the main anatomical difference between the digenean and monogenean parasites shown in Figures 3.2a and 3.3?

● The monogenean has a huge posterior attachment organ (opisthaptor) which may be armed with several hooks and suckers that aid attachment. Digeneans attach themselves by means of suckers, with most species having two, as shown in Figure 3.2a.

Monogenean flukes, living normally on the outside of their first host, are able to pass an infective stage through the water to another host. Figure 3.3 illustrates this simple direct life cycle in which a fertilized egg is released from the adult fluke, hatches in the water into a free-swimming ciliated larval stage known as an **oncomiracidium**, and attaches by means of its opisthaptor to a new host of the same species. You may also notice in Figure 3.3 that both testes and ovary are present, i.e. the animal is **hermaphrodite.** Although self-fertilization can occur, if worms are in close proximity on a host, cross-fertilization is believed to be common. However, digenean flukes, which also prefer cross-fertilization, have a problem — how can they transfer from the inside (often the gut) of one vertebrate definitive host to another?

○ How is a similar problem solved by protoctistan parasites, e.g. malaria parasites, living in the bloodstream of vertebrate hosts?

● The parasites rely on blood-sucking vectors to transfer their infective stages (Section 2.3.2).

In digenean life cycles (Figure 3.4), a more elaborate transfer system has evolved, involving one or more intermediate hosts. Larval forms actively seek another host or are dependent on a host passively transferring a stage to the next host during feeding, which is not unlike vector transfer mentioned above. The adult fluke in the definitive host is, like monogeneans, hermaphrodite. Some remain in the body and pass eggs out into the gut lumen which are subsequently voided with the faeces, but many gut digeneans are voided whole and full of fertilized eggs often within a few days of their arrival in the gut. Outside, the fluke body disintegrates releasing the eggs. The eggs hatch in water to release minute free-swimming **miracidium** larvae which penetrate into the soft tissues of the **primary** (= first) **host**, usually a gastropod or bivalve mollusc. From here they migrate to the haemolymph (blood spaces) of the digestive gland and gonads, developing into a bag-like **mother sporocyst**.

○ Why should these particular sites be chosen by the parasite?

● Nutritional dependency is an important aspect of parasitism. The blood spaces surrounding the digestive gland and gonads of the primary host and the gut of the definitive host contain accessible nutrients.

The forms that develop within the primary host differ between species but all undergo massive proliferation in the mollusc. The mother sporocyst, a rather structureless sac, undergoes reproduction producing *either* large numbers of **rediae** (singular, **redia**, pronounced 'reed-e-a') *or* **daughter sporocysts**, both of which produce large numbers of free-swimming **cercariae** (singular, **cercaria**, pronounced 'ser-care-ee-a'). These reproductive stages are often imprecisely described as 'asexual' but this type of reproduction is more precisely known as apomictic parthenogenesis or apomixis (Dyson, 2001). Motile cercariae with muscular lashing tails are released from the mollusc, and encyst either on vegetation or in the tissues of an invertebrate or vertebrate intermediate host. In either case these larval stages, known as **metacercariae**, suspend development

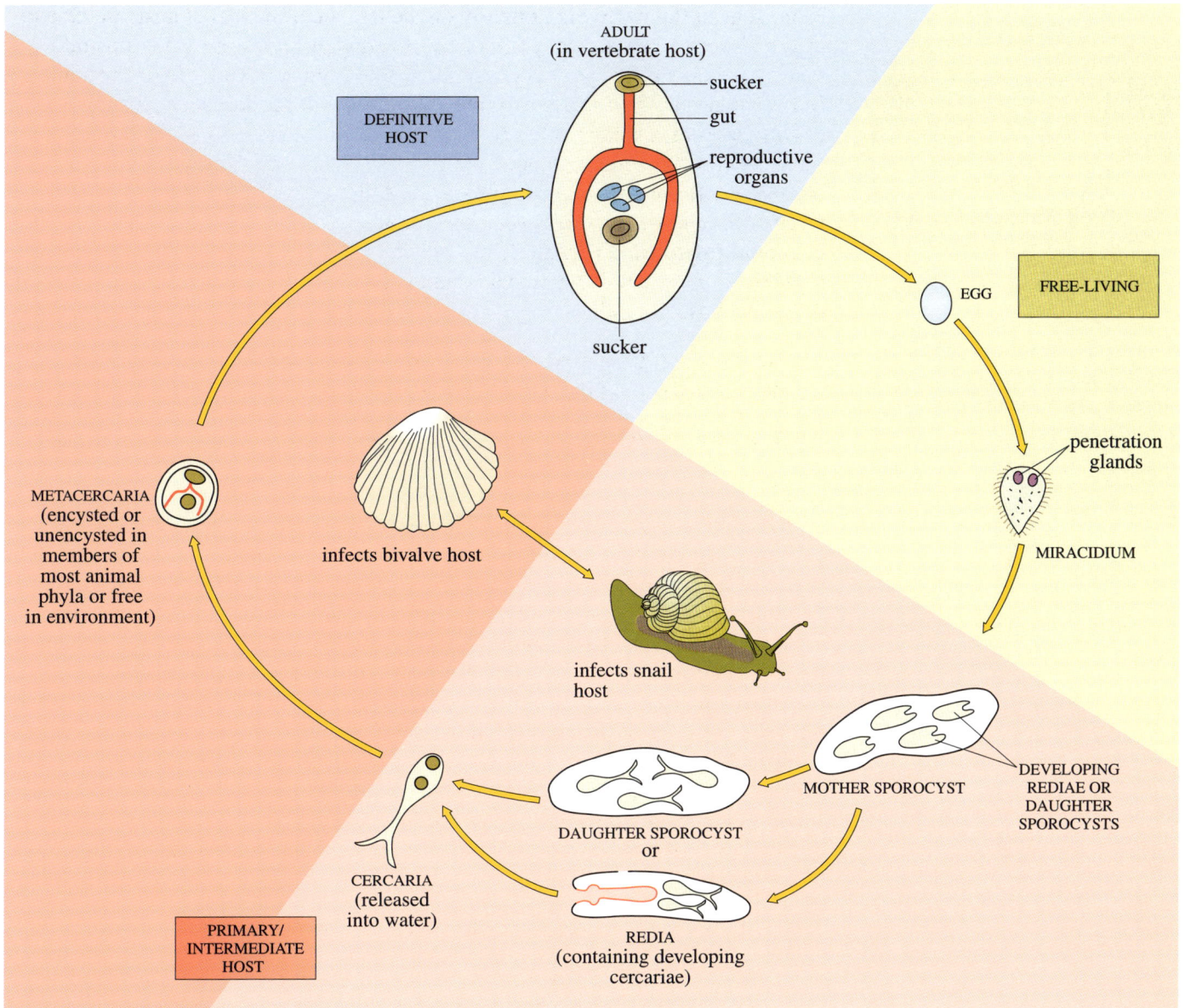

Figure 3.4 General digenean life cycle (not to scale). The definitive host is a vertebrate. Primary hosts can be snails or bivalve molluscs.

until eaten by the appropriate definitive host. There, they are released from the cyst (triggered by the pH changes along the gut), attach between the villi of the intestine, grow into the adult fluke and usually browse on the gut mucosa or, as in *Schistosoma*, penetrate directly into the tissues (see below).

This life cycle (Figure 3.4) can thus be described as an indirect life cycle. It is extremely adaptable and different species show many different variations on the basic pattern. One such group of digenean parasites with some unusual features are the blood flukes (*Schistosoma* spp.) which are of enormous medical importance. A brief outline of their main features serves to illustrate their beautiful adaptations.

penetration
glands

developing
miracidium
larva

lateral
spine

terminal
spine

(a)   50 μm   (b)

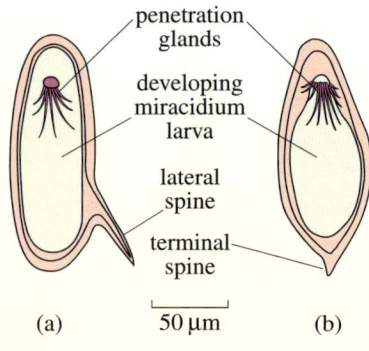

Figure 3.5 Fertilized eggs of
blood flukes: (a) *Schistosoma
mansoni*; (b) *Schistosoma
haematobium.*

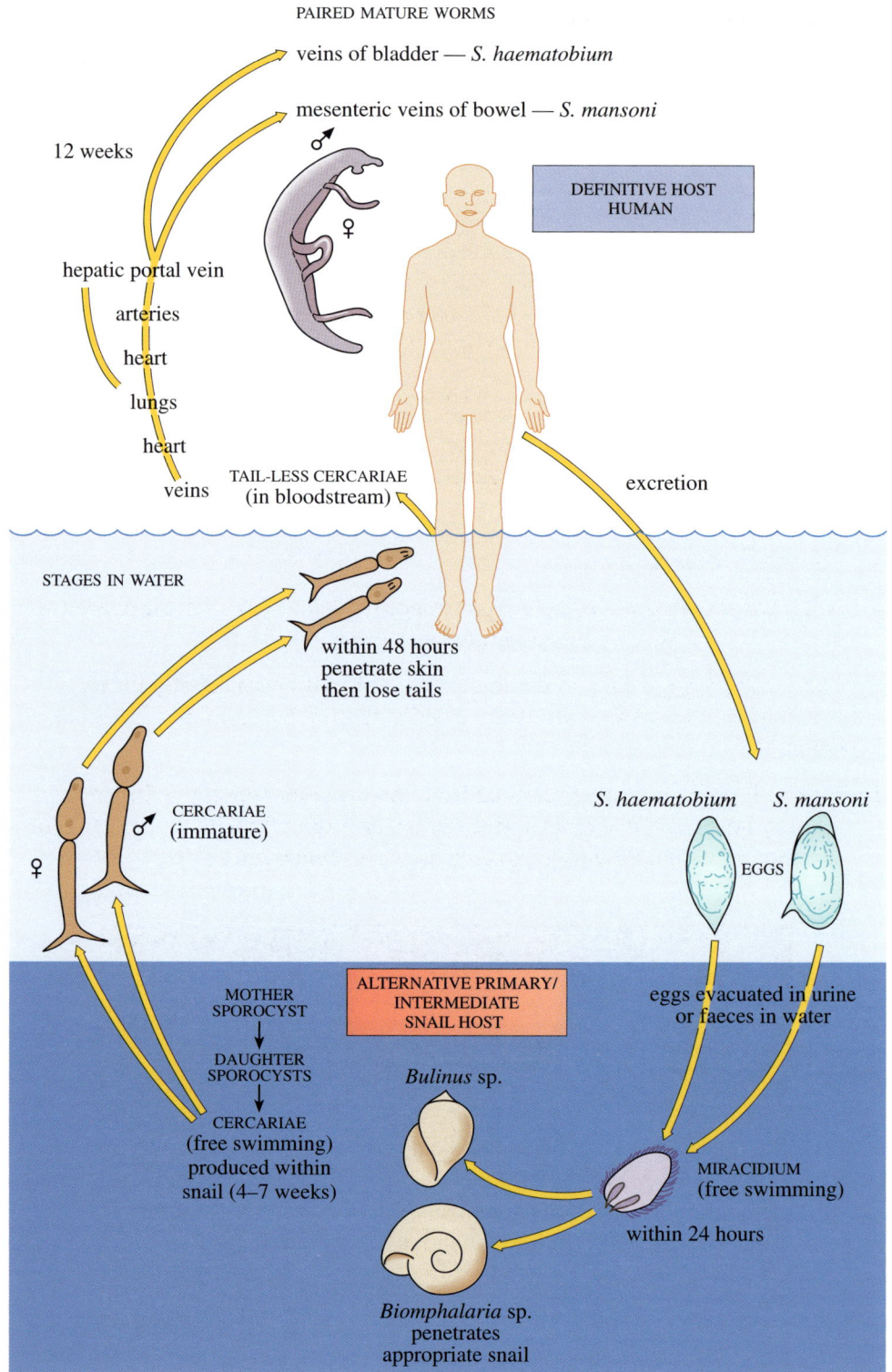

PAIRED MATURE WORMS

veins of bladder — *S. haematobium*

mesenteric veins of bowel — *S. mansoni*

12 weeks

DEFINITIVE HOST
HUMAN

hepatic portal vein

arteries

heart

lungs

heart

veins

TAIL-LESS CERCARIAE
(in bloodstream)

excretion

STAGES IN WATER

within 48 hours
penetrate skin
then lose tails

S. haematobium   *S. mansoni*

EGGS

CERCARIAE
(immature)

♀   ♂

MOTHER
SPOROCYST

↓

DAUGHTER
SPOROCYSTS

↓

CERCARIAE
(free swimming)
produced within
snail (4–7 weeks)

ALTERNATIVE PRIMARY/
INTERMEDIATE
SNAIL HOST

Bulinus sp.

eggs evacuated in urine
or faeces in water

MIRACIDIUM
(free swimming)

within 24 hours

Biomphalaria sp.
penetrates
appropriate snail

Figure 3.6 Life cycles of two species of *Schistosoma.*

3.2.3 THE BLOOD FLUKES

The first modern record of a blood fluke in humans was recorded during a post-mortem examination in Cairo by Theodor Bilharz in 1851 when he found the fluke in the mesenteric veins in the abdomen. He also demonstrated spiny eggs in the urine and faeces of many other patients, noticing that such eggs possessed either a lateral or a terminal spine (Figure 3.5). Subsequently, it was discovered that two species were present, one (*Schistosoma mansoni*) inhabiting the mesenteric veins around the gut and the other (*S. haematobium*) the veins of the bladder. Although other species of schistosomes are now known, these two remain the most important species infecting humans in the Middle East, Central and South-east Africa and parts of South America.

The life cycles of these parasites are shown in Figure 3.6.

○ Examine these life cycles (Figure 3.6) and Figure 3.7 carefully and list any differences you notice from the generalized digenean life cycle shown in Figure 3.4.

● The main differences are:

(i) Adult worms are thread-like in appearance (an adaptation to their existence in the narrow veins of the bowel or bladder).

(ii) The sexes are separate and the female worm is held permanently in an extensive ventral groove along the usually shorter, but more massive, male worm.

(iii) Free-swimming cercariae released from the daughter sporocysts in the snail primary host reinfect the definitive human host directly without the need for encystment within an intermediate host or on vegetation.

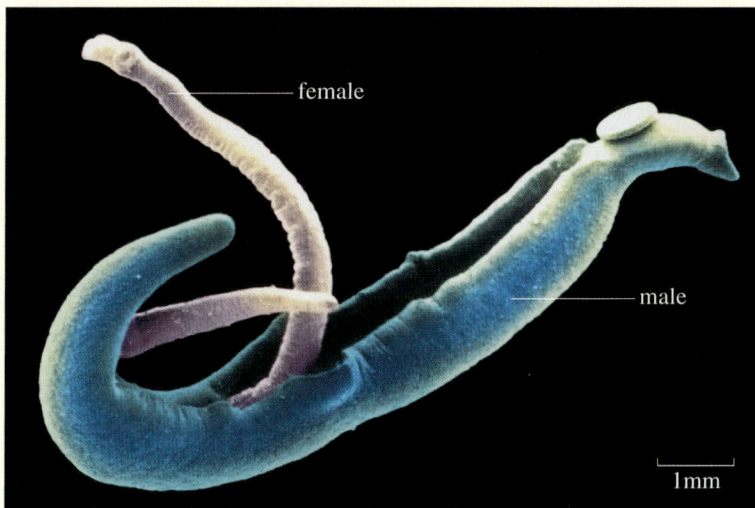

Figure 3.7 Coloured SEM of paired adult female (pink) and male (blue) *Schistosoma mansoni* worms.

Thus, schistosomes are bizarre digeneans which have become adapted to an unusual habitat and, as such, show unique morphological, physiological and life-cycle strategies which are quite different from all other members of their taxon.

Paired worms (Figure 3.7) produce large numbers of eggs which are prevented from being swept away in the bloodstream by the spines (Figure 3.5) which lodge them in the innermost layer of the venule. As these eggs pack the blood vessel, it eventually ruptures releasing the eggs into adjoining connective tissue. Many of the eggs tear through this tissue into the lumen of the gut or bladder from which they are passed to the exterior. However, many remain embedded in the tissues, especially in the liver, spleen and intestinal wall in the case of *S. mansoni* or in the bladder wall in the case of *S. haematobium* (Figure 3.6). These eggs cause lesions which often calcify causing a number of debilitating and painful symptoms. Medical reports suggest that carcinomas may occur as a result of prolonged infection. Schistosomiasis (also known as bilharziasis) is widespread in many parts of the tropical and subtropical world and, in terms of the human misery it causes, is second only to the devastation caused by the protoctist malaria parasite.

○ What essential environmental factor and human habit is needed for the successful completion of the parasites' life cycle?

● Rainfall, and an adequate habitat for the survival of one or more species of the freshwater snail as primary host. When raw human faeces and urine containing the parasite's fertilized eggs are deposited, released larvae can enter the snail. Reinfection of humans occurs when such pools are used for washing and bathing.

On hatching from the eggs, the minute miracidia larvae actively swim in the water. Exactly how they find their snail primary host is not fully known but it is believed to involve a combination of phototaxis, which attracts the larva to the surface waters containing the snails, and chemotaxis toward specific chemicals exuded by the snail. This chemical stimulus may contain soluble fatty acids and/or amino acids, the neural transmitter serotonin and specific glycoproteins in mucus. On contact, the miracidium attaches by a secretion from apical gland cells. Penetration occurs as a result of a secretion of lytic enzymes aided by a boring movement of its attached end. In the correct host, the miracidium loses its ciliated surface and metamorphoses into a mother sporocyst which reproduces to form many daughter sporocysts which are released into the haemolymph of the digestive gland tissue in the snail. Each of these daughter sporocysts is capable of parthenogenetic reproduction resulting in hundreds of cercariae. Miracidia are genetically either male or female so that all cercariae produced from the one invasive miracidium eventually develop into adults all of the same sex. Cercariae are shed from the snail host under particular light stimuli, with somewhere between 1000 and 3000 being emitted from each snail over a brief period during each day. Snails are usually infected with large numbers of daughter sporocysts arising from each miracidium.

The non-feeding cercariae have large glycogen reserves upon which they depend for energy, but very few survive for longer than 48 hours. They swim in bursts, periodically rising and sinking just below the surface of the water and await a chance encounter with human skin. Sets of glands at the anterior end produce enzymes to digest their way through snail tissue, adhesive secretions to assist attachment to the skin of the new host, and lytic substances to aid penetration through the human skin. Penetration occurs rapidly in a few minutes (Figure 3.8). After the tail is lost, the larval worms proceed through the subcutaneous tissue into the peripheral lymphatic or venous vessels and thence via the circulation through heart and lungs to the liver, where they mature. Maturation, pairing and often mating occur in the deeper portal veins of the liver. The male then carries the female via the portal veins to the mesenteric veins around the intestine or urinary capillaries around the bladder where egg laying occurs. The time from entry to establishment of matured worms in the portal veins is about 25–30 days but the adult worms can survive for several years producing hundreds of fertilized eggs each day!

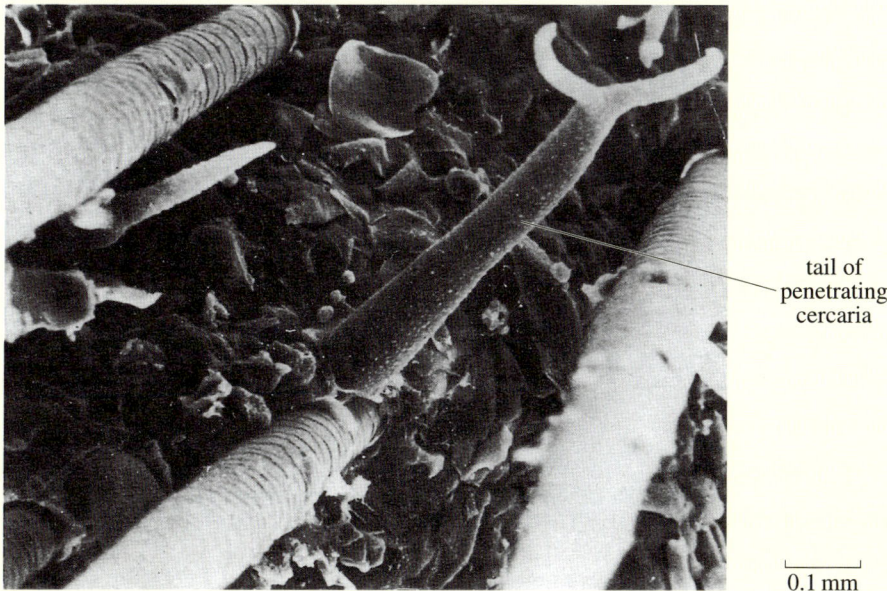

tail of penetrating cercaria

0.1 mm

Figure 3.8   SEM of a cercaria of *Schistosoma mansoni* penetrating the skin of an experimental mouse host down the side of a hair shaft. The head of the cercaria has already disappeared beneath the skin epidermal cells, leaving only the forked tail, which is subsequently lost.

In spite of the precarious nature of the life-cycle strategy, it is obvious from their extensive geographical distribution and the vast numbers of humans infected that these creatures are amazingly successful.

○   How would the human immune system react to schistosome infection?

●   Schistosomes are invasive organisms and the host immune system should recognize them as 'foreign', thus triggering the immune response.

○   What differences would there be in the likely immune responses to parasitic microbes compared with invasion of a digenean fluke?

●   Because microbes multiply rapidly in the host and therefore constitute a life-threatening situation, the host tries to reduce the numbers as quickly as possible and antibodies, with or without complement, and phagocytic cells can cause their destruction. Invasive flukes do not multiply within the definitive host and are much bigger organisms, too large to be destroyed in this way, so that the immune response, which is probably directed against the eggs and cercariae, is geared more to preventing reinfection by these stages rather than destruction of established flukes.

What evidence is there that such a mechanism occurs in blood fluke infections?

It is very difficult to produce conclusive data from field studies although epidemiological evidence showing that the symptoms of schistosomiasis are often less severe in older infected persons suggests that some kind of immunity to reinfection from cercariae may occur, even though the adult worms present are still producing eggs. However, experimental laboratory work has provided convincing data. Figure 3.9 shows the results of experiments in which established adult schistosome worms were transferred surgically from infected mice and monkeys into other monkeys. With this transfer of adults from these donor hosts into the experimental monkeys, no further blood flukes became established when these monkeys were exposed to schistosome cercariae, indicating that the transferred flukes had conferred some degree of immunity on their new host. However, in most cases, these adult worms continued to produce eggs as they would have done within their original host.

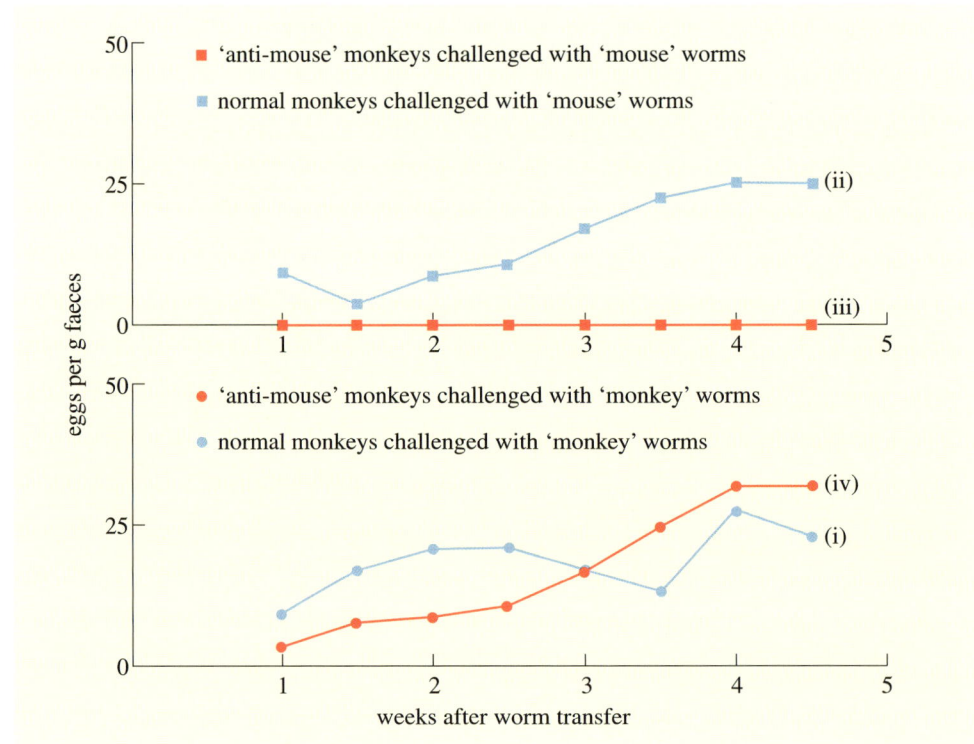

Figure 3.9 Numbers of eggs of *Schistosoma mansoni* in the faeces of normal monkeys and monkeys immunized against mouse antigens, to which pairs of 'mouse' or 'monkey' worms were transferred. The number in brackets represents the experiment number (see text). Data adapted from Smithers (1968).

○ By reference to Figure 3.9 what was the fate of schistosome egg production when:

(i) normal monkeys were challenged with 'monkey' worms;

(ii) normal monkeys were challenged with 'mouse' worms;

(iii) 'anti-mouse' monkeys (monkeys previously immunized against mice antigen) were challenged with 'mouse' worms;

(iv) 'anti-mouse' monkeys were challenged with 'monkey' worms?

● When normal monkeys were exposed to 'monkey' worms (i), egg production began early and reached a peak quickly after four weeks. When parasites from mice were transferred to normal monkeys (ii), egg production was initially checked but after 4–5 weeks, rose to levels comparable with (i). Egg production was non-existent in (iii) where 'anti-mouse' monkeys were challenged with 'mouse' worms but when challenged with 'monkey' worms (iv), eggs rose to levels comparable to (i) and (ii) after 4–5 weeks.

○ What conclusions can you draw from these results about the nature of the immune reaction between the parasite and its host?

● Experiment (i) shows that worms transferred from a first donor monkey to a second monkey produced eggs normally, thus indicating that whatever resistance the worms had to the first host immune response continued in the second host. Experiment (ii) suggests that mouse worms, on transfer to a monkey host, initially show a drop in egg production but re-adapt antigenically to survive the new host's immune system and eventually produce eggs normally. Experiment (iii) shows that it was possible to immunize monkeys against mouse donor worms by 'pre-immunizing' them with mouse antigen. On transfer, the worms carrying mouse antigen acquired when in the donor host, were thus exposed to anti-mouse antibody and died before they could re-adapt antigenically. However, if these 'pre-immunized' monkeys were challenged with monkey donor worms (experiment (iv)) they were unable to mount an effective immune response against the worms before the latter adapted antigenically to their new host antibodies. Thus in experiments (i), (ii) and (iv), transferred worms, on acquiring monkey antigen, in some way become protected from the immune response of the second host and achieve normal egg production.

Most helminths, as we have seen here with blood flukes, have larval stages that migrate through the tissues of the host before settling in a final site. Thus, the immune response is elicited against these stages and may subsequently be effective against further invasion. The situation described in this experiment shows that the host may tolerate and harbour the present infection but may be immune to re-infection, a response known as **concomitant immunity**.

Such experiments show that adult schistosomes may be subjected to an immune response in the bloodstream. We would expect the worms to be coated with IgG or IgE antibody which binds cells that should at least debilitate the parasite.

This immune response is ineffective as the parasite can survive in the bloodstream for a number of years. How do they avoid destruction? As indicated by the above experiment, the evidence points to the fact that blood flukes, and possibly most other parasites, can evade the host response. For example, it is believed that the protoctistan *Trypanosoma* sp. varies its surface antigens to simulate the host's proteins whilst schistosomes are even more sophisticated: they can coat themselves with host antigens, including glycoproteins or glycolipids which are identical to those of host membranes, thus disguising themselves from the host's immune system. It is uncertain whether they are produced by the schistosome or are acquired from host blood cells. This antigenic disguise has been investigated mostly in schistosomes but it is likely that other helminths have also developed immunological devices for maintaining their integrity. Indeed, all parasites may have such mechanisms. Schistosomes, for example, are also believed to inhibit eosinophil attachment whilst invading larvae can break up specific antibodies and inactivate macrophages (Hagen *et al.*, 1993).

We return to discuss immune aspects of host–parasite relationships more fully in Section 3.4.2.

3.2.4 TAPEWORMS

These ribbon-like helminths are almost invariably found as adults in the intestine of vertebrates. Some (order Pseudophyllidea) infect marine fish, birds and mammals and have life cycles similar to those described for digeneans, with shelled eggs that hatch into a free-swimming first larval stage (coracidium) and further larval stages in two intermediate hosts, usually crustaceans and fish, before the adult arrives in the gut of the definitive host.

○ Would you describe this life cycle as 'direct' or 'indirect'?

● Indirect, as intermediate hosts are involved in transfer between definitive hosts.

The cestode order Cyclophyllidea includes some dangerous, but interesting, parasites of humans. The cyclophyllidean life cycle is shown in Figure 3.10.

○ What are the differences between the life cycle shown in Figure 3.10 and the fluke life cycle (Figure 3.4)?

● There is no free-swimming first larval stage, the larval stage being ingested directly by the sole intermediate host.

An **oncosphere** is the name given to the first stage in the cyclophyllidean life cycle, that replaces the shelled egg in the pseudophyllidea. It consists of a living membrane (tegument) that surrounds and feeds a **hexacanth** larva (homologous to the pseudophyllidean coracidium) which has three pairs of hooks (= 6) at one end. In cyclophyllidean tapeworms, as shown in Figure 3.10, the embryo develops to the oncosphere stage within the uterus of the gravid segment.

On ingestion by the vertebrate intermediate host, for example a ruminant mammal, the oncospheres hatch after being subjected to digestive enzymes

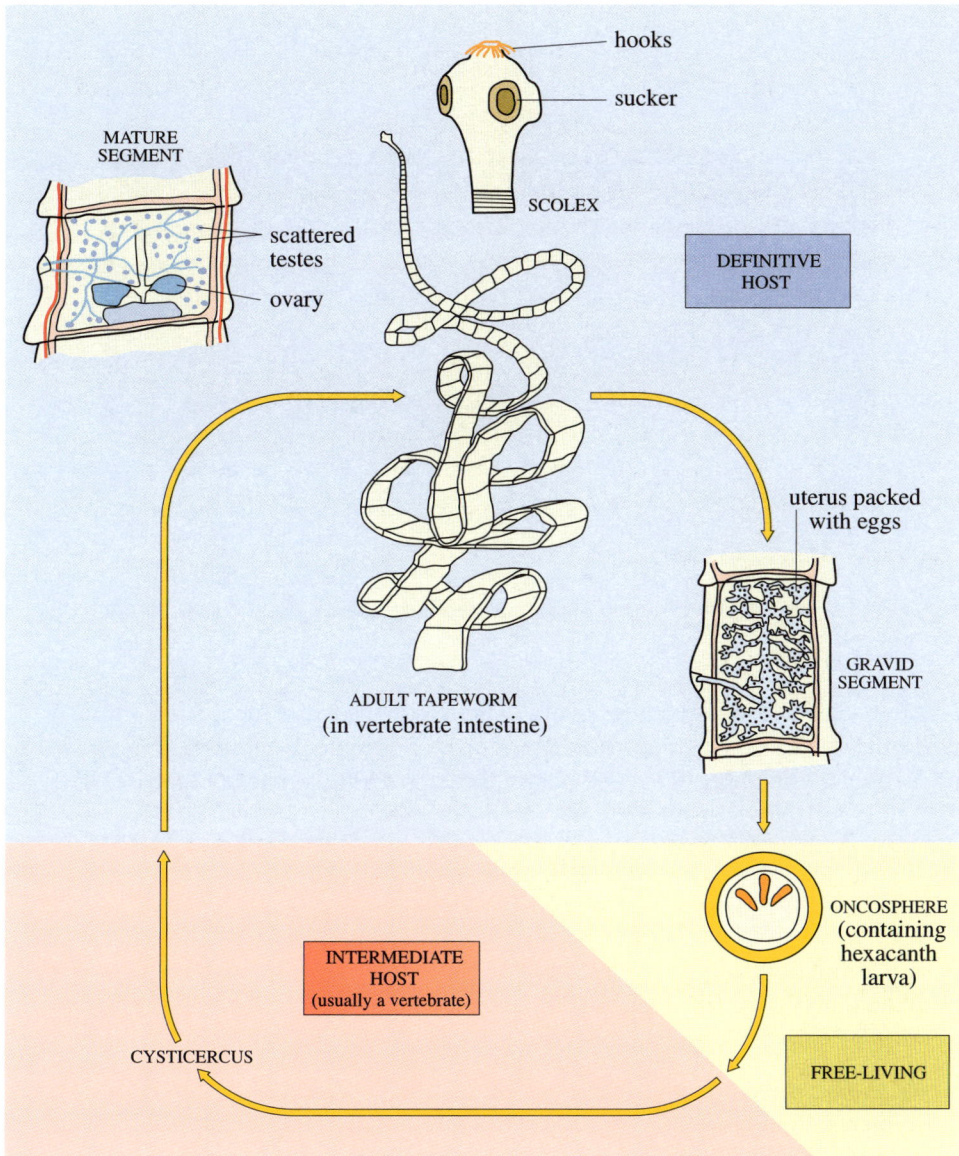

Figure 3.10 Generalized life cycle of a cyclophyllidean tapeworm (not to scale).

within the abomasum and duodenum, releasing the hexacanth larva. Proteolytic secretions from glands in the larva allow it to bore into the mucosa and from there enter the blood circulation. On reaching the host's muscles, the larva develops into a ball-like **cysticercus** which has an invaginated scolex (Figure 3.11). Should the muscle be eaten undercooked, the cysticercus ruptures in the duodenum of the final host, the scolex evaginates and the four suckers attach to the lining of the ileum.

At the rear of the 'neck' region of the scolex, proglottids are formed, one after another, to form a long tape or **strobila**. Each proglottid has a complete set of male and female reproductive organs when mature (Figures 3.10 and 3.12) and although self-fertilization is possible, it is likely that cross-fertilization occurs with ripe proglottids of an adjacent worm. The length of the strobila, which may possess between 1000 and 2000 proglottids, is usually 3–8 m but some specimens

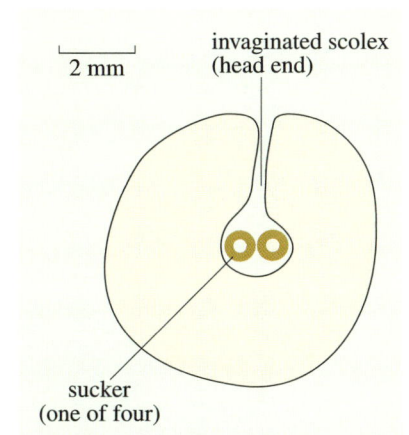

Figure 3.11 Cysticercus of *Taenia saginata,* the beef tapeworm.

proglottids, each with mature
reproductive system

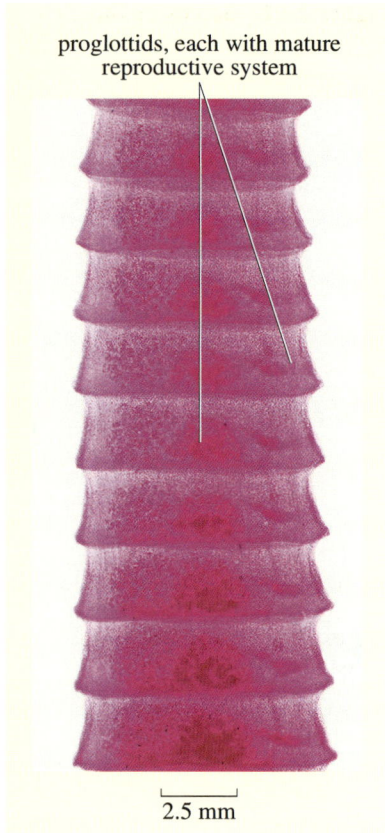

2.5 mm

Figure 3.12 *Taenia* sp. Mature
proglottids showing reproductive
system, producing both male and
female gametes.

uterus packed with oncospheres

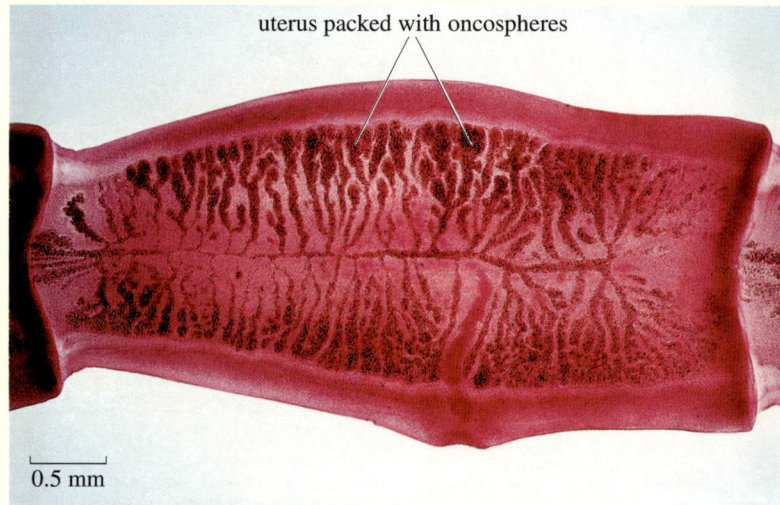

0.5 mm

Figure 3.13 Gravid proglottid of *Taenia* sp. showing extensive uterus containing
oncospheres.

of 12 m have been reported after being expelled by drug treatment from some
unfortunate victims! The tapeworm sheds up to 10 gravid proglottids (Figures 3.10
and 3.13) a day with host faeces. Each of these proglottids may contain 100 000
eggs, each capable of developing into an oncosphere.

The gravid proglottids become detached and shed with the faeces, decaying to
release the oncospheres. In certain countries, where raw sewage is used as
fertilizer, millions of oncospheres may be spread on pasture from infected faeces
but only a fraction find their way into the cattle intermediate host. Little is known
about the viability of the oncospheres but they have been found to be still viable
after one month at a temperature of 30 °C.

Taenia saginata is known as the 'beef' tapeworm, as cattle and other ruminants
act as intermediate hosts, the final host being humans. The scolex of this
tapeworm bears four saucer-like suckers with which it embeds itself between the
villi in the gut of the vertebrate host. Other taeniids, including *T. solium*, possess
fearsome-looking hooks as well as suckers (Figures 3.2d and 3.14).

Although infections with 'beef' tapeworm infections are not unknown in Britain,
modern abattoir inspections and more refined sewage disposal methods have all
but eliminated it. However, be warned should you fancy rare steak, that tapeworm
infection still occurs abundantly in some Mediterranean and African countries.

Echinococcus granulosus (Figure 3.15) is a small tapeworm, with different
strains being found in domestic dogs, foxes and other related Canidae throughout
the world.

○ What is the major difference in body structure of the tapeworm in Figure 3.15
   compared with the *Taenia* species shown in Figure 3.10?

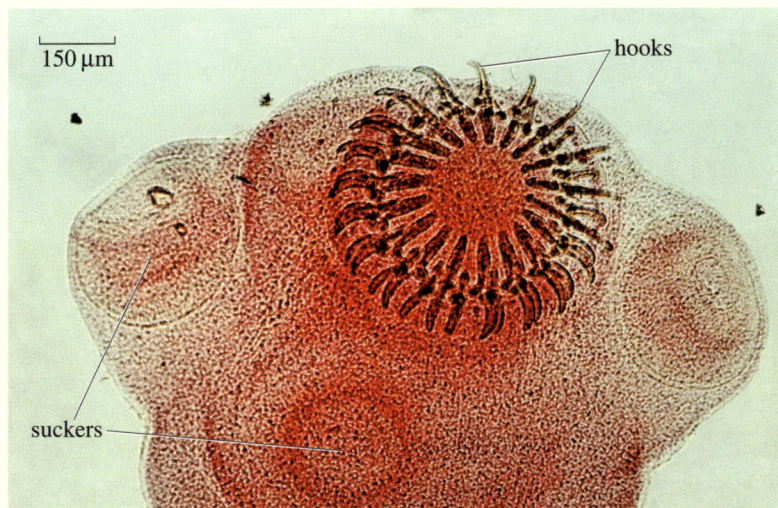

Figure 3.14 Scolex of *Taenia solium* showing suckers and hooks for attachment to gut villi.

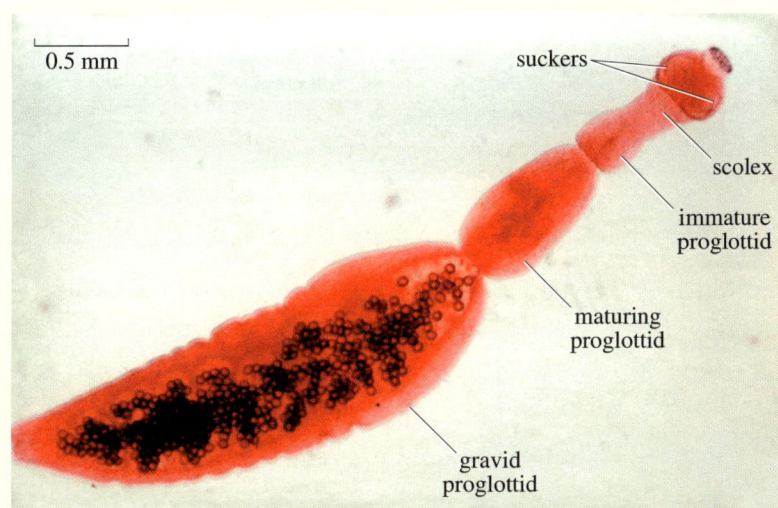

Figure 3.15 An adult *Echinococcus granulosus*.

● *Echinococcus* has only three proglottids behind its scolex and neck regions, with the end one containing the gravid reproductive system.

Having only one mature proglottid at a time, fewer oncospheres are shed by this tapeworm than by typical tapeworms. After being deposited with the faeces of the canid host, the oncospheres are usually ingested by browsing herbivores, such as sheep, but also frequently by humans (Figure 3.16). However, what it lacks in oncosphere numbers, it certainly makes up for by massive production of infective stages within the intermediate host. Hexacanth larvae released from the oncospheres in the herbivore gut bore into the tissues, and probably via the bloodstream, end up in organs, such as the brain, liver and kidneys, or even bone. A common location in sheep and humans is the liver. There the larva (bladderworm) develops into a massive cyst, called a **hydatid cyst** which can contain several litres of fluid.

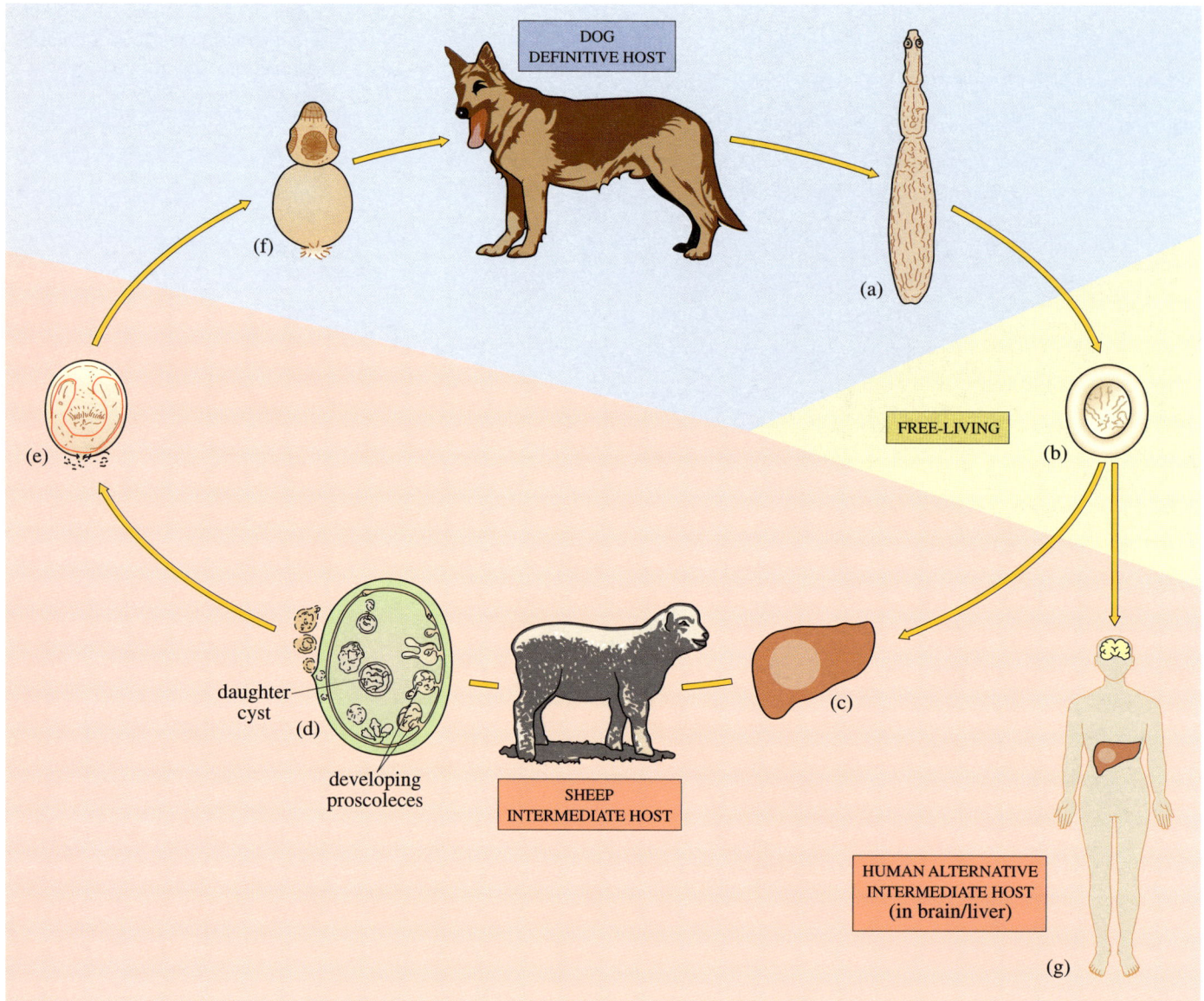

Figure 3.16 Life cycle of *Echinococcus granulosus* (not to scale). Stages in the dog definitive host: (a) adult tapeworm with the final proglottid gravid; (b) oncosphere shed in faeces. Stages in the sheep intermediate host: (c) bladderworm stage in liver; (d) hydatid cyst with daughter cysts and developing proscoleces; (e) one invaginated proscolex released from dead host; (f) one everted scolex in gut of definite host. (g) The human alternative intermediate host.

Each hydatid cyst (Figure 3.17) can produce, by budding, hundreds of thousands of individual **proscoleces** (singular, **proscolex**) which are released and evert when contaminated tissue is eaten by the carnivore final host. As mentioned earlier it is possible for oncospheres from infected dogs to be taken in by humans, in which case a hydatid cyst often forms in the liver, occasionally the brain. Untreatable, it can eventually result in death, the cyst being identified only after autopsy or by brain scans. In Britain it is found to occur from time to time in humans in a few farming communities. Regular dosing of domestic dogs with antihelminthic drugs, especially in sheep-rearing areas, is desirable to eliminate the risk. Sadly, this tapeworm and others, including *Taenia multiceps,* have been found not only in a large number of farm dogs in Wales (Table 3.1) but also in many foxhound packs (Table 3.2) in spite of a high-profile campaign by the veterinary service to highlight the danger (Williams, 1992). *Taenia multiceps* produces cysts similar to those of *E. granulosus* in the brains of sheep, causing symptoms called 'gid' (or 'staggers').

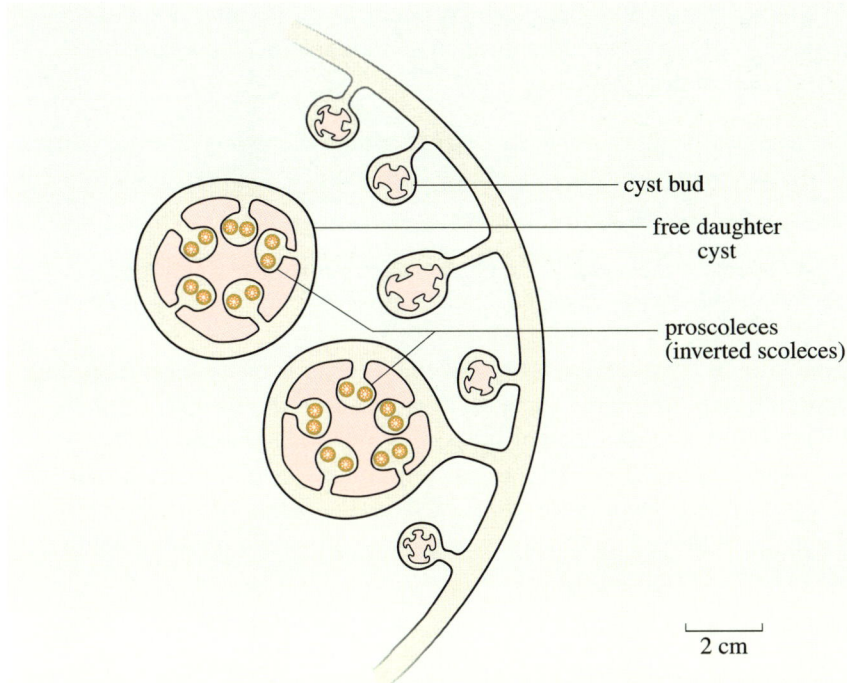

Figure 3.17   Part of a hydatid cyst of *E. granulosus*. The cyst can grow up to 15 cm in diameter.

cyst bud

free daughter cyst

proscoleces (inverted scoleces)

2 cm

Table 3.1   The prevalence of cestodes in farm dogs on 100 sheep-rearing farms in Wales and the number of farms with dogs infected with each cestode species. Data from Williams (1992).

Species of cestode	Number (%) of infected dogs	Number of infected farms
Echinococcus granulosus[*]	48 (15)	17
Taenia multiceps[*]	37 (11.5)	20
T. ovis[*]	22 (70)	7
T. hydatigena[*]	147 (46)	72
T. serialis[†]	14 (4)	3
T. pisiformis[†]	45 (14)	23
Dipylidium caninum[†]	26 (8)	12

[*]Intermediate stages in sheep.

[†]Intermediate stages not in sheep, e.g. in rodents, fleas.

Table 3.2   The prevalence of cestodes in 552 hounds from 12 packs of foxhounds in Wales. (*Note*: some hounds harboured more than one species of parasite.) Data from Williams (1992).

Species of cestode	Number (%) of infected hounds	Number of infected packs
Echinococcus granulosus	162 (29)	8
Taenia multiceps	41 (7.5)	4
T. ovis	32 (6)	6
T. hydatigena	270 (49)	10
T. serialis	6 (1)	2
T. pisiformis	53 (10)	3
Dipylidium caninum	24 (4.3)	2

Table 3.1 shows that certain of these cestodes occur as cysts (intermediate stages) in sheep but some occur in other animals such as rodents, or, in the case of *Dipylidium caninum*, in dog fleas.

○ What do these data suggest about the way in which these dogs, both in farms and in foxhound packs (Table 3.2), become infected with those cysts found in sheep?

● As the cysts occur in the liver of sheep, the dogs must have eaten these organs from the carcasses of recently dead sheep.

Indeed it was common practice at the time of this study for many farm dogs and foxhounds to be thrown offal from sheep that had died on the farms, and some would have been infected with tapeworm cysts. In the dogs, they would have developed into adult tapeworms. The danger to humans lies in the chance of transferring eggs to their own mouths, derived from the saliva of a dog when licked, who may have picked up the tapeworm eggs in its own mouth from the contaminated faeces of other dogs.

The data in Table 3.3 give an indication of the longevity of the infection in dogs.

○ From the data in Table 3.3 what can be indicated about:

(i) the time taken for tapeworm egg production;

(ii) tapeworm survival;

(iii) host resistance?

Table 3.3 Rate and duration of proglottid release of *T. multiceps* in five dogs experimentally infected by feeding with scoleces of *T. multiceps*. Data from Williams (1992).

Dog no.	Number of T. multiceps scoleces fed	Day on which gravid proglottids first detected	Number of proglottids detected per day		Number of days on which no proglottids detected	Last day on which proglottids detected	Number of T. multiceps recovered
			Range	Average			
1	2	42	0–7	1.9	32	140	0
2	5	51	0–8	2.1	16	180[†]	2
3	5	61	0–4	1.7	19	180[†]	2
4	10	47	0–16	2.8	8	121	0
5	5	49	0–11	3.4	11	180[†]	3
5[*]	30	56	1–21	4.3	2	147	19

[*]Reinfected 20 days after artificial termination of the first experimental infection.

[†]Infection terminated.

● (i) Development of proglottids to the gravid condition (see Figure 3.10) takes around 40–60 days.

(ii) Gravid proglottids were still being detected at 180 days when the infection was terminated by administration of a drug. Thus the tapeworm probably survives in the gut considerably longer than this time.

(iii) When dog 5 was reinfected 20 days after the artificial termination of the first infection, tapeworm development and egg release showed no significant difference from the preceding infection, indicating, in this instance at least, that no immunity had been achieved by the host.

We have already mentioned aspects of host immunity to blood flukes in Section 3.2.3, and will continue the discussion further in Section 3.4.2. There is little evidence for host immunity against cestodes. The host could kill the cestode or expel it from the intestine or break up the strobila so that it did not produce gravid proglottids. However, the experimental infections summarized in Table 3.3 are light compared with some of the field infections from which it is much more difficult to obtain evidence.

The fact that adult tapeworms survive in their final host for months, often years, contrasts with the situation mentioned earlier, for digenean parasites, whose lifespan as an egg-producing adult is usually very much shorter.

3.2.5   ROUNDWORMS

Helminths of the phylum Nematoda — roundworms — are long and slender (Figure 3.18), round in cross-section, and move by whip-like lashings of the body.

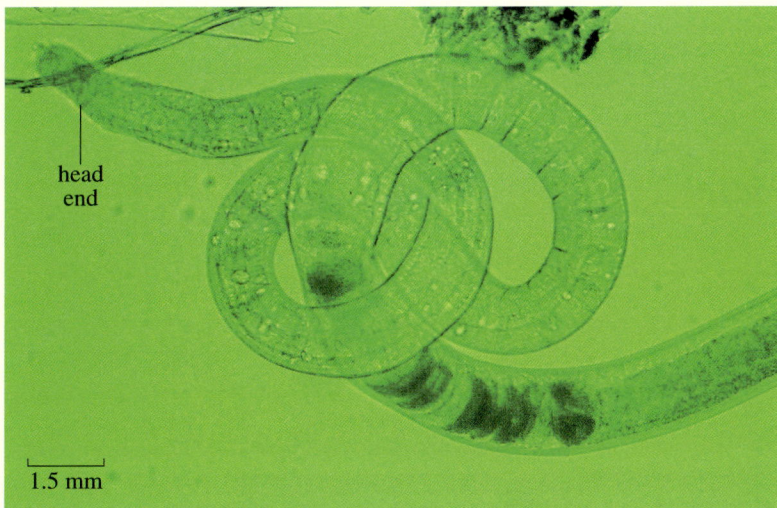

Figure 3.18   An adult nematode worm, *Trichuris trichiura*. This species is a parasite of the colon, found particularly in children living in overcrowded, unsanitary conditions.

Roundworms are distributed widely in all terrestrial and aquatic habitats, being particularly abundant in soil and freshwater. The parasite morphology is very little changed from free-living nematodes whose life cycle involves a series of larval stages and an adult stage (Figures 3.19 and 3.20). One of these larvae, usually, but not always, the third stage, is the infective stage of parasitic species. The larvae grow and pass from one stage to the other by a series of moults which cast off the cuticle that surrounds the ectodermal layer of the body wall and lines the buccal cavity and the rectum.

In comparison with flukes and tapeworms, where the larval stages are quite different morphologically (in flukes: mother sporocyst, redia, cercaria, metacercaria; in tapeworms oncosphere, hexacanth larva, cysticercus, hydatid cyst), all nematode larvae resemble immature adults. Indeed they are more correctly referred to as 'juvenile' stages.

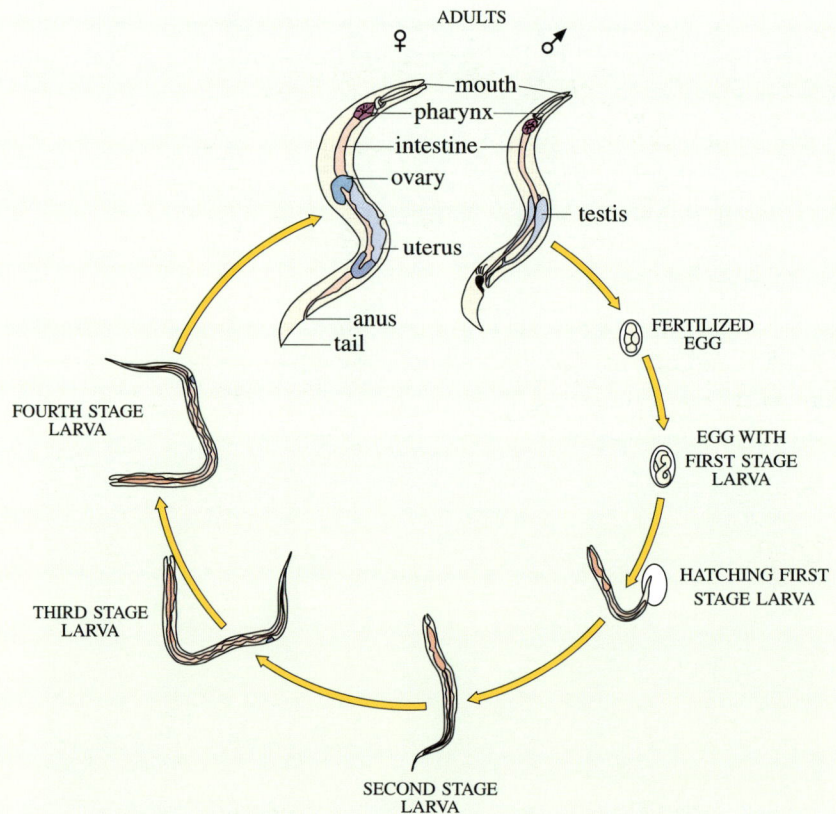

Figure 3.19 Generalized roundworm life cycle (not to scale). The adults produce fertilized eggs that hatch to release larvae. There are four larval stages, each followed by a moult. The infective stage is often, but not invariably, the third stage larva. *Note*: some species of roundworm are hermaphrodite.

As parasites, nematodes occur either as adults or juveniles in many different kinds of tissues and systems of both vertebrates and invertebrates and also in plants. In humans, intestinal nematodes can transfer directly from one host to another via invasive juvenile stages, whilst in some nematodes that live in blood and other tissues, the juvenile is carried from one definitive host to another by insect vectors, or by passive intermediate paratenic hosts.

An example of roundworms with a direct life cycle are the hookworms (Figure 3.20). Two different species infect humans, namely *Ancylostoma duodenale* in southern Europe, Africa and Asia and *Necator americanus* in the tropics and southern parts of North America. In both species, eggs hatching from shed faeces release a juvenile stage, called a rhabditiform larva, which eventually moults twice into an infective form. This third stage juvenile or filariform larva is capable of penetrating bare human skin, should a shoeless victim happen to walk by. The male and female adult worms end up in the intestine where they cut into a plug of mucosa of the intestinal wall by means of fearsome cutting plates in their buccal capsule (Figure 3.21) causing bleeding and interference with nutrient absorption. An infection by these worms causes severe anaemia and undernourishment.

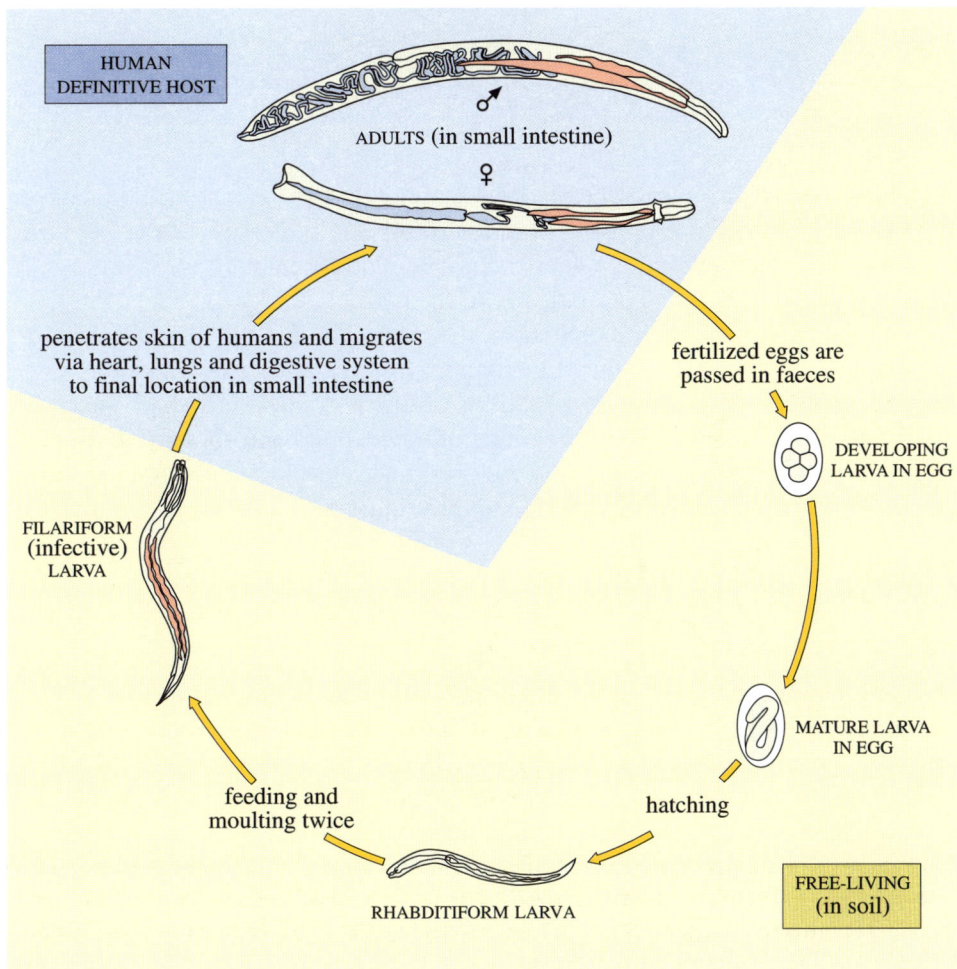

Figure 3.20 Life cycle of hookworms (not to scale).

Figure 3.21 Anterior of the hookworm *Ancylostoma duodenale* showing the buccal capsule. The adult sucks a plug of mucosa into this capsule and feeds on the blood that escapes from the cuts made by the cutting plates.

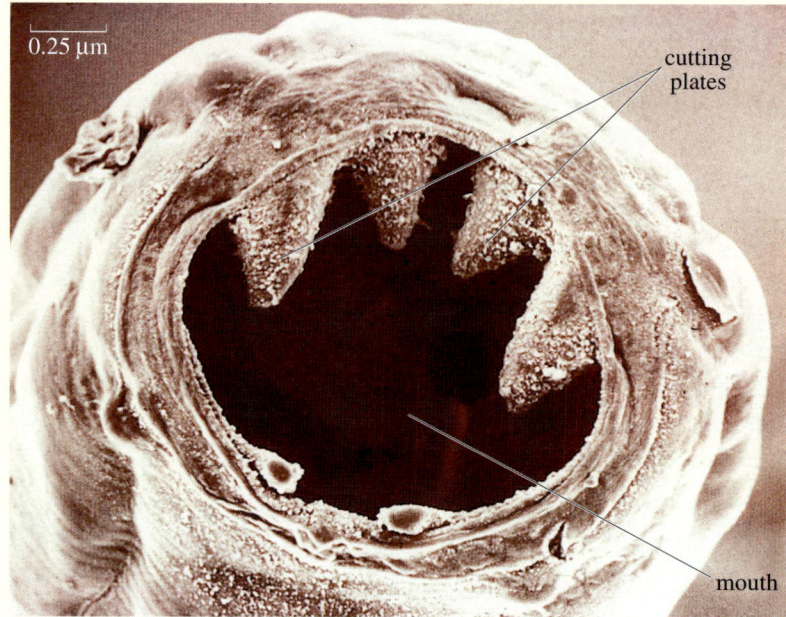

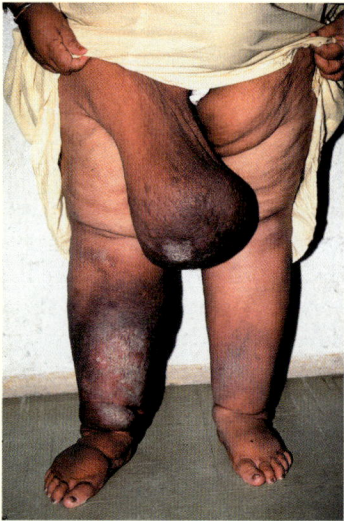

Figure 3.22 Elephantiasis of the leg caused by the presence of the nematode worm *Wuchereria bancrofti* in the lymph system.

Wuchereria bancrofti is a roundworm with an indirect life cycle. The adults occur coiled in the lymph glands or ducts of humans, especially in parts of the Far East. Eggs hatch and release minute juveniles into the bloodstream. These larvae, called microfilariae, migrate periodically to the peripheral blood vessels where they are taken up by an intermediate host, the mosquito, in a blood meal. Within the mosquito, they migrate to the thoracic muscles and, after growth and moulting, eventually pass into the proboscis of the mosquito to be shed onto the skin during blood sucking. They penetrate into the wound and are carried to the lymph system where they mature. *Wuchereria* causes inflammation and blockage of the lymph system, eventually producing grotesque fleshy deformity in their hosts called 'elephantiasis' (Figure 3.22), for which there are few effective treatments.

3.2.6 PARASITIC ARTHROPODS

Whether all arthropods that associate closely with other organisms can be regarded as parasites is a debatable point. Crustaceans such as *Lernaea* (class Copepoda) and *Sacculina* (class Cirripedia) (Figure 3.23) are almost unrecognizable as arthropods in the parasitic adult form in which they are found in fish or other crustaceans. Their taxonomic status can only be identified by their larval stages, for example the nauplius stage or the cypris stage (see Figure 2.14c, d). Similarly, the Pentastomida (Figure 3.24) is a small group of animals (about 90 species) which, as adults, are parasites in the respiratory tract of vertebrates, especially reptiles. It is thought that this very ancient group of parasites has been associated with their reptile hosts since the Mesozoic. The selection pressure arising from such a long association in a very specialized habitat has resulted in a body structure that makes it impossible to be certain about their origins or taxonomic relationships. Adults are worm-like with a blood-filled body cavity in

(i)

head rooting processes
in flesh of gill

body   (b)

two egg-sacs
coiled together

gill filaments of fish

(c)

(a)

♀

(d)   ♂

(ii)

(b)

rooting processes
in crab

(c)

external sac
produces eggs

(a)

parasite

crab gut

(d)

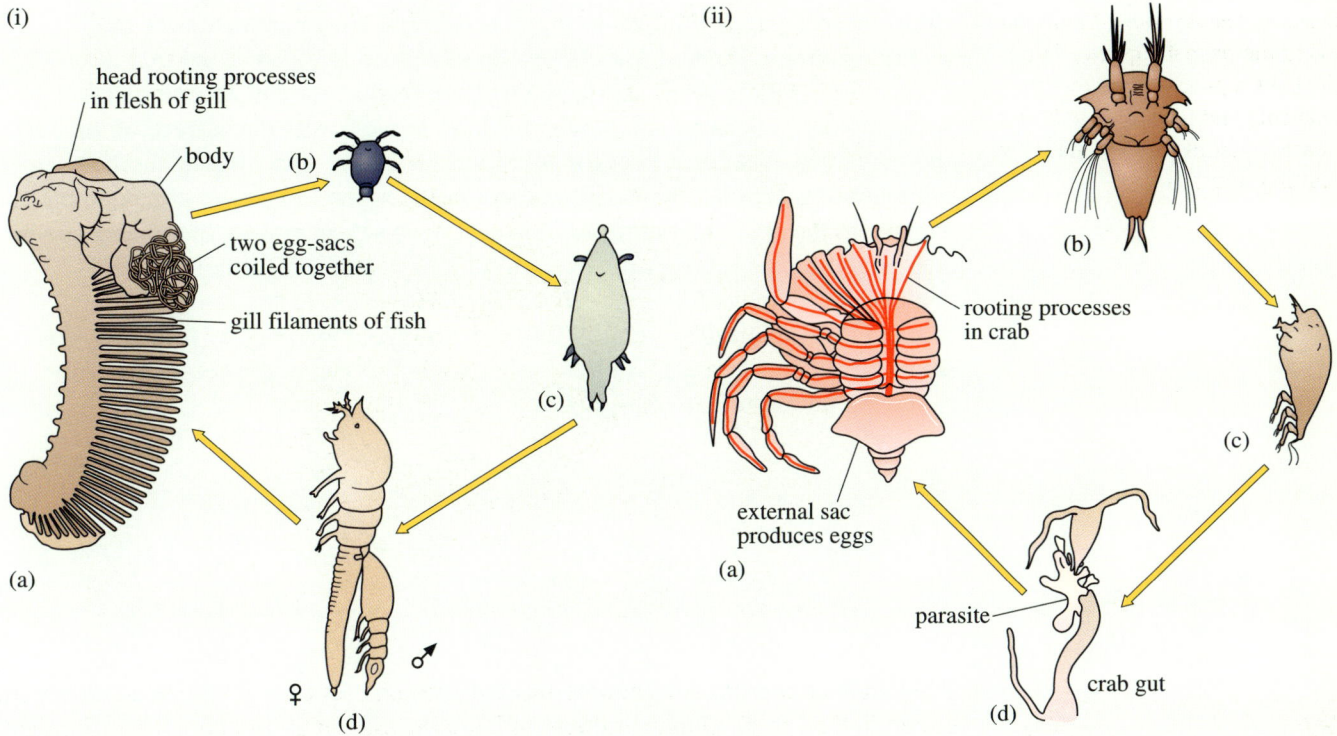

which simple organs 'float'. Only the larval stages have some resemblance to arthropods. Once placed in a separate phylum, they are now considered to be related to crustaceans and are placed in a class of that subphylum. They usually have an indirect life cycle via invertebrate intermediate hosts but some transfer directly when taken in with food by a reptile. Several larval moults occur, the hooked appendages of the larva assisting in migration through host tissue.

Figure 3.23   Parasitic crustaceans. (i) Stages in the life cycle of *Lernaea* (Copepoda): (a) adult female attached to gill arch of codfish; (b) larva hatches from egg and swims to flatfish; (c) stage that attaches to flatfish gill; (d) male and female become free-living and mate. The males die and females attach to the gills of codfish as in (a). (ii) Stages in the life cycle of *Sacculina* (Cirripedia): (a) a crab *Carcinus meanus*, infected with rooting processes and carrying sac with gonads; (b) newly-hatched larva; (c) larva that attaches to crab; (d) adult parasite inside the crab.

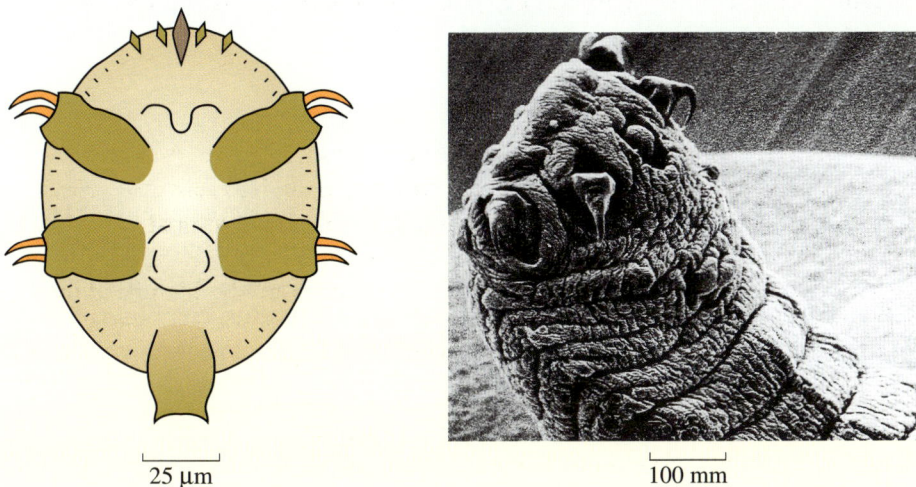

25 μm

100 mm

Figure 3.24   (a) Larval pentastomid bearing four double-hooked appendages and penetration structures at anterior end for tissue migration in host. (b) SEM of a young adult pentastomid. Note the four prominent hooks at the anterior end and the worm-like body.

Most crustacean parasites, other than *Lernaea* and *Sacculina*, are only slightly modified compared with free-living relatives and are ectoparasites of fish, occurring on the skin and sometimes the gills. Fish lice, *Argulus* spp. (class Branchiura) (Figure 3.25), are found on both freshwater and marine fishes. Although obligate parasites, they are able to move across the surface of the host and may even abandon a host for short periods before attaching to another. The salmon louse, *Lepeophtheirus salmonis* (class Copepoda) causes considerable damage to the skin of Atlantic salmon as it browses on mucus, blood and skin cells. In wild populations of salmon, the incidence of infection is low and pathological damage has not been reported. However, in sea-farmed salmon, it results in high mortality. Salmon find their zooplankton prey by following steep salinity gradients and the fish lice aggregate in the vicinity of these gradients following odour trails to their hosts.

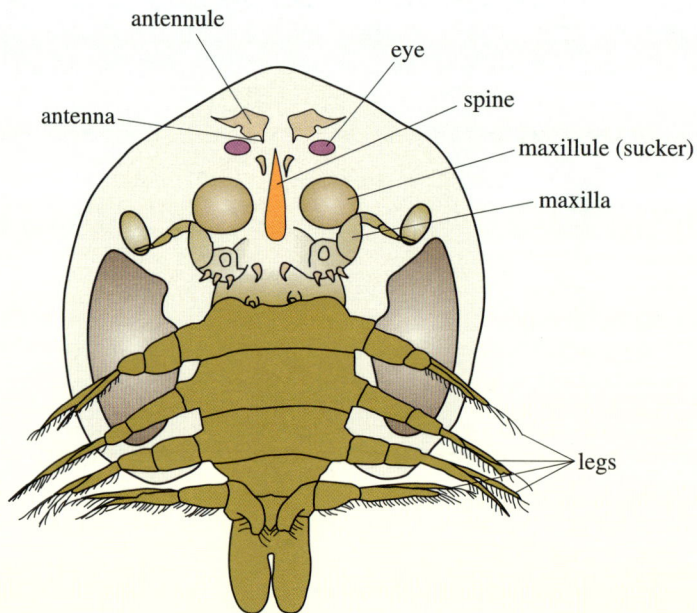

Figure 3.25 Ventral view of *Argulus foliaceus*, a branchiuran ectoparasite of fish (not to scale).

Bloodsucking flies such as mosquitoes and tsetse flies (class Insecta, order Diptera) are sometimes described as parasites and do indeed fall within our definition but their association with the host is of short duration. Fleas (class Insecta, order Siphonaptera) (Figure 3.26), lice that infect humans (class Insecta, order Phthiraptera) (Figure 3.27) and, among arachnid parasites (class Arachnida, order Acarina) ticks (Figure 3.28) all suck blood from their hosts, whereas parasitic mites (Figure 3.29) feed on skin cells. All these ectoparasitic arthropods have an association with their hosts which probably extends to the period shortly after the hosts' evolutionary origins. They are perfect examples of parallel evolution of the parasitic habit on hosts with contrasting life-styles.

Figure 3.26 Adult flea, *Pulex irritans*.

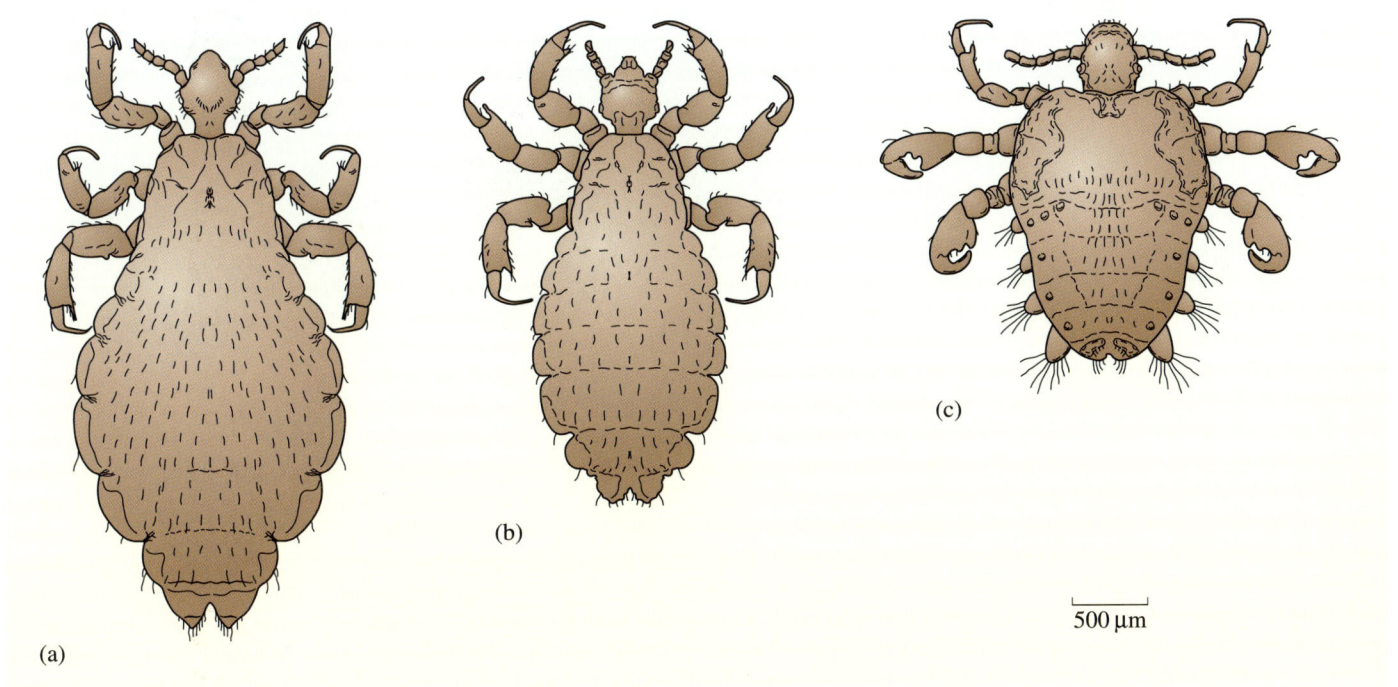

Figure 3.27 Adults of human-sucking lice: (a) *Pediculus humanus* (body louse); (b) *Pediculus capitis* (head louse); (c) *Phthirus pubis* (pubic or crab louse).

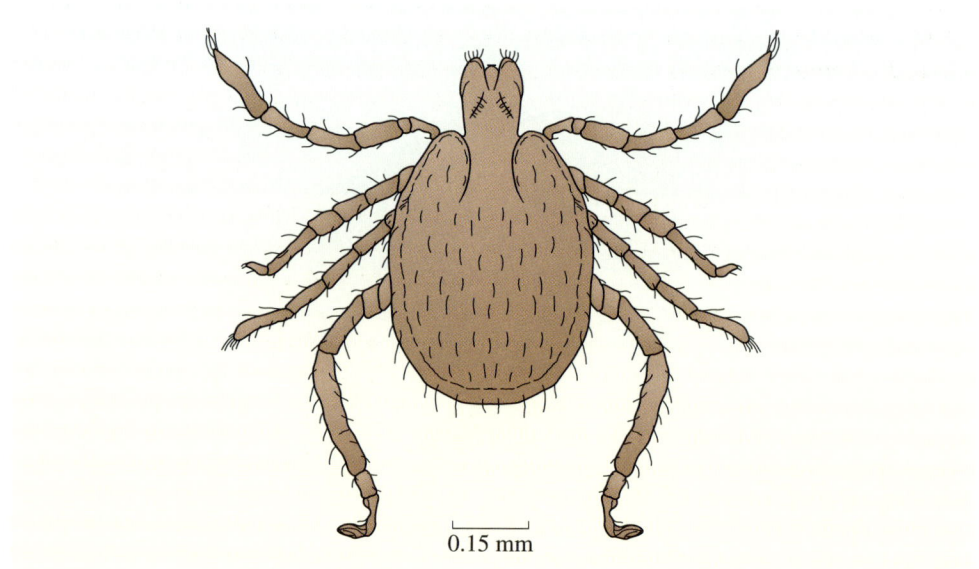

0.15 mm

Figure 3.28 *Ixodes ricinus*, a tick found on cattle and other ungulates, including deer, throughout the UK.

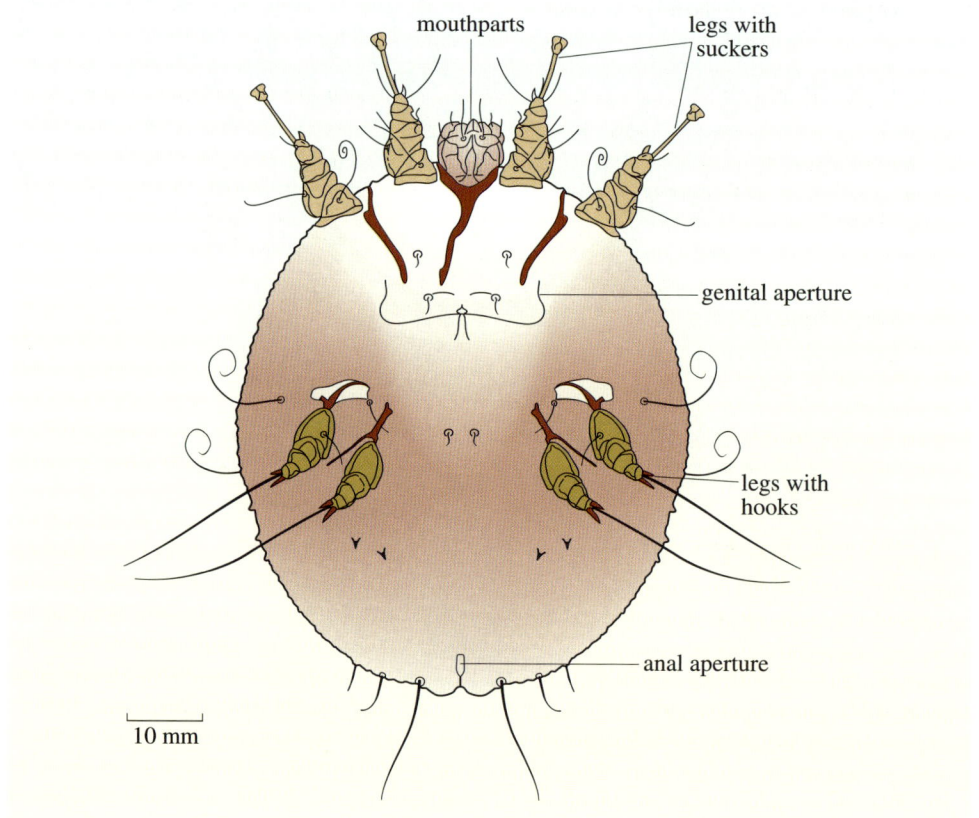

Figure 3.29 Female itch mite, *Sarcoptes* sp.

These parasites are often specific to one species of terrestrial host, or at least to closely related species. Human lice, for example, feed exclusively on human blood. *Pediculus capitis* is the head louse, *P. humanus* the body (or clothes) louse and *Phthirus pubis* the pubic (or crab) louse. All are transmitted from host to host by close bodily contact. Lice cause skin irritations and sometimes allow secondary bacterial infections to develop. Fleas, such as *Pulex irritans* (Figure 3.26), on the other hand, can show less specificity to a host and, as a result, transfer readily between human and rodents. Unfortunately, fleas feeding on rats may ingest the bacillus, *Yersinia pestis* with their blood meal and, should their next meal be on a human, to which they can jump if in close proximity, the bacilli are regurgitated with saliva into that host. They cause septicaemia and the disease known as bubonic plague, a scourge of past populations when both host species become crowded together such as in time of war. Dog and cat fleas, *Ctenocephalides canis* and *C. felis* respectively, can be penetrated by the larvae of the tapeworm *Dipylidium caninum* (see Tables 3.1 and 3.2) hatched from eggs in the host faeces. These larvae penetrate the gut wall of the larval flea and mature into cysticerci (Figure 3.11) when the flea matures. The definitive hosts become infected by swallowing the flea whilst grooming, the cysticerci everting into tapeworms which attach to the gut lining.

Skin infestations by mites do not have such serious consequences but the mange mite *Sarcoptes* sp., which can excavate tunnels in the epidermis of its host as it metamorphoses from nymph to adult stage, causes scabies, an intense itch, often common in young children who spread the mite by skin contact.

Ticks (Figure 3.28) are mostly intermittent ectoparasites of terrestrial vertebrates. They may feed occasionally on humans to whom they attach from vegetation on which they spend some of their time between blood meals. If they do so, they sometimes transmit harmful rickettsial microbes, such as those that cause Lyme disease in country walkers, an infection normally restricted to deer.

Many wasps and wasp-like insects (class Insecta, order Hymenoptera) and a few flies (order Diptera) often lay eggs directly into the body of another insect where they hatch and develop into a larva which completely devours the body juices and tissues of the host.

Because their life cycles are not clear-cut cases of parasitism, these creatures are referred to as 'parasitoids'. Free-living adults (Figure 3.30) generally resemble their non-parasitoid relatives. Adult females locate their hosts by systematic searching and lay eggs either directly into the tissues of the host or in its immediate vicinity from a specialized ovipositor which may also secrete a toxin to paralyse the host. On hatching, larval endoparasitoids develop within the host body whereas ectoparasitoids live externally but with their mouthparts buried in the body of their host. Eventually, both types pupate within the host, finally emerging as a new adult. Most parasitoids are solitary and compete with other species whose eggs may be laid in the same host but, occasionally, a female lays her eggs directly into the pupa of a developing parasitoid of another species already in the host, a case of hyperparasitism (Section 3.1). Parasitoids are of great importance in natural and agricultural ecosystems as they may regulate the population density of their hosts.

Figure 3.30 Adults of some main groups of insect parasitoids (not to scale). (a)–(d) Hymenoptera; (e) a dipteran. The figures illustrate the position taken up by the female when ovipositing her eggs into a host.

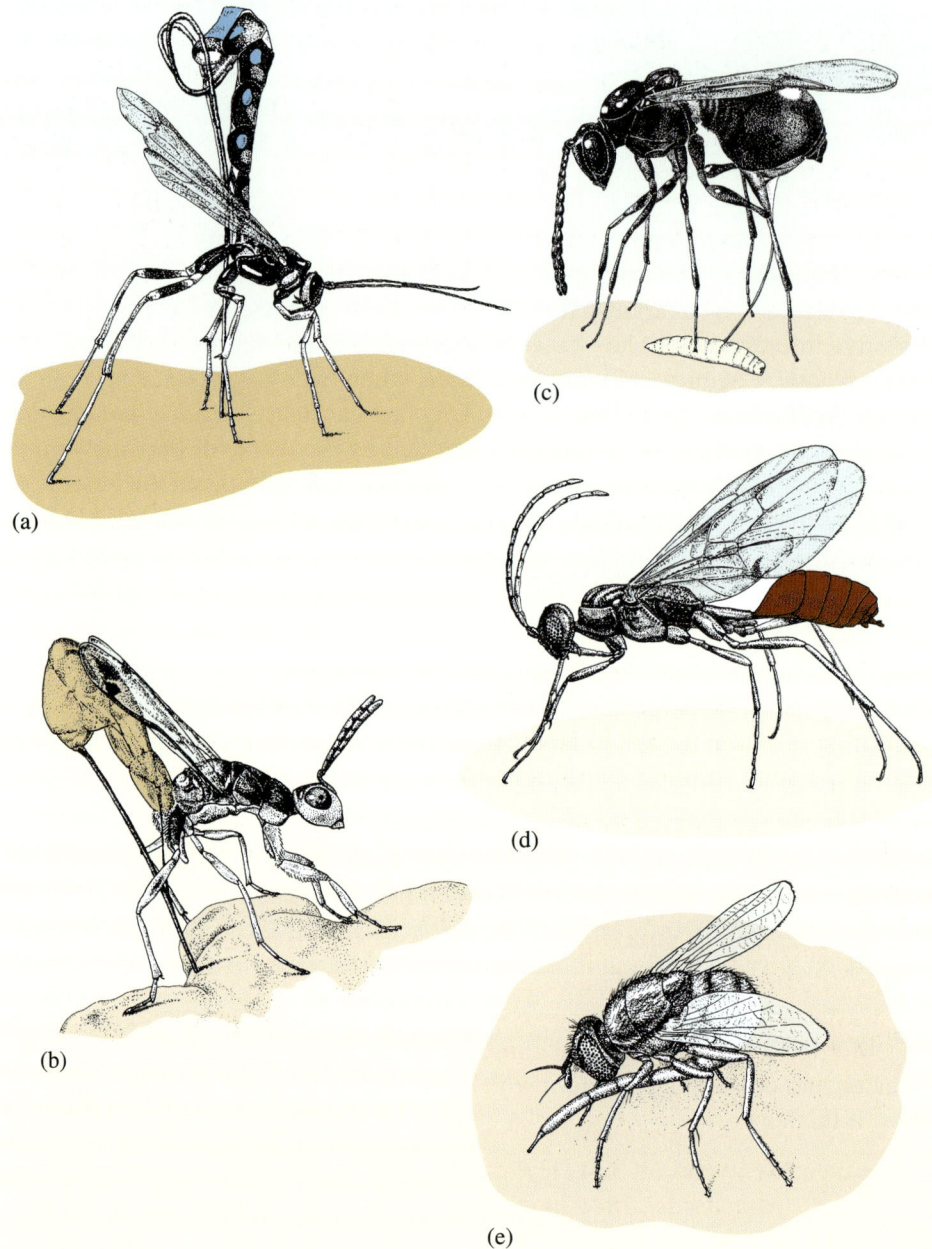

3.2.7 PARASITIC MOLLUSCS

Parasitism is not common in the phylum Mollusca but has evolved amongst some groups of gastropods (Figure 3.31). Ectoparasites such as *Brachystomia* (Figure 3.31b) have a sucking pharynx and chitinous jaws which clamp on to the tissues of their bivalve host.

Endoparasites are found only in two closely related families of primitive gastropods. *Stilifer* sp. (Figure 3.31a) occurs embedded in the body wall of echinoderms. These snails possess a shell and a sucking proboscis but have a

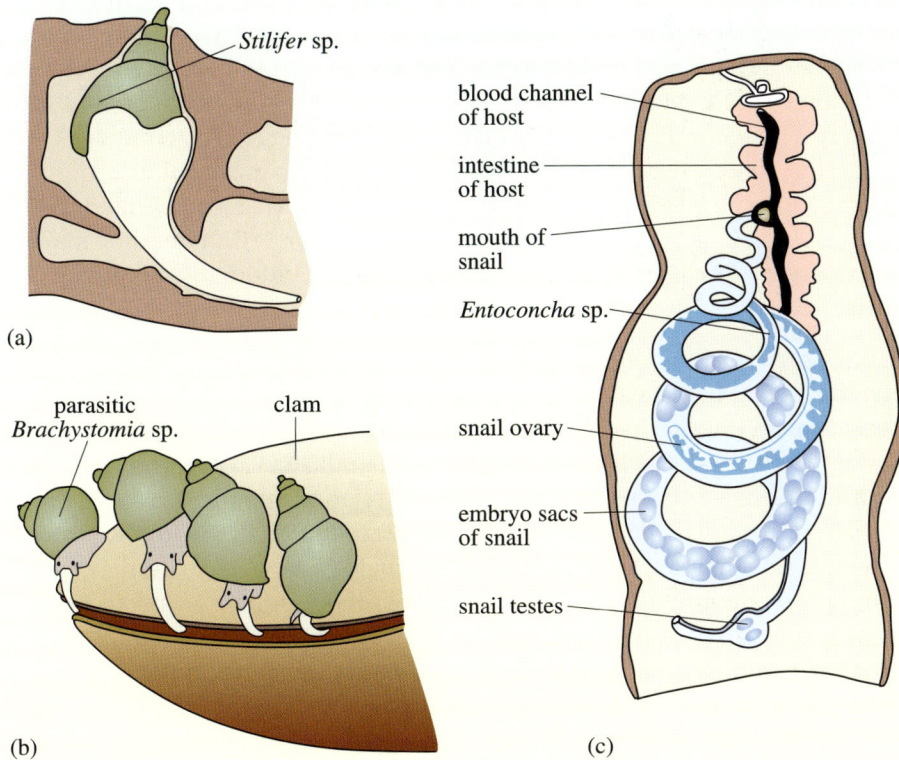

Figure 3.31 Parasitic molluscs (not to scale). (a) *Stilifer* sp. embedded in body wall of a sea-star (Echinodermata, Asteroidea). (b) *Brachystomia* sp., ectoparasitic on clam (Mollusca, Bivalvia), shown feeding on body fluids. (c) *Entoconcha* sp., an endoparasite within the body cavity of sea-cucumber (Echinodermata, Holothuria).

vestigial foot. *Entoconcha* sp. (Figure 3.31c) shows the greatest modification of the basic body plan, being worm-like and only possessing a shell during their larval development. These snails clamp their mouth region on to a blood vessel of their echinoderm host (see Section 1.4.2), but, having lost their gut, take up nutrients through the body wall.

○ Why do they use their mouth in this way?

● For attachment to host tissue.

Interestingly, a few species of freshwater bivalve molluscs such as *Anodonta* and *Unio* have larvae, known as glochidia, which attach to the fins of a fish host by hinged spines on their shell. They become encysted within host tissue, which they digest and absorb nutrient from, before metamorphosing into a young mussel which drops off the fish to live independently as an adult.

3.2.8 CONCLUSION

We began this section by stating that parasitism cannot be regarded as an unusual existence. The life cycles given in this section should have shown you that parasites are wonderfully adapted to their existence and have evolved a number of intricate life-cycle strategies. These strategies are based on the requirements of direct or indirect transfer between definitive hosts. We have seen that such transfer modes may be *passive* where one or more intermediate hosts are eaten by others further up the food chain, each host passing on a stage of the life cycle. An example is the tapeworm, where an oncosphere is taken in by an intermediate host, hatches into a hexacanth larva that metamorphoses into a cysticercus (which may, in some species, form a hydatid cyst that multiplies by budding) and which ends up in a predator of that intermediate host. This category also includes vector transmission, direct transfer between definitive hosts by means of a carrier organism. Other strategies involve active transmission via free-moving larval stages which seek out new hosts, either intermediate or definitive (they may have indirect or direct life cycles), such as digenean flukes with free-swimming cercariae.

This brief review of the different kinds of parasites found in the major animal phyla, shows that parasites do not form a homogeneous group. Some groups of animals contain only parasitic species (e.g. flukes and tapeworms) whilst others have only a few representatives (e.g. gastropod molluscs). The exact body forms of parasitic species reflect the type of intimate relationship they have with their host. Indeed, we made the point early in this section that the use of the word 'degenerate' is not one which describes such highly specialized animals. Certainly they are modified compared with free-living relatives and every group illustrates a diversity of modifications for what essentially are similar functions, those of maintaining a position in or on another organism, feeding off it and making sure that its offspring reach a new host.

All organisms show a structural division of labour. Parasites go further, in having evolved different structural stages within their life cycles, e.g. the miracidium, cercaria, redia, sporocyst and adult forms found in monogeneans and digeneans are different morphological phases of one individual, reflecting the complexity of a life-style involving a number of intermediate hosts. Nematodes, on the other hand, have juvenile transfer stages which, to a large extent, are small adults with a similar morphology. The larval stages of parasitic molluscs and crustaceans are indistinguishable from those of their free-living relatives, yet the adults are totally modified for their different way of life.

Although different groups of parasites can be quite diverse in structure, within taxonomic groups evolutionary pressure means that sibling species often show identical morphology even though they may inhabit different microhabitats. Modern tools of molecular biology, such as DNA profiling, have aided taxonomists in separating such species and show that there may be many more species of parasites within each phylum than can be separated on morphological grounds.

○ What do the body forms of the various kinds of parasites illustrated in Figures 3.1–3.4, 3.6, 3.14, 3.15, 3.18, 3.23, 3.24 and 3.31c have in common?

● They all tend to be worm-like in appearance.

This body shape adapts them to occupy tissues and organs such as the gut, blood system, bile duct, etc. and especially to taking up nutrients across the body wall. Endoparasites lie amongst their food source, e.g. in the blood or gut so a long and flattened body shape allows a larger surface area with more contact with the food.

We turn to a closer examination of this feature of parasites in the following section.

SUMMARY OF SECTION 3.2

1   Helminth parasites include the flukes and tapeworms (Platyhelminthes), the roundworms (Nematoda) and spiny-headed worms (Acanthocephala).

2   Monogenean flukes with direct life cycles are ectoparasites of vertebrates. Digenean flukes, which may have one or more intermediate hosts between the molluscan primary host and the vertebrate definitive host, are endoparasites.

3   Blood flukes show extreme adaptation of morphology and adaptive responses to the immune system of their hosts.

4   Tapeworms with indirect life cycles occur in the gut and are characterized by having a scolex with suckers and hooks and a segmented body with repeating units or proglottids, the sexually mature in the middle and the gravid at the end. They occur in the gut.

5   Roundworms may have direct or indirect life cycles.

6   Parasitic arthropods range from the highly modified forms such as the pentastomids and *Sacculina* to those which barely differ from free-living forms.

7   Fleas, lice, mites and ticks are ectoparasites, relatively unmodified compared with their free-living relatives, and often act as vectors of disease-causing microbes.

8   Parasitoids are insect larvae which always kill their host by feeding on its internal tissues.

9   A few gastropod molluscs are ecto- or endoparasites.

3.3   PARASITE NUTRITION

Broadly speaking, parasites lie surrounded by a food source or on the surface of it. Two kinds of nutritional mechanism can be identified: parasites that have a functional gut obtain most of their nutrients by digestion and absorption in the same way as free-living forms; those without a gut have to rely on uptake of digested nutrients across the body wall. However, the situation is more complicated and both types of nutrient uptake often occur in those with a mouth and gut.

○   Of those parasites covered in Section 3.2, which types would utilize only the body wall routes for absorbing nutrients?

●   Digenean mother and daughter sporocysts and tapeworms.

Tapeworms lack a mouth or gut and thus the only route for absorption of nutrients is across the body wall or tegument.

Figure 3.32 Diagrammatic representation of a section through the tegument of a tapeworm.

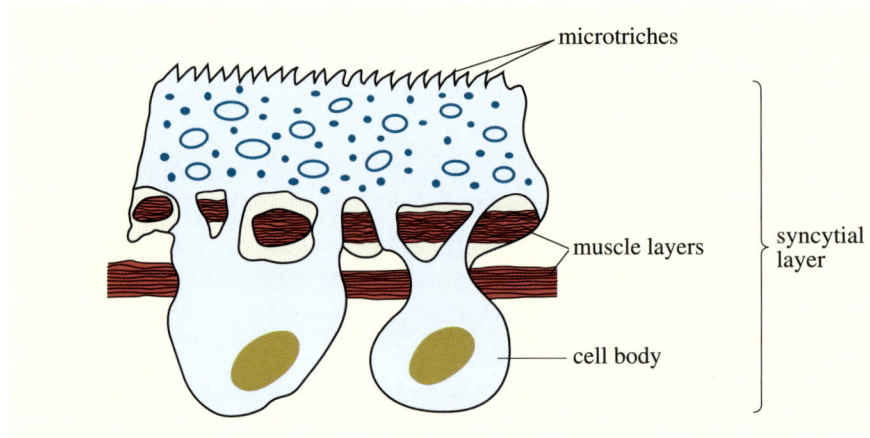

The tegument (Figure 3.32) consists of a syncytial layer that forms a thin absorptive surface. Small microvilli-like extensions, called microtriches (pronounced 'micro-trikes'), increase the surface area for absorption.

Only limited information is available about the transport systems in these parasites but it is likely that they share many of the features of vertebrate intestinal cells. It is not entirely clear how the tapeworm avoids being digested by the host, but the apical plasma membrane of the tapeworm tegument is coated with mucopolysaccharides and glycoproteins forming a very thin glycocalyx, which is continuously replaced and may well protect against host digestive enzymes. Knowles and Oaks (1979) showed that the protease-sensitive peptide sequences of the proteins making up this layer are protected from enzyme attack. It is also known that this layer is capable of binding high molecular mass substances such as host digestive enzymes to its surface. Thus, it may confer a competitive advantage on the worm, allowing it to carry out 'membrane digestion' and thus take up nutrients more efficiently than the host intestinal cells.

Adult monogenean and digenean parasites feed mainly via the gut, taking up blood, mucus, host tissue or host gut contents depending on their location. However, the structure of the tegument in both groups is very similar to that of tapeworms and there is evidence that the tegument, particularly in the Digenea, is permeable to sugars and amino acids which may be absorbed by facilitated diffusion. Pinocytosis, endocytosis and phagocytosis are also known to be involved in some Digenea.

With very few exceptions such as in filarial worms, the cuticle of nematodes (Figure 3.33) does not have an absorptive function, instead being protective as well as important in the whip-like movement brought about by antagonistic action between its fibrous layer and the underlying muscle blocks. Fluid food is pumped into the intestine by the oesophageal muscles and, after digestion, wastes are voided via the anus. Some blood-feeding nematodes are able to secrete anticoagulants from pharyngeal glands, and solid tissue feeders sometimes secrete enzymes which break down the host's tissue.

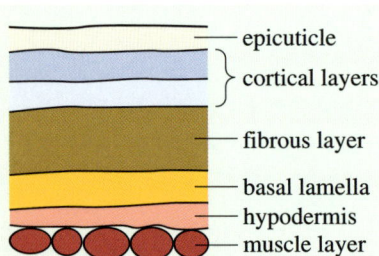

Figure 3.33 Section through the body wall of a roundworm.

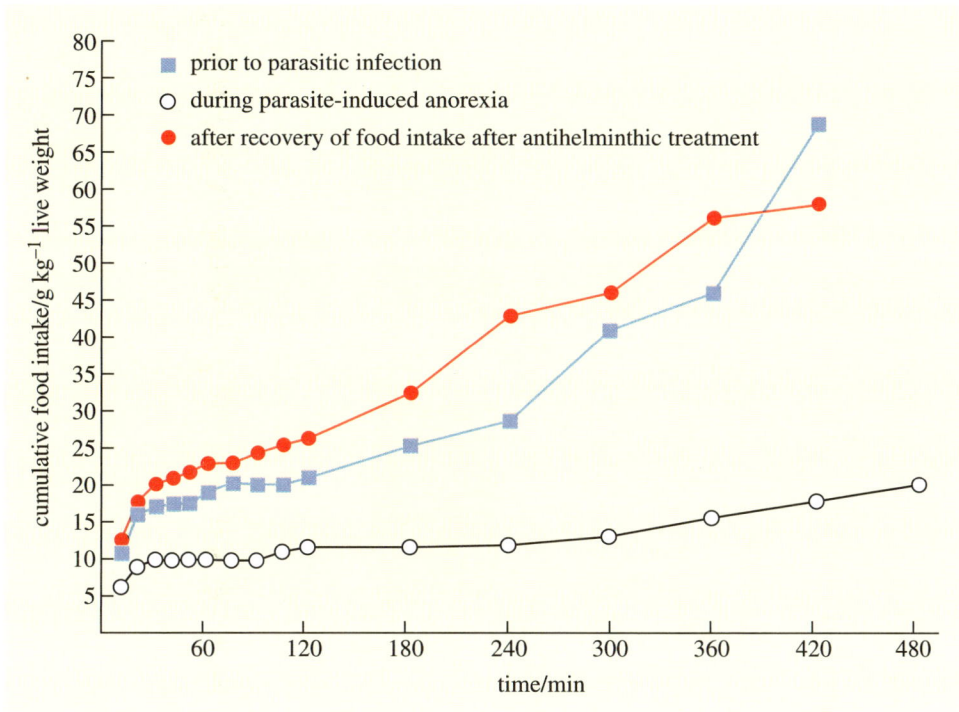

Figure 3.34 The short-term (over 8 hours) cumulative food intake in lambs prior to parasite infection, during parasite-induced anorexia and after recovery of food intake after antihelminthic treatment which is followed by expulsion of parasites (data from Dynes, 1993). The lambs were fed once a day and the cumulative food intake lines can be seen as indicative of their feeding rates.

The traditional view of the effect of parasitism on the host is that it deprives the host of essential nutrients but infection has more complicated effects. Figure 3.34 shows the pattern of food intake in lambs prior to and during parasite infection and after expulsion of parasites.

○ What can you deduce from the results shown in Figure 3.34?

● During parasitic infection, food intake itself is greatly reduced.

Thus the response of the host to infection by parasites seems to be a voluntary reduction in the amount of food taken in. These results show the food intake for short 8-hour periods during the course of a longer infection. If anorexia continued for a long period, the lambs would become very underweight. This phenomenon of parasite-induced anorexia has been recognized for some time. Indeed, in Victorian times in the 19th century early 'weight watchers' attempted to sell 'tapeworm tablets' as a cure for obesity! Is this anorexia merely a pathological response of the host to the presence of the parasite or, as some workers believe, does it have a more important functional basis in the host–parasite relationship?

Food intake and appetite are known to be affected by the release of neuropeptides from the hypothalamus in the brain, some of which increase food intake and reduce energy expenditure while others reduce food intake and increase energy expenditure. Although the mechanisms are very complicated, messenger molecules circulating in the blood, such as leptin, insulin and corticosterone, are known to be components of feedback mechanisms to the hypothalamus. Increased amounts of these hormones cause the release of the neuropeptides that reduce food intake and reduced amounts cause the release of neuropeptides that stimulate food intake.

Figure 3.35 Plasma concentration of leptin, insulin and corticosterone in rats infected with *Nippostrongylus brasiliensis* and their matched uninfected control and pair-fed groups. Control animals were allowed unlimited food, pair-fed animals were uninfected but fed the same amount of food as the infected animals. Data from Mercer and Chappell (2000).

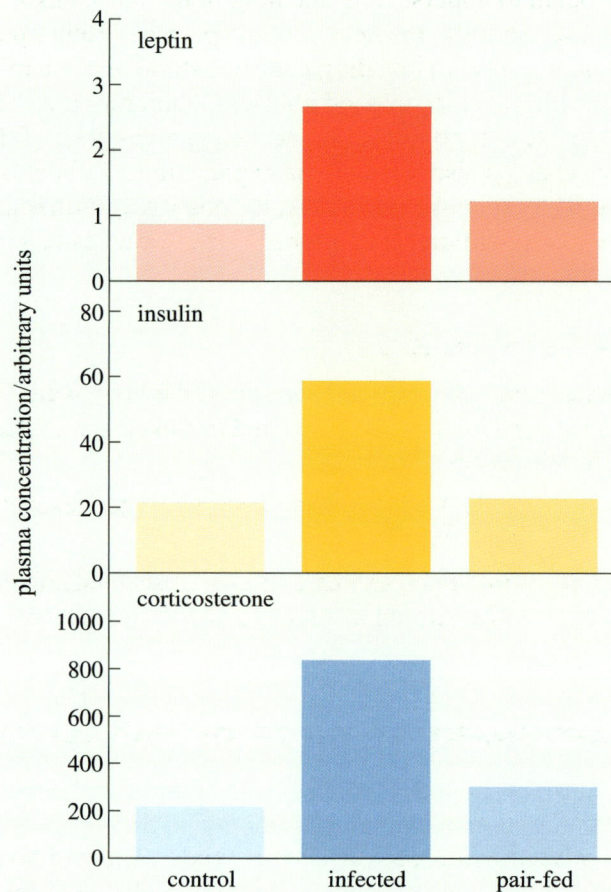

Figure 3.35 shows the results of measuring these blood proteins in experimental rats infected with the gut nematode worm *Nippostrongylus brasiliensis*.

○ What do these results show?

● Leptin, insulin and corticosterone levels are all raised in rats infected with the nematode worm. Both control and pair-fed groups show similar reduced levels.

○ What is the purpose of both pair-fed and control groups?

● Pair-fed groups allow a direct comparison of hormones or messenger molecules in the blood for the same food intake. The control group is to eliminate the possibility that the amount of food eaten determines the levels of hormones in the blood.

So, in parasitized individuals the results are consistent with the hypothesis that the increased levels of these hormones stimulate the release of neuropeptides from the hypothalamus that reduce appetite. The continuous high levels, as a result of infection, result in anorexia.

Although we can begin to understand something of the mechanism underlying the response, is it of cost-benefit to the host or to the parasite? Either possibility exists. The immune response to invading parasites stimulates the release of various cytokines which also lower food intake in laboratory rats. Thus, it may well be that a reduction in food intake is a necessary side-effect of the host's immune response. There is also a functional explanation: if a host is induced by its parasites to cut its food intake drastically, the chance of further parasitic infections, and thus competition for the already established parasites, are minimized.

SUMMARY OF SECTION 3.3

1   Internal parasites often take up nutrients across the body wall. The tegument is important in absorption in tapeworms and in flukes, particularly for the larval stages in molluscan hosts.

2   Roundworms (nematodes) have a gut and, with very few exceptions, cannot absorb nutrients through the cuticle of the body wall.

3   Parasites are known to induce anorexia in their hosts which may be a consequence of the host's immune response or help the parasite to maintain its infection.

3.4   EVOLUTIONARY ASPECTS OF PARASITISM

3.4.1   ORIGINS

As we have seen in Section 3.2, parasites have a wide variety of host niches and many different kinds of life-cycle strategies for moving between hosts. Thus, there is no obvious single origin of parasitism even within one phylum.

Although there are differences in interpretation, most helminthologists believe the origin of flukes and tapeworms to have been from a turbellarian-like ancestor.

The most likely origin for platyhelminth parasites, for example, could have been from the muddy ooze at the bottom of oceans. Flatworms living in the anoxic mud of the ocean bottom and being surrounded by bacteria, could have evolved a body tegument that gave protection against bacterial invasion and bacterial enzymes. Some of these flatworms, perhaps accidentally, found their way on to fish where they browsed on mucus and epithelial cells of the skin and gills. Various suckers and hooks evolved as an adaptation allowing a more permanent contact with their food source and thus gave rise to the monogenean parasites. Many monogeneans are highly specific parasites of fish, particularly of cyprinoids, an ancient group of teleost fish. It is likely that some monogeneans moved into the cloaca of fish and from there, some successfully migrated into the intestine, giving rise to the cestodes.

○   Which habits of a monogenean mentioned in Section 3.2.2 might support this scenario?

●   *Polystoma intergerrimum* which occurs in the urinary bladder and cloaca of frogs could represent this scenario.

Another example is *Fecampia* sp., a free-living turbellarian, the larvae of which sometimes end up in the haemocoel of marine crustaceans where they lose most of their body organs except the intestine and reproductive system. When sexually mature, the flatworm leaves the host, produces eggs and dies.

○ What other feature of these turbellarians would adapt them to living in host tissues?

● The body wall structure mentioned above, the tegument of which has many of the same features as those of monogeneans, digeneans and cestodes.

○ What kind of host is almost universally found in digenean life cycles (Figure 3.4)?

● Molluscs.

Thus the most likely route of origin of parasitic digeneans was via molluscs. In all probability, larval or adult benthic turbellarian flatworms were accidentally taken into the mantle cavity of marine molluscs. There, some survived and penetrated into the gut. *Paravortex* sp. (Figure 3.36) are flatworms found browsing along the gut of bivalve molluscs such as the cockle *Cerastoderma edule*. They reproduce within the gut, giving rise to live young which are swept out of the host's anus. Later in digenean evolution, the parasite penetrated through the gut wall and developed into parthenogenetic rediae or daughter sporocysts in the nutrient-rich haemolymph of the gonads and digestive gland. At this point in evolution, probably as far back as in the Cambrian, the parthenogenetic stages in the mollusc alternated with a free-living sexual stage. Thus, as in all other known examples in the animal kingdom, apomictic parthenogenesis in sporocysts and rediae has evolved from sexual reproduction. The life cycle of *Parvatrema homoeotecnum,* which has parthenogenetic stages with many adult features, in the haemolymph of the ridged periwinkle *Littorina saxatilis* supports this theory (James, 1964). Such sporocyst and redia stages are much simpler structures compared with sexual adults, but these reproductive stages probably evolved to maximize the chance of transfer to a new host. Much later in evolution, the free-living sexual stages were swallowed by a fish and henceforth the adult developed in the gut. Larval stages (miracidia or cercariae) were retained or evolved respectively to connect the parasitic stages in the primary molluscan and definitive vertebrate hosts. Later still in evolution, more intermediate hosts and stages such as encysted metacercariae appeared which facilitated transmission.

○ Why might encysted metacercariae help transmission?

● The metacercariae, encysted on vegetation, or in or on an intermediate host protect the larvae and prolong their lives thereby increasing the chance of transmission to the final host.

We mentioned, in Section 3.2, the universal nature of the roundworm body plan.

○ Which features of roundworms permitted the evolution of a parasitic way of life?

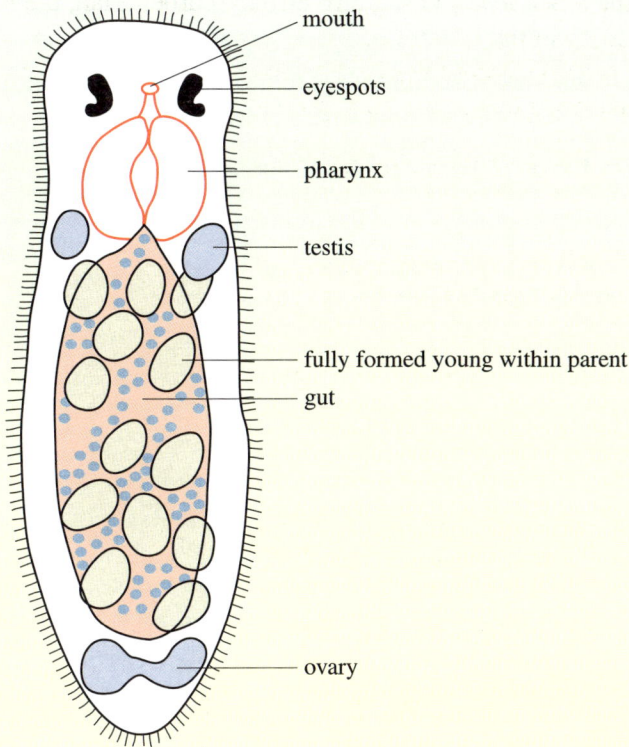

Figure 3.36   *Paravortex cardii*, a sexually mature symbiont in the gut of the cockle *Cerastoderma edule*.

- mouth
- eyespots
- pharynx
- testis
- fully formed young within parent
- gut
- ovary

● The cuticle, because of its resistance to chemical and physical attack, would certainly have helped the roundworms to resist digestive juices when they were first accidentally taken into the gut. Also the spindle-like shape and substrate-burrowing habit of many roundworms would have aided penetration of the host integument or gut into the internal tissues.

○ If you compare the body structure of the parasites in the animal phyla mentioned in Section 3.2 with the features of their free-living relatives, what strikes you as the main difference in their body plans?

● With the exception of the roundworms, which are a very specialized group, parasitic members tend to have a simple body plan.

Amongst crustaceans, parasitic isopods also show very little modification apart from prehensile appendages and small changes in their mouthparts and gut. However, the more primitive copepods and cirripedes may show considerable modification of body structure. *Lernaea* sp. (Figure 3.23a), for example, lives in the confined space of the branchial chamber of fish and has thus evolved an elaborate folding of their body and of the exposed coiled egg sacs, which has resulted in them achieving the greatest possible reproductive capacity in that space. Pentastomids are worm-like with extremely simplified organs.

The evolution of parasites is, of course, inextricably linked to the evolution of their hosts. For example, many parasites have evolved features which manipulate the behaviour and/or life-cycle patterns of their hosts to their own benefit.

3.4.2 HOST–PARASITE COEVOLUTION

Coevolution has often been described as a kind of evolutionary 'arms race' between parasite and host. Reciprocal genetic changes in the interacting populations give rise to adaptations which at first give one of a pair of species an advantage, only to be countered by neutralizing adaptations from the other. An example is antigenic variation in *Schistosoma* sp., already discussed in Section 3.2, in which species show avoidance of the immune response of their hosts by coating themselves with the host antigen. Similarly the protoctist, *Trypanosoma* sp., is known to produce antigenic variants.

Coevolution between host and parasite could explain why polymorphism occurs in natural populations. Extreme polymorphisms where different structural and antigenic variants occur, as seen in *Trypanosoma* sp., allow the parasite to survive and avoid the immune response of the host. Each variant possesses a different range of antigens and as the host antibodies kill one form, the survivors are able to change morphologically and antigenically. Thus, although the infection is reduced, the host is unable to eliminate it completely. As you are aware, the immune systems of animals are sophisticated and complex and it has been suggested then that such systems evolved, and have been maintained, because of the selective pressure of the challenge from parasites (Wakelin, 1997).

○   What life-cycle process enhances the spread of variations on which natural selection acts?

●   Sexual reproduction.

Indeed we may well have parasites to thank for the origin of sex. Although there are a number of theories about the origin of sex, that of an adaptation such as the 'evolutionary arms race' between species and their environment is a strong contender. Increasing variability caused by the redistribution of genes during crossing over and meiosis increase the chance of offspring adaptation to varying microhabitats (the tangled bank hypothesis) or to future environmental changes (the Red Queen hypothesis).

The tangled bank hypothesis suggests that parasites reproducing sexually, and therefore producing different genotypes, would compete more effectively for the myriad of microhabitats available to them in any one species of host. Because of small differences amongst individuals their competition for these restricted habitats would be less severe than that amongst clones produced asexually which, all possessing the same genotype, would compete for each microhabitat. However, as we have already discussed, parasitic invasion is a constantly changing challenge in which the host's immune system must keep pace with the changing antigenic properties of the parasite. Thus sexual reproduction increases antibody and antigen variation in host and parasite populations. The parasite version of the Red Queen hypothesis suggests that sex is the response of hosts to the invasion by the parasites and vice versa.

Anderson and May (1982) have argued that a 'well balanced' host–parasite association is not necessarily one in which the parasite does little harm to the host.

○   What advantage might destructive effects on the host have for the parasite?

●   If the parasite relies on the next host to devour its present host for transmission to occur, then there is some advantage in causing debility or deformation that might make this event more likely. Thus debilitating the host or changing the host's behaviour may enhance the chance of the parasite being passed on to the next host in its life cycle and/or other definitive hosts.

In addition, virulent parasites may reduce numbers of the dominant host species, allowing greater diversity within habitats and thus a better environment for the host species. Moreover, parasite individuals are not randomly distributed throughout the host population, being overdispersed or clumped (aggregated) where some hosts are more heavily infected than others.

Figure 3.37 shows the distribution of two species of parasite on their fish hosts.

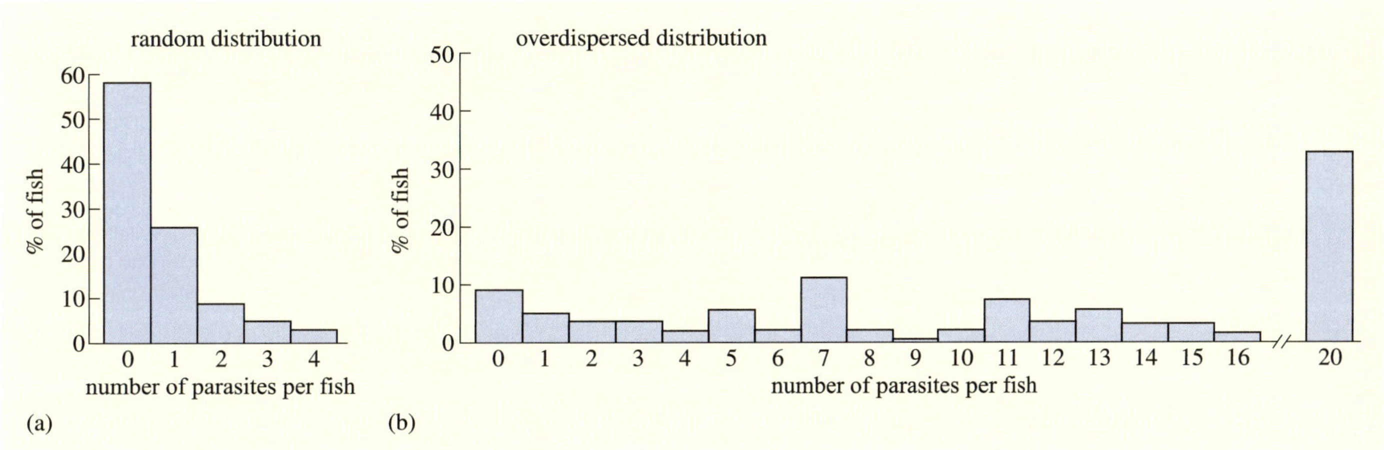

(a)   (b)

○   From the data in Figure 3.37 suggest two possible reasons why the distribution of most parasites in their host is more likely to follow the overdispersed pattern than the random pattern.

Figure 3.37   Frequency distribution of two species of parasite in their fish host. (a) Random distribution of *Proteocephalus filicollis*, a tapeworm; (b) overdispersed (clumped) distribution of *Diplostomum gasterostei*, a digenean fluke. Data from Kennedy (1975).

●   (i) Clumping (overdispersion) may be due to greater susceptibility because of lowered immunity; the presence of some parasites weaken the host and may make them more susceptible to repeated infection. (ii) The individual host's proximity to the source of infection may vary.

Naturally it is beneficial to host and parasite that it kills only a few, weaker hosts. Thus the coexistence of different strains of host and parasite may depend on such factors as virulence and transmission. This coexistence is also, almost certainly, based on the balance that exists between the effects of the host's immune system on the invading parasite and the countermeasures set up by the parasite (see Section 3.3). How effective is the host's immune system in its own countermeasures against the parasite? The fact that human AIDS patients being treated with immunosuppressant drugs suffer more greatly from opportunistic infection with a range of parasites suggests that the immune system does play an important role in normal individuals' control of invading organisms. It is likely that this principle is true in most other animals (Davey and Gillman, 2001).

It is possible to regard the host–parasite immune response as one of a conflict between the two in which the host seeks supremacy. On the other hand, as we have suggested above, it may be a device to allow both partners to compromise so that an uneasy equilibrium is established. Producing an immune response is a costly metabolic exercise for the host and thus complete elimination of the parasite may not be the aim. It may be sufficient to reduce the parasite's capacity to feed on host tissue and/or to replicate.

Most protoctist parasites replicate within the host and thus the host's response is geared towards stopping reproduction. If the parasites live extracellularly, their plasma membrane is subjected to damage by immune-mediated effector mechanisms involving antibody, complement and cytotoxic cells. Intracellular parasitic protoctists are more protected (hence the difficulty in producing a successful vaccine against the malaria parasite) but can be attacked by cell-mediated defence mechanisms that involve the production of reactive oxygen species or toxic nitrogen metabolites.

Parasitic worms such as flukes, tapeworms and roundworms, as well as generally being much larger, have a complex body surface structure (Section 3.3). However, the plasma membrane is still potentially vulnerable to immune-mediated damage, especially in the case of the larval stages of some nematodes that live in the tissues before migrating to mature in the gut, and the invading cercariae of blood flukes. Little is known about the responses to large mature worms in the gut. However, the feeding behaviour of the worms and ectoparasites on the outer surface of hosts may initiate a response. Close proximity with host blood or tissue fluids results in the parasite coming into contact with host antibody which can activate cell-mediated toxicity and the complement cascade. The most likely effect, though, is the production of antibody that is directed against the invading larval stages of subsequent infections: the response is geared to preventing further infection rather than eliminating the present one, i.e. concomitant immunity (see Section 3.2.3).

We mentioned earlier that combating parasitic invasion has a cost to the host. In addition to requiring metabolic resource, cytotoxic cells often release potent biologically active mediators such as histamine, serotonin and leukotrienes which initiate hypersensitivity responses in local tissue. Thus, the host must strike a balance between the potentially harmful effects of the immune response combined with the diversion of resources from growth and reproduction and the beneficial one of destroying or reducing the parasite's effect. Host 'trade-off' is mirrored by the parasite where it also must balance its own capacity to grow and reproduce against causing undue damage to the host and losing, too soon, its vehicle for survival and transmission. A certain amount of effective host immunity may actually be beneficial to the parasite in preventing overinfection and premature host death. Indeed, there is some evidence that the fecundity of female schistosomes and outward movement of their fertilized eggs through tissues are promoted by the host's immune response.

There is also evidence to suggest that this 'trade-off' might involve other physiological mechanisms, e.g. in some bird–parasite systems there may be a link between male secondary sexual characters and parasitism (Hillgarth and

Wingfield, 1997). Those males heavily infected by parasites divert resources away from production of these characters with the consequence that females are more likely to choose males that are parasite-resistant and thus possess fully developed sexual attractant structures and behaviour. It is possible that enhanced testosterone levels, necessary for the development of secondary sexual characters, result in a reduction in immunocompetence. In humans, there is some evidence that stress lowers immunocompetence (Mitchison and Oliveira, 1986) and thus the link between hormone levels and susceptibility to parasitic infection may be a feature of many kinds of vertebrate.

Further evidence that there is a 'trade-off' between the relative growth of certain body characters comes from a study on developing barn swallows which can be infected in the nest by the larvae of the dipteran fly *Ornithomyia biloba* (Saino *et al.*, 1998). Nestlings exposed to many such parasites invest more resource in feather growth than in general body growth, which enables their earlier departure from the nest and thus a reduction in the time exposed to the blood-sucking fly larvae, which are left behind in the nest.

Therefore the complex association between host and parasite is often a compromise, rather than a conflict, between the competing demands of fitness of the host and parasite. It has been suggested that the immune response may thus be partial and limited and that the dangers to the host of immunopathology (caused by the hypersensitive inflammatory response) balanced against the need to protect against parasitic invasion may actually have led to selection of genes in the host species that downregulate, rather than upregulate, immune responsiveness.

As fascinating as host–parasite immune compromise is, it is only one of a number of evolutionary devices that increase parasite fitness. Another mechanism, for which there is some circumstantial evidence, is one in which the parasite affects host behaviour in such a way as to enhance transmission from host to host.

3.4.3 PARASITE-INDUCED TRANSMISSION STRATEGIES

Changes to the host's morphology and behaviour often occur as a result of parasitism (Moore, 1995). For example, *Leptothorax nylanderi*, an ant inhabiting rotten wood, tends to be reddish-yellow or brown in colour. However, ants parasitized by the larval stage of a tapeworm of the spotted woodpecker are larger, bright yellow in colour, and move very sluggishly.

○ What disadvantage to the ant would such coloration and behaviour have?

● They could be selectively preyed upon by birds including woodpeckers.

Such selection therefore enhances transmission of the parasite. In fact, enlarged size, or gigantism, is a common feature of many parasitic infections especially of marine molluscs infected with digenean and nematode larvae. This gigantism, probably due to metabolic or hormonal effects caused by the presence of the parasite, not only makes the mollusc more vulnerable to predation (and therefore to being eaten by a new host) but creates more space for parasite growth and reproduction. Parasitic castration is a common feature of molluscs infected with the larval stages of flukes. The destruction of gonadal tissue may be the result of the secretion of 'castration' hormones by the parasite but is often due to mechanical damage inflicted by the growing sporocysts or rediae. The use of released host body space is illustrated by the crustacean cirripede, *Sacculina* (Figure 3.23ii). This parasite occupies the external location normally taken up by the host crab's eggs. The crab behaves as if it is carrying eggs and thus protects the parasite as it would its own eggs.

Altered host behaviour can also be associated with enhanced transmission. For example, digenean metacercariae which occur in cockles (Mollusca, Bivalvia) as larvae and as adults in oystercatchers have been shown to interfere with the cockles' shell-closing mechanism. Thousands of metacercariae are found along the growing edge of the shell valves or enveloped in host tissue below the hinge mechanism (Bowers *et al.*, 1996). They prevent complete closure of the valves, forcing the cockles to turn around just under the mud surface to operate their siphons successfully (Figure 3.38). In doing so, the gape is exposed to the long bill of the oystercatcher which easily scours out the parasite-infested flesh. Transmission in this case depends on the intermediate host (the cockle) being eaten by the definitive host (the oystercatcher).

Figure 3.38 Orientation of parasitized and unparasitized specimens of cockles.

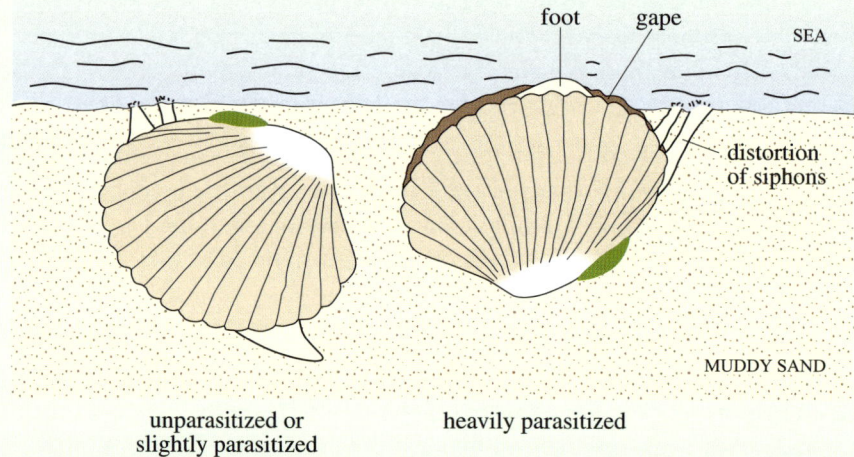

unparasitized or slightly parasitized   heavily parasitized

Some protoctistan parasites, such as *Plasmodium*, are transmitted when the intermediate host or vector, the mosquito, feeds on the definitive host, the human. Mosquitoes infected with *Plasmodium* are more likely to continuously seek a blood meal — a kind of 'insect Count Dracula'. The parasite appears to alter the neurochemistry controlling the mosquitoes' abdominal stretch receptors so that it bites insatiably. Other studies have suggested that mosquitoes containing infective sporozoites in their salivary glands carry out more probing on more people before being able to suck blood. This change in habits could arise from a biochemical reaction to the presence of the sporozoites which reduces the anti-blood-clotting enzyme in the gland, thus forcing the mosquito to take shorter but more frequent meals, and enhancing the parasite's distribution by infecting more victims with its sporozoites.

In tsetse flies, mechanoreceptors that monitor blood flow can be physically blocked by the parasite, thus disrupting satiety signals and ensuring continued feeding on several hosts.

The speed of replication of the parasite is linked to its virulence. The parasite faces an evolutionary trade-off between being transmitted quickly and killing the host. If replication is too fast, the host may die from released toxins, if too slow, transmission chances may be reduced. A central premise of coevolution is that the 'best' state for the parasite would be an intermediate state of virulence — sufficiently harmful to the host to promote the likelihood of transmission but not nasty enough to kill.

Changes in host behaviour can only be regarded as adaptive if they confer fitness benefits on the parasite and increase host vulnerability to predation. Although the above examples are suggestive, it is not clear if the modifications are the result of manipulation by the parasite or just a lucky side-effect of the pathology of the infection. For example, the parasitized mosquitoes may just be more hungry than the healthy ones. Much further research is required before such questions are resolved.

SUMMARY OF SECTION 3.4

1   Platyhelminth parasites probably arose from free-living ancestors in the oceanic ooze.

2   Those attaching to the external surface of fish became monogenean parasites.

3   Those invading the gut of molluscs evolved into digenean parasites.

4   The likely ancestors of tapeworms were monogeneans which invaded the cloaca and then the intestine of fish.

5   Both parasite and host evolve together in stepwise fashion, i.e. coevolution.

6   The evolution of sexual reproduction, which enhances variation, may have been a response to parasitism.

7   Parasites often induce morphological or behavioural changes in their hosts that enhance their chances of transmission to the next host.

3.5   CONCLUSION

We have seen that parasitism is so widespread throughout the animal kingdom and parasites have developed such beautiful adaptations and complicated life cycles that we cannot regard it as an abnormal way of life. We have also seen how the parasite can affect the behaviour and possibly the morphology of the host to enhance transmission.

Finally, parasite virulence may reduce overcrowding in host populations, particularly as many parasites are often clumped within weaker hosts. The death of heavily infected hosts reduces the chance of an overcrowded population destroying its own habitat, and, incidentally benefits the parasite because potential hosts are maintained. Thus our final thought is that parasitism, although often destructive to individuals, actually maintains and increases biodiversity.

REFERENCES

Anderson, R. M. and May, R. M. (1982) Co-evolution of hosts and their parasites, *Parasitology*, **83**, pp. 411–426.

Bowers, E. A., Bartoli, P., Russell-Pinto, F. and James, B. L. (1996) The metacercariae of sibling species of *Meiogymnophallus minutus* including *M. rebecqui* comb. nov. (Digenea: Gymnophallidae) and their effects on closely related *Cerastoderma* host species (Mollusca: Bivalvia), *Parasitology Research*, **82**, pp. 505–510.

Davey, B. and Gillman, M. (2001) Defence, in *Generating Diversity*, M. Gillman (ed.), The Open University, Milton Keynes, pp. 151–200.

Dynes, R. A. (1993) Factors causing feed intake depression in lambs infected by gastrointestinal parasites. Unpublished Ph.D. Thesis, Lincoln University, New Zealand.

Dyson, M. (2001) Reproduction in *Generating Diversity*, M. Gillman (ed.), The Open University, Milton Keynes, pp. 111–150.

Hagen, P., Garaside, P. and Kusel, J. R. (1993) Is tumour necrosis factor the molecular basis of concomitant immunity in schistosomiasis? *Parasite Immunology*, **15**, pp. 553–557.

Hillgarth, N. and Wingfield, J. C. (1997) Parasite-mediated sexual selection; endocrine aspects, in *Host-Parasite Evolution; General Principles and Avian Models*, D. H. Clayon and J. Moore (eds), Oxford University Press, Oxford, pp. 78–104.

James, B. L. (1964) The life cycle of *Parvatrema homoeotecnum* sp. nov. (Trematoda: Digenea) and a review of the family Gymnophallidae Morozov, 1955, *Parasitology*, **54**, pp. 1–41.

Kennedy, C. R. (1975) *Ecological Animal Parasitology*, Blackwell Scientific Publications, Oxford.

Knowles, W. J. and Oaks, J. A. (1979) Isolation and partial biochemical characterization of the brush border plasma membrane from the cestode *Hymenolepis diminuta*, *Journal of Parasitology*, **65**, pp. 715–731.

Mercer, J. G. and Chappell, L. H. (2000) Appetite and parasites, *Biologist*, **47**, pp. 35–40.

Mitchison, N. A. and Oliveira, D. B. G. (1986) Chronic infection as a major force in the evolution of the suppressor T-cell system, *Parasitology Today*, **2**, pp. 312–313.

Moore, J. (1995) The behaviour of parasitized animals, *Bioscience*, **45**, pp. 89–96.

Ridge, I. (2001) Diversity in protoctists, in *Introduction to Diversity*, I. Ridge and C. M. Pond (eds), The Open University, Milton Keynes, pp. 55–92.

Saino, N., Calza, S. and Muller, A. P. (1998) Effect of a dipteran ectoparasite on immune response and growth trade-offs in barn swallow, *Hirundo rustica*, nestlings, *OIKOS*, **81**, pp. 217–228.

Smithers, S. R. (1968) Immunity to blood helminths, *Symposium of the British Society for Parasitology*, **6**, pp. 55–66.

Wakelin, D. (1997) Parasites and the immune system, *BioScience*, **47**, pp. 32–40.

Williams, B. M. (1992) Studies on the epizootology of canine cestodes particularly *Taenia multiceps* and bovine coenurosis. Unpublished Ph.D Thesis, The Open University, Milton Keynes, 359 pp.

FURTHER READING

Bush, A. O., Fernandez, J. C., Esh, G. W. and Seed, J. R. (2000) *Parasitism. The Diversity and Ecology of Animal Parasites*, Cambridge University Press, Cambridge. [A text describing the diversity of the major eukaryotic parasites of animals. It considers both pathology and regulation and the impact parasites have on their hosts.]

VERTEBRATES

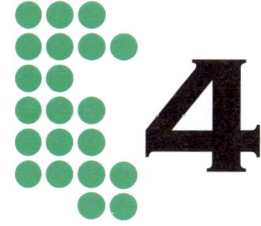

4.1   INTRODUCTION

The largest, most powerful and most familiar animals are vertebrates: fish, amphibians, reptiles, birds, mammals. They swim, walk or fly very well and their senses, especially vision, hearing and smelling are acute; many groups can subdue powerful prey in their strong jaws and swallow large objects. This chapter is about the origins and functioning of the special features that have enabled vertebrates to become such dominant animals. Vertebrates have a wider variety of kinds of cells than invertebrates, and the group has several unique types of tissue. Most large animals are vertebrates, but paradoxically their cells are mostly smaller than those of invertebrates, though very numerous. Having tissues, especially the central nervous system, composed of fewer, larger cells is one of the reasons why physiologists prefer invertebrates to vertebrates as experimental subjects.

4.2   THE CHORDATE BODY PLAN

Chordates derive their name from the skeletal rod in the dorsal half of the body called the **notochord** (*not*- back + *chord*- gut, string of musical instrument). As well as the vertebrates (subphylum Craniata), chordates include two groups with similar basic body plans (Figure 4.1), the salps and sea-squirts (subphylum Urochordata), and the lancelets (subphylum Cephalochordata). Although vertebrates themselves are conspicuous and well known, these **invertebrate chordates** that share some of their characteristics, and may resemble their ancestors of long ago, are much less familiar.

All chordates have a notochord: the outer sheath is formed from collagen fibres secreted by connective tissue cells, and the core consists of special cells containing vacuoles that can swell osmotically to two or three times their original volume, thereby stiffening the whole structure, as in other hydrostatic skeletons (see Section 1.2.2). A notochord of similar structure appears early in vertebrate development, where it plays a central role in the elongation of the embryo and the arrangement of the other tissues. It is similar in many vertebrate larvae (Figure 4.1a), but in most adults, the notochord is partially or entirely replaced with cartilage or mineralized skeletal tissues. But the notochord persists throughout life in a few groups, notably crossopterygians (Figure 4.1c).

The notochord is usually surrounded by articulated vertebrae and blocks of strong muscles (seen in Figure 4.1a) that bend it from side to side in swimming or burrowing. They are controlled by spinal nerves, so called because they emerge in pairs from the main dorsal nerve cord, also called the *spinal cord*, between the vertebrae. The vertebral column, back muscles and spinal cord thus form serially arranged segments. Chordate segmentation involves only the dorsal half of the body, and is in other ways very different from that of annelids and arthropods.

Figure 4.1 (a) Cross-section through the pharynx of a larva (ammocoete) of the lamprey (*Petromyzon*), a primitive agnathan vertebrate. The numerous gills serve in filter-feeding as well as respiration: water enters the gut through the mouth and passes over the gills, which, in this specialized larva, are covered by a fold of tissue. The arrangement of the notochord, nerve cord, swimming muscles and major blood vessels is similar to that of other vertebrates. (b) The main features of the body plan of a vertebrate. (c) A coelacanth, *Latimeria chalumnae* dissected to show the notochord. Much of the rest of the skeleton, including part of the skull, pectoral girdle and fin rays visible here, is cartilage.

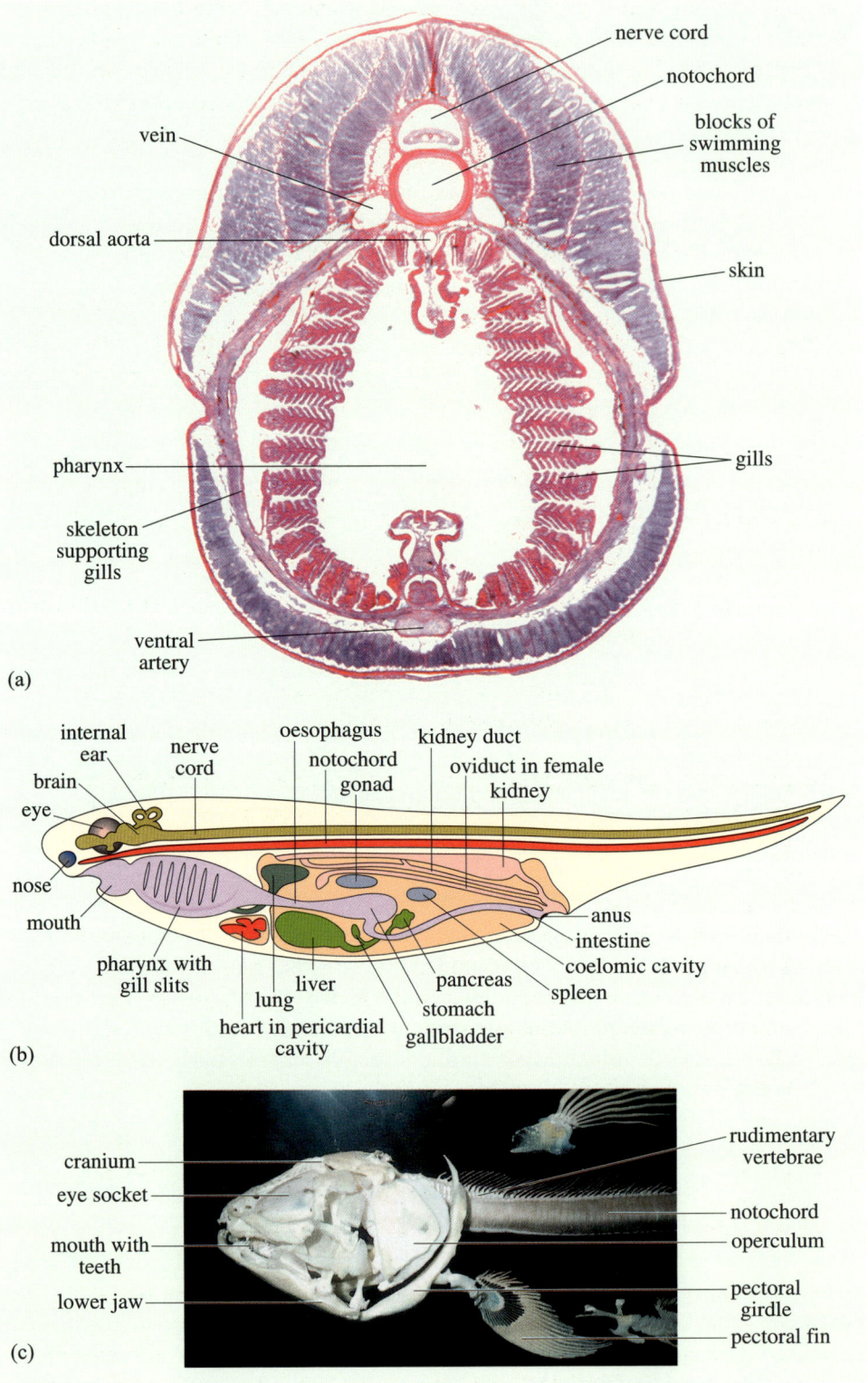

(a)

Labels: nerve cord, notochord, blocks of swimming muscles, vein, dorsal aorta, skin, pharynx, gills, skeleton supporting gills, ventral artery

(b)

Labels: internal ear, nerve cord, oesophagus, kidney duct, brain, notochord, oviduct in female, gonad, kidney, eye, nose, mouth, anus, intestine, coelomic cavity, pharynx with gill slits, liver, pancreas, spleen, lung, stomach, heart in pericardial cavity, gallbladder

(c)

Labels: cranium, rudimentary vertebrae, eye socket, notochord, operculum, mouth with teeth, lower jaw, pectoral girdle, pectoral fin

The main coelom containing the intestines, liver and other abdominal viscera is ventral, not partitioned into segments, and does not extend into the head or the tail (Figure 4.1b). The segmental arrangement of the nerves, muscles and skeleton is much more conspicuous in early embryos, and genes that control their number and internal arrangement prove to be important in evolutionary changes, as discussed in Chapter 5.

○ How does the relationship between the nervous system and the gut in Figure 4.1a and b differ from that of the invertebrate phyla discussed in Chapter 1?

● The nerve cord is dorsal to the gut and it does not form a ring of ganglia around the pharynx, as there is in annelids, arthropods and many other phyla. In fact, a third structure, the notochord lies between the nerve cord and the gut.

The chordate nerve cord differs from that of other phyla in other ways: it is a thick-walled hollow tube, formed by rolling up a flat sheet, not a solid thread as in most invertebrate phyla. The separation of the dorsal nerve cord from the gut allows both structures to become large and elaborate.

○ What features of the gut shown in Figure 4.1 (a) resemble and (b) differ from that of invertebrates discussed in Chapter 1?

● (a) As in most invertebrates, the gut has both a mouth and an anus and lies in the coelom but (b) it is perforated by gill slits.

These **pharyngeal gill slits** extend from the gut right across the body wall forming bilaterally symmetrical perforations on the sides of the 'neck' (Figure 4.1a and b). Water (and anything else) that enters the mouth and reaches the pharynx can *either* pass out over the gills, *or* continue along the gut until it reaches the anus and a second opportunity for expulsion. You can see these alternatives clearly by feeding a fish (e.g. a goldfish or guppy) a pellet that releases a coloured dye or ink or opaque particles when crushed: some of the marker is expelled almost at once through the gills behind the head, while the rest takes some time to emerge from the anus.

○ Are respiratory structures usually derived from the gut?

● No. In most animal phyla, the gills are extensions of the body wall or of the limbs. For example, crustacean gills are parts of limbs, molluscan gills extend from the body wall, as do the tracheae found in most uniramians and echinoderm tube-feet, etc.

With a few exceptions (notably frogs and other amphibians), gas exchange is confined to the pharyngeal gills (Figure 4.1b), or in air-breathing fish and all tetrapods, to lungs which also form as an extension of the gut. In all vertebrates (and most other adult chordates), the blood is actively pumped through these localized respiratory surfaces, where it picks up oxygen, and on to the rest of the body. The heart is derived from a muscular blood vessel folded to form several chambers separated by valves. It is situated ventral to the gut near to the pharyngeal gills that are perfused with the blood that it pumps.

These features characterize all chordates, including the invertebrate chordates. In salps, sea-squirts and lancelets, and a fair number of aquatic vertebrate chordates (i.e. some fish and amphibian larvae), the gills serve not only for respiration and for excretion of some metabolic waste products, but also as a filter on which to collect fine particles of food from the water.

Figure 4.2 shows two such filter-feeders; tunicates (sea-squirts) attach themselves to any surface exposed to food-bearing water currents and are often seen on rocks and piers, etc. just below the high-water mark. They draw in water through the inhalant canal (which corresponds to the mouth). It passes over the very numerous and extensive gills and out through the exhalant canal, aided by a squirting action of the muscles in the body. The pharynx and its many gill slits are the largest component of the body of adult sea-squirts (Figure 4.2a), with the other viscera and the nerve cord and notochord being relatively small. The scores of gill slits of cephalochordates (Figure 4.2b) also serve in filter feeding: ciliated oral tentacles drive a current of water into the mouth and out across the gills. These animals resemble small, transparent fish and are common in certain areas of warm, shallow seas or estuaries. They swim only weakly but use the muscles and stiff notochord, which extends the entire length of the body, to wriggle into soft sand, partially burying themselves with their anterior ends exposed.

Figure 4.2 The three subphyla of Chordata: (a) Urochordata: An adult sea-squirt (*Ciona* sp.) attached to rocks in shallow coastal waters. Water enters the inhalant (buccal) siphon and flows over the numerous gills in the greatly enlarged pharynx. Particles of food are extracted and pass to the stomach and on to the intestine. The water current and the faeces are expelled through the exhalant siphon. (b) Cephalohordata: a stained whole-mount of amphioxus. They take in water through the mouth and collect particulate food on the many pharyngeal gills. (c) Craniata: a dogfish (*Squalus* sp. class Chondrichthyes), a jawed fish with only 5 pairs of gill slits. Gills serve only for respiration in this active predator, that seizes whole animals or bites off morsels with its large, jawed mouth.

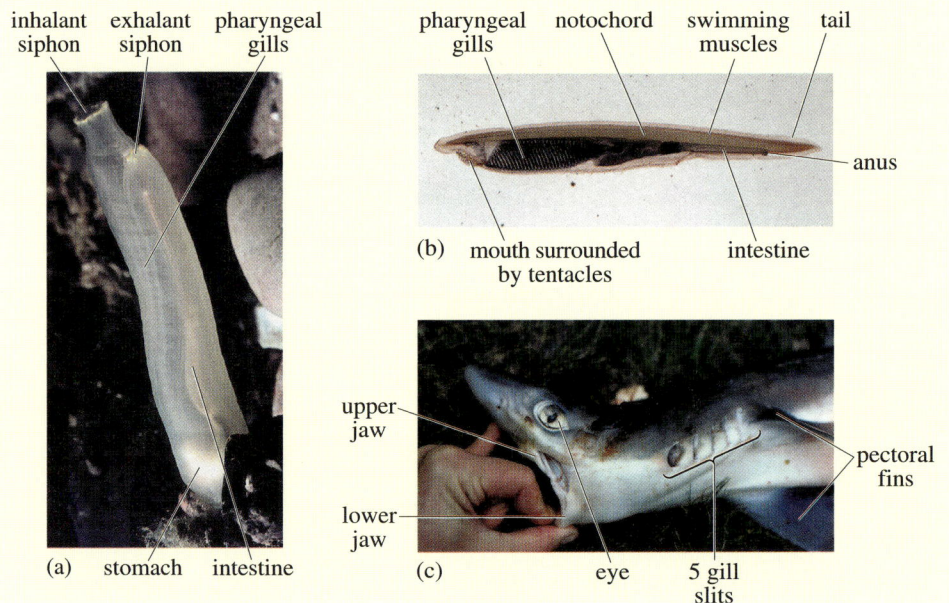

These ancient lineages, although locally abundant, are not nearly as large or diverse as the craniate chordates, generally known as vertebrates. The most primitive fishes, the Agnatha, have seven pairs of gills and gill arches, but in all other (gnathostome) fishes, the two anterior arches form the jaws and their attachments to the rest of the skull, so (with a few obscure exceptions), they have only five pairs of gills (Figure 4.2c). Reinforced with jaws and armed with teeth, the mouth is large in most vertebrates, and is the principal feeding organ. The pharyngeal gills and associated structures disappear in terrestrial vertebrates (though rudimentary clefts can be seen transiently in their early embryos), but the jaws, teeth and skull become adapted to exploit land-based foods.

The steps by which the chordate body plan evolved into that seen in craniates are much discussed using evidence from comparative anatomy, palaeontology and, increasingly, genetic analysis and molecular biology. In this chapter, we concentrate on three major groups of tissues and organs that are major factors in the success of vertebrates as large, powerful animals: the blood and cardiovascular systems, the skeletal and skin tissues and sense organs.

SUMMARY OF SECTION 4.2

1   The chordate body plan includes a tubular nerve cord dorsal to a central notochord, and the pharynx perforated to form gill slits. The dorsal musculature, muscles and nerve cord are segmented. A ventral heart pumps blood through the gills and the rest of the body.

2   Vertebrates are the most abundant and diverse of three subphyla of chordates. The other groups are mostly marine filter-feeders with numerous gill slits.

3   Primitive jawless vertebrates have seven pairs of pharyngeal gill slits each supported by a skeletal tissue that forms the gill arch. The two anterior arches form the jaws and their attachment to the braincase in other fishes, leaving five functional pairs of gills. Gills disappear entirely in terrestrial vertebrates.

4.3   THE VERTEBRATE BLOOD SYSTEM

More efficient blood, and more elaborate vessels for conveying it around the body under high pressure, are among the most important features that enabled vertebrates to become very large and powerful. The basic plan of the vertebrate circulatory system and the tissues that form the vessels were established in early fishes. Many bony fishes, including some of the most ancient, had lungs as well as gills, which enabled them to breathe air from the surface as well as, or instead of, water. Further modifications are associated with the evolution of endothermy in birds and mammals. Endothermy is the production of so much body heat that the core temperature is substantially higher than ambient.

○   Why would endothermy entail more efficient blood and cardiovascular systems?

●   Almost all metabolic pathways proceed faster at higher temperatures. Faster metabolism requires more oxygen and generates more waste products that must be removed.

In birds and mammals, all the blood passes twice through the heart each time it circulates round the entire body, once to and from the lungs, where it is oxygenated and releases wastes, and again before perfusing the rest of the body. The heart is 'divided' (because the blood on these two routes is not mixed) and the circulation is described as 'double' because its pressure is raised again after being dissipated by passing through the lungs, thus hastening its passage to the brain, abdomen, muscles and other tissues.

The following account relates mainly to tetrapod vertebrates (i.e. amphibians, reptiles, birds and mammals).

4.3.1 THE BLOOD VESSELS

In vertebrates, the blood is always contained within vessels, albeit sometimes very fine vessels (diameter about 1 to 10 µm) with permeable walls only one cell thick, called *capillaries*. Gases, dissolved nutrients and small cells migrate across the cells that form the capillaries, called endothelial cells, in both directions, supplying the perfused tissue and removing its wastes. Networks of fine capillaries (called capillary beds) also take up nutrients from the epithelial cells in the gut lining, and exchange gases, taking up oxygen and releasing carbon dioxide and other waste products in lungs and gills. At least in mammals, the endothelium of organs such as the brain, bone and muscles differs in form and biochemical properties, allowing selective uptake of certain substances from the blood to the tissues.

Vertebrate tissues are permeated by capillaries containing blood, rather than being bathed in blood, as happens in many invertebrates. Because the cardiovascular system is closed (i.e. blood is contained in specialized vessels), blood circulates under pressure on (at least as far as the larger vessels are concerned) a regular route.

The muscular heart generates the pressure that drives the blood through capillaries and the larger vessels, called *arteries* if they carry blood away from the heart, or *veins* if they convey it back (Figure 4.3). The absolute size of these vessels depends upon the species (the diameter of the largest is around 1 mm in mice, 1 cm in humans, and over 20 cm in large whales) and in general, arteries are smaller with relatively much thicker walls than veins. Arterioles and venules are narrower, thinner-walled versions of arteries and veins. The diameter of typical arterioles is around 1% that of large arteries of the same animal. With the exception of very fine venules, all these vessels have walls that are too thick for materials to move across them. Although they serve only as conduits, their mechanical properties are intricate and very important to the functioning of the blood system as a whole.

No plumber would attempt to drive a viscous fluid such as blood through narrow channels by an intermittent pumping action like that produced by the heart because the flow might become turbulent and waste a great deal of the energy used to propel it. * Smooth, energetically efficient flow is due mainly to the unusual mechanical properties of the walls of the arteries (shown in Figure 4.3), which stretch sufficiently to accommodate the bulge of blood produced by the heartbeats but also sustain the pressure that drives the flow.

○ As well as the pumping of the heart itself, what other factors could cause kinks and bulges in arteries?

● Animals change posture by bending their back and limbs and open their jaws, which compresses or bends blood vessels. Wounds and bruises also alter blood volume and pressure.

* Artificial hearts currently being developed to treat severe cardiovascular disease produce continuous flow, not an intermittent pumping action.

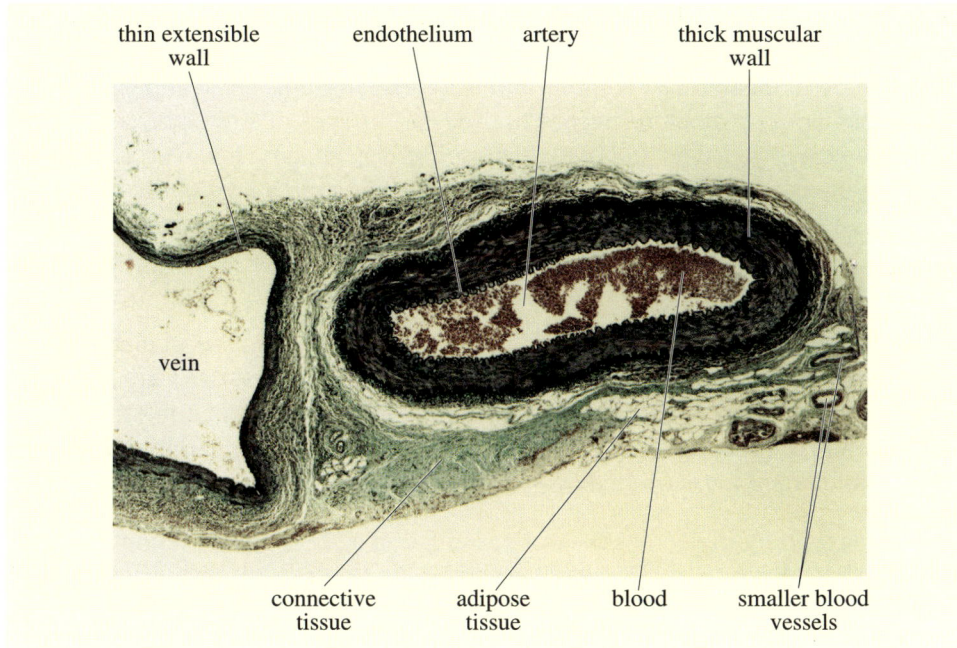

Figure 4.3 Stained cross-section of a mammalian vein, left, only partially in the picture, and artery, right. This image is about 3 mm across.

Clearly, a kink-resistant and bulge-resistant vascular system is essential in animals that swim or burrow using large changes in body shape. Most pipes with extensible walls, such as rubber balloons, inflate more readily when they are already stretched by internal pressure: blowing up a balloon is difficult to start with, then gets easier. Arterial walls do the opposite: they become stiffer as they are stretched wider. These unusual properties enable arteries to accommodate the pumping of the heart while avoiding bulges (called aneurisms), which are both inefficient (the blood sits there, doing nothing) and potentially dangerous if the vessel stretches far enough to burst, spilling its contents.

○ Are arteries mainly cellular, actively metabolizing tissue, or extracellular? (*Hint*: Think about the fragments of major blood vessels that are often visible in fresh meat, especially liver and heart.)

● Large blood vessels must be mainly extracellular material, because they always appear creamy white (i.e. not themselves perfused with blood that brings the nutrients required by living cells) against the dark red, richly perfused liver, muscle or other forms of 'meat'.

Large blood vessels are indeed mainly extracellular material, and are composed of collagen and a uniquely vertebrate protein called **elastin**. Elastin is found in the major vessels of all gnathostomes and is also a major component of ligaments around joints. It is not known in invertebrates (or in agnathans) though structurally different proteins with similar capacity to store energy have been described in other groups of animals.

Elastin can be stretched to about double its normal length without breaking, and it is quite resilient, storing over 90% of the energy used to deform it and recoiling to its former shape when released. Collagen is about a thousand times stiffer than elastin and less extensible, being irreversibly deformed by stretches of more than about 10%.

The arrangement of these two materials in the walls of arteries (and smaller arterioles) gives them their crucial properties. The elastin fibres are close to their resting length in unstretched arteries, and the collagen fibres are crimped, so the artery doesn't put up much resistance to their early stages of stretch because the former are very extensible, and the latter are just uncrimped. However, once the crimp is pulled out, the stiffer collagen resists further stretch and makes the artery very difficult to burst. Further pressure drives blood flow, not expansion of the arteries. These properties are precisely matched to those of the pumping heart to produce smooth, energetically efficient flow.

The maximum volume of all the blood vessels put together is always much more than that of the blood itself, because only a fraction of the capillaries are fully perfused at once. Smooth muscle in the vessel walls can constrict arterioles and small arteries, re-routeing the blood flow to other vessels, and restricting the perfusion of certain areas of the body.

○ What signalling molecules control vasodilation and vasoconstriction?

● The release of nitric oxide promotes vasodilation and the neurotransmitter noradrenalin causes vasoconstriction. (See Clements and Saffrey, 2001.)

Conservation of body heat by shock-induced pallor, erection of the penis, and reducing perfusion of the extremities, all arise from re-routeing of blood flow by selective constriction and/or opening of blood vessels. These changes are coordinated by the nervous system, which continuously monitors blood flow and adjusts the pressure and degree of perfusion to posture and the energy requirements of different tissues during activities such as exercise, digesting a large meal, etc.

Veins and venules operate on similar principles but have thinner walls that allow much more swelling. Even large veins have strict limits on how much they can expand: distorting blood distribution by adopting abnormal postures (e.g. being upside down) or applying a tight tourniquet may cause veins to bulge uncomfortably, but they very rarely burst. Should a vessel rupture by extensive bruising or by the kind of tearing or biting wound that large animals inflict on each other, the smooth muscle in the walls of the large blood vessels quickly contract and close off the damaged areas.

○ What other property of the blood system helps to limit the harm caused by such wounds?

● The blood itself clots as well as the vessels sealing themselves.

These mechanisms together curtail bleeding very efficiently. High-pressure blood systems such as those of mammals and birds would be impractical without these safety systems that reduce spillage. The arrangement of the major arteries and veins is established during foetal development, but the smaller vessels grow 'on demand', increasing or reducing blood perfusion to a tissue according to its needs, and can regenerate after injury.

Most of the pressure imparted to the blood by pumping of the heart is dissipated as it passes through the capillary bed, leaving little to propel the blood back to the heart. However, since the vein walls are thin and flexible, movements and changes in shape of the tissues surrounding them make a major contribution to squeezing the blood along them. Many veins also contain small flaps of connective tissue that act as valves, promoting the one-way flow of blood.

○   Why would valves be important in the limb veins of terrestrial animals?

●   Because the blood in such veins would usually be moving upwards, against gravity, and would tend to flow backwards unless restrained by valves.

The fluid and some larger molecules squeezed out of the capillaries and into the tissue by the high blood pressure is collected in another set of vessels, the lymphatic system. Lymph vessels resemble small veins in many ways and transport fluid, cells of the immune system and other materials back to the blood. The lymphatic system of mammals is more elaborate than that of lower vertebrates, and includes many lymph nodes embedded in adipose tissue that contain high concentrations of leukocytes and other immune cells. (See Davey and Gillman, 2001.)

4.3.2   BLOOD

Vertebrate blood and respiratory systems are much more complex than those of invertebrates, involving several different kinds of cells, many of them present in large numbers. Because it contains so much cellular material, vertebrate blood is usually regarded as a tissue rather than a body fluid.

○   What is the principal mechanism by which oxygen reaches invertebrate tissues and waste products are removed from them? How does the anatomical arrangement of respiratory surfaces match this mechanism?

●   Simple diffusion of dissolved gases is sufficient for all but the largest and most active invertebrates such as cephalopods and large arthropods. Respiratory surfaces including gills and tracheae are often widely distributed around the body, so the gases never have to diffuse far to reach active tissues.

○   Would similar arrangements be adequate for craniates (Figure 4.2c)?

●   No. The respiratory surfaces are usually confined to the pharyngeal gills and/ or to the lungs, which are remote from most of the muscles.

Invertebrate blood is usually watery with only a few cells, most of them involved in clearing up wastes and removing foreign materials and microbes, and only rarely contains blood pigments.

○   What kinds of invertebrates have blood pigments?

●   Animals that live in environments such as mud, sand or stagnant water where oxygen is scarce (e.g. earthworms, lugworms, sea-cucumbers, mosquito larvae), and large, powerful invertebrates (such as the large crustaceans and cephalopods). (See Sections 1.2.2, 1.3.2 and 2.2.2.)

These invertebrate blood pigments that transport the respiratory gases to the tissues often consist of many subunits that associate to form large aggregates of very high molecular weight. For example, the haemoglobin of *Daphnia,* a water flea, has approximately 24 subunits; one species of *Arenicola,* the lugworm, has 96 subunits (and an M_r of 2.85 million) and some other polychaetes have haemoglobin with over 200 subunits.

In vertebrates, the haemoglobin is in the form of relatively small molecules (M_r of approximately 68 000), each of which can bind up to four molecules of oxygen, and is concentrated inside numerous tiny cells (erythrocytes) (Figure 4.4). Free haemoglobin molecules are small enough to be filtered out of the blood in the kidney. Being locked inside cells prevents the pigment molecules from being lost in the urine and protects them from damage in the fast-flowing blood. Red pigment in the urine is a sure sign of damage to red blood cells somewhere in the body, or of kidney malfunction.

The proportion of cells to the fluid plasma (which has no cells or other structures visible under the microscope, but contains many dissolved nutrients and hormones) varies from 20% to over 45% of the total volume. It is generally higher in physically active species, such as most birds and mammals including rats (Figure 4.4). The white cells (leukocytes) are larger than red blood cells but far less abundant.

Figure 4.4 Blood of an adult rat. The very numerous erythrocytes are self-coloured red, but the 'white' cells, leukocytes, have been stained with a blue dye to reveal the nuclei.

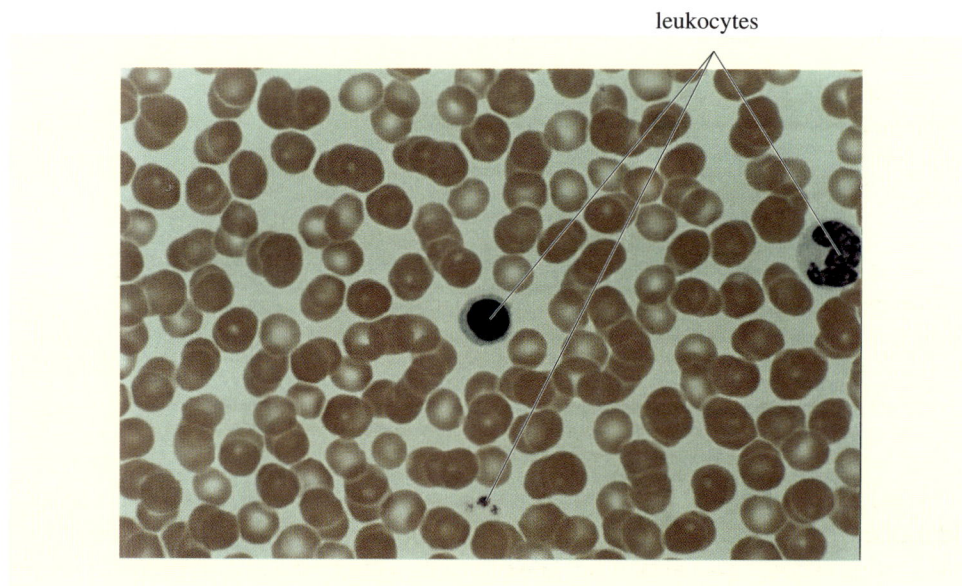
leukocytes

Erythrocytes enormously increase the amount of oxygen that can be carried: typical mammalian blood plasma at 37 °C holds only about 3 ml oxygen per litre (usually expressed as 0.3% volume) in simple solution, but fresh whole blood can carry up to 200 ml per litre (20% volume). The oxygen capacity of the blood depends upon the haematocrit (the volume of erythrocytes per unit volume of blood) and upon the concentration of haemoglobin in each erythrocyte. A greater proportion of the latter is the main reason why the blood of diving mammals such as seals can take up even more oxygen (up to 26.4% volume) than that of most terrestrial mammals.

The cellular composition of the blood of many vertebrates changes as the animals mature. Most tortoises and terrapins are quite small, but all living species of sea turtles grow to at least 100 kg, some species much more, before they return to the shore to lay their eggs. Green turtles (*Chelonia mydas*) which reach a mass of about 300 kg, swim long distances in warm seas throughout the tropics, where they eat mainly large algae. Table 4.1 shows some measurements from the blood of adult turtles, and from that of hatchlings that weighed about 100 g when they were collected as they emerged from their nest above the high-water mark of warm sandy beaches.

Table 4.1   The composition of the blood of hatchling and adult green turtles (*Chelonia mydas*) collected near their nesting grounds on beaches near the Australian Barrier Reef. Data from Wells and Baldwin (1994).

	Haemoglobin/ g l^{-1} of whole blood	Millions of erythrocytes per µl	Mean erythrocyte volume/pl	Mean cell haemoglobin /10^{-12} g	Mean cell haemoglobin concentration/g l^{-1}
Hatchlings	77	0.81	0.34	96	275
Adults	86	0.36	0.75	245	327

○ What is the haematocrit of whole blood in hatchling and adult turtles?

● In hatchlings, there are 0.81×10^6 erythrocytes in each µl of blood (equivalent to 0.81×10^{12} in a litre), each of which occupies a volume of 0.34×10^{-12} l, so the blood is $(0.34 \times 0.81) \times 100 = 28\%$ red cells. A similar calculation from the data for adults yields 27%.

○ What is the single major difference between the blood of hatchling turtles and adults of the same species?

● The hatchlings' erythrocytes are half the size of those of adults but they are twice as numerous in the blood.

In spite of this difference, the haematocrit and the haemoglobin concentration in the blood as a whole, and inside the erythrocytes, are about the same in hatchling and mature turtles. Further studies showed that the capacity for binding and transporting oxygen was about the same in both sets of samples. The change may be related to the mechanics of blood flow rather than its contribution to tissue respiration. Vertebrate blood is literally thicker than water, and in turtles and many other active animals is so packed with cells that its viscosity approaches the limit for its efficient circulation through small vessels. The hatchlings have to run several hundred metres across a predator-infested beach before reaching the sea, so their capacity for strenuous exercise must be nearly as good as that of the adults.

The largest known vertebrate erythrocytes are hundreds of times greater in volume than the smallest. The exact reason for such a wide range has not been firmly established, but the explanation that best fits the comparative data is that they match the capillary size. Erythrocytes are about 25% wider than the diameter of the capillaries they pass through and modern techniques for visualizing living cells *in situ* show that they are bent and partly squashed as they flow through. This deformation actually improves their efficiency for gas exchange, reducing the diffusion path between the blood pigment and the cell surface. Smaller, denser capillaries deliver oxygen to active tissues more efficiently, but they do present higher resistances to blood flow, which requires higher blood pressures to maintain the circulation. Larger animals need higher blood pressure than smaller ones of similar habits and body shape, because the blood has further to go before getting back to the heart.

○ What explanation for the data in Table 4.1 do these concepts suggest?

● Numerous tiny erythrocytes may be better suited to fast passage through the small blood vessels of hatchlings. If adults' erythrocytes and capillaries were equally small, their blood pressure would have to be very high.

The change with life history is noticeable in green turtles because the hatchlings are so small compared to the adults, but equally active. This example illustrates the matching between the structure and properties of the blood, the vessels it flows through, and the habits and habitats of the animals it serves.

Blood vessels as well as the blood itself can adapt to changes in physiological demand. Endothelial cells can detect the forces generated as the blood moves past them. Prolonged or frequent increases stimulate their proliferation, thus enlarging the capillary and enabling the blood to flow through it faster.

○ What changes in an individual's habits or circumstances might stimulate this effect?

● Greater forces on the endothelial cells may arise from increases in blood pressure, either in the system as a whole (e.g during strenuous exercise) or locally, as a result of occlusion of adjacent vessels, or swelling following an injury or infection.

A large proportion of the volume of all mature erythrocytes is haemoglobin, and these cells seem to be capable of little else than carrying gases. Those of mammals are so highly specialized that even the nucleus is absent. The cells are thus incapable of any form of growth or repair; instead, worn or inactivated cells are replaced continuously, new ones being formed from the recycled remains of 'old' ones in haemopoietic* tissues, which include the liver, kidneys and gut wall. The contribution of these sites differs between species and with stage of development: in adult mammals, erythrocytes are produced almost exclusively in the red bone marrow.

* This term is sometimes spelled hemopoietic, or hematopoietic or haematopoietic.

4.3.3 THE MECHANISM OF OXYGEN TRANSPORT

Just increasing the oxygen content of blood does not by itself equip the blood to supply oxygen to tissues remote from respiratory surfaces. The pigments must *reversibly* combine with oxygen, so that they easily pick up oxygen at the lungs, and, since the 'residence time' of erythrocytes in the capillaries of oxygen-using tissues is short, usually 0.2 to 2.0 s in mammals, give it up readily when and where required.

○ What other form of gas transport is necessary for prolonged, efficient aerobic respiration?

● The waste product, carbon dioxide, must be conveyed from the respiring tissues where it is produced, to the respiratory surfaces, where it is excreted. (Water is also formed, and may be excreted as water vapour, or through the kidneys, or retained for other physiological roles.)

Blood pigments also contribute to the important function of transporting carbon dioxide. The amount of oxygen available depends upon its partial pressure, a concept explained in Box 4.1.

BOX 4.1   PARTIAL PRESSURES OF GASES

The total pressure of a mixture of gases is the sum of the partial pressures that each component would exert if it alone were present.

Table 4.2 shows the composition of artificially dried air from the normal atmosphere and the partial pressures of its major components at sea-level. Although most forms of biological activity involve exchanging gases with the atmosphere, wind and thermal currents mix the air very well, and the proportions of these gases are remarkably constant except in small spaces such as inside burrows and termite mounds, near fires, etc. The amount of water vapour is much more variable, depending upon the temperature of the air and its proximity to bodies of liquid water. The proportion of water vapour is often more than 5% by volume, exerting a partial pressure of around 5 to 10 kPa and reducing proportionately the partial pressures of other gases.

Table 4.2 The composition and the partial pressures of each major gas in dry air at sea-level under the standard conditions of 1 atmosphere = 760 mm Hg = 101.3 kPa. The inert gases (mainly argon) and various synthetic gases together comprise about 1%.

Gas	% in dry air	Partial pressure/mm Hg	Partial pressure/kPa
Oxygen	20.95	159.2	21.2
Carbon dioxide	0.03	0.2	0.03
Nitrogen	79.0	600.6	79.6

○ What is the partial pressure of oxygen in air containing 5% water vapour (typical for a lowland garden on a summer day) that exerts a partial pressure of 5 kPa?

● The partial pressures of the other gases are reduced to (101 − 5) kPa = 96 kPa, of which, from Table 4.2, 0.2095 × (100 − 5) = 19.9% is oxygen, so the partial pressure of oxygen is 96 kPa × 0.199 = 19.1 kPa.

○ From Figure 4.5, would the presence of this quantity of water vapour make any noticeable difference to the ability of your blood to bind oxygen?

● No. These pressures are off the scale of the horizontal axis of Figure 4.5, far into the region at which the haemoglobin is almost 100% saturated.

Even the amount of water evaporated into the air during uncomfortably hot, damp weather (or in confined spaces like kitchens or bathrooms) makes no significant difference to oxygen uptake. However, the total air pressure, and hence the partial pressures of its component gases, decreases substantially with altitude: for example, at 6000 m above sea-level *, the proportion of oxygen is the same as at sea-level, but its partial pressure is halved.

○ From Figure 4.5 and Table 4.2, would the ability of human blood to take up oxygen be impaired at this altitude?

● Yes. At partial pressures of around 10 kPa, human haemoglobin is at most 95% saturated.

At this altitude, most people feel ill and are incapable of exercise.

The blood pigments of birds such as bar-headed geese (*Anser indicus*), whose annual migration involves flying over the Himalayas, are adapted to function at lower partial pressures of oxygen.

Different pigments are often compared by measuring the amount of oxygen they bind at a particular partial pressure of oxygen (P_{O_2}) expressed as a percentage of the amount of oxygen bound when the pigment is fully saturated. A respiratory pigment's *percentage saturation* plotted over a range of partial pressures produces a curve called the **dissociation curve** for oxygen (or sometimes the oxygen equilibrium curve) as shown in Figure 4.5 for haemoglobin. In the capillaries that perfuse the tissues, the partial pressure of oxygen is lower, around 4 kPa, so haemoglobin releases its oxygen, whereas at respiratory surfaces, where the partial pressures of oxygen are usually much higher, the pigment picks up oxygen (in which state it is called oxyhaemoglobin).

○ Is the oxygen dissociation curve for haemoglobin (Figure 4.5) linear? What are the implications of the curve's shape for the effects of a change in oxygen pressure on the amounts that are released from or remain bound to the pigment?

* The summit of Mont Blanc in the Alps is 4807 m above sea-level; Mount Everest is 8848 m high.

● No, the curve is not linear. The amounts of oxygen bound or released depend on where on the curve the change in the partial pressure of oxygen occurs. If the change took place over the very steep part of the curve (for example, the portion arrowed in Figure 4.5), a small change in partial pressure could induce a large change in the amount of oxygen bound to the pigment.

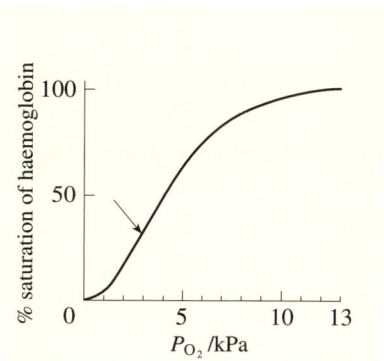

Nearer to the top and bottom of the curve (i.e. at partial pressures of oxygen below 2 kPa or above 8 kPa), the saturation of the pigment changes less rapidly, i.e. the *affinity* of the haemoglobin for oxygen is low. The shape of the curve is crucial to how efficiently a pigment can act as an oxygen carrier.

Figure 4.5   Oxygen dissociation curve for the blood of an adult terrestrial mammal such as a human.

Vertebrate haemoglobin is unusual in that, as shown by the part of the curve to the left of the arrow in Figure 4.5, its affinity for oxygen increases as more of the pigment is oxygenated. This particular property is displayed only when the four subunits, each with a discrete haem group, combine to form a tetramer. Changes in affinity for oxygen arise from alterations in the interactions between the haems of each molecule. The acquisition of a single oxygen molecule by one subunit increases the affinity of the neighbouring haems for oxygen, so that once a haemoglobin molecule takes up one oxygen molecule, it is easier for it to acquire the other three. Conversely, the loss of oxygen from haem in a saturated molecule of oxyhaemoglobin lowers the affinity of the remaining haems and two or three more oxygen molecules usually become detached. We now know that the binding (or release) of a single oxygen molecule to (or from) a haem influences the affinity of distant haems by changing the shape of the whole tetramer.

Haemoglobin is among the most thoroughly studied of all biological molecules. The combination of detailed knowledge of its primary and tertiary structure, and of allosteric effects in enzymes, led to a very convincing molecular explanation of how cooperation between the monomers can modify the binding of oxygen at individual sites. The haemoglobin tetramer can exist in two interchangeable forms, one called the *T (tense) structure*, the haems of which have a comparatively low affinity for oxygen, and the other the *R (relaxed) structure*, in which the haems have a high affinity for oxygen. The two states differ only in a slight rotation of one pair of subunits with respect to the other pair.

The uptake of oxygen by one subunit in the T structure promotes a small shift in the arrangement of the whole complex to form the R structure. The transition from T to R is not induced by the binding of a set number of oxygen molecules but becomes more probable with each successive oxygen molecule that is bound.

○ Are similar mechanisms known in enzymatic proteins?

● Yes. Cooperativity of substrate binding (Walker and Swithenby, 2001) involves conformational changes in an enzyme following binding of a substrate molecule to one active site that increases the affinity of the other active sites for their substrate.

The transition from T to R states of haemoglobin changes the bonds that link the subunits, which shifts the spatial relationship between the haem iron and the plane of the porphyrin ring, so altering the binding affinity of the iron for molecular oxygen. The switch from T to R alters the binding characteristics so effectively that the last haem group has an affinity for oxygen that is much like that of a free subunit, and is one hundred times the affinity of the first haem group.

○ How could such cooperative interactions between subunits help to maintain the conditions under which oxidative tissues work best, and minimize the energy consumed by pumping the blood around the body?

● The cooperative interactions between subunits ensure that a large proportion of the oxygen carried in the blood bound to the haemoglobin is delivered to the tissues, even without the need for a large drop in their oxygen content.

The mitochondria only function optimally when oxygen diffuses into them at a steady rate; their supplies would be impaired by a steep drop in the oxygen pressure around them. 'Futile' passages of the blood through the capillaries are reduced because at least some oxygen can be off-loaded into the tissues each time, even if only a little more is required.

CARBON DIOXIDE

As well as influencing oxygen binding, carbon dioxide also binds directly to vertebrate haemoglobin though not at the sites occupied by oxygen molecules. Carbon dioxide (CO_2) is about 30 times more soluble in water (and thus in plasma) than oxygen, but even so, its binding to haemoglobin increases by up to 100-fold the CO_2 held in the blood over that present in simple solution in the plasma. Like that of oxygen, the affinity between CO_2 and haemoglobin is affected by pH (i.e. the interaction has its own Bohr effect, see below), but to a much smaller extent. Unbound CO_2 readily diffuses across the respiratory surfaces from which it is excreted if the concentration gradient favours its passage from the blood to the external medium.

○ Would much CO_2 pass into animals from air or from water?

● Almost none is taken up from air, because the concentration of CO_2 in air (see Table 4.2) is so low compared to that of the blood. However, water, especially stagnant pools rich in decaying organic matter, sometimes contains large quantities of dissolved CO_2 so the gas passes into fish, tadpoles and other water-breathing vertebrates.

Although chemical conditions seem to favour the passage of CO_2 from the blood into the air, in mammals, only a small amount manages to escape through the lungs because it is bound to the blood pigments. The rest is excreted in solution through the kidneys as bicarbonate ions.

Perhaps surprisingly, neurons in the brain that control the rhythm of breathing movements in mammals and birds are much more sensitive to small changes in the concentration of carbon dioxide in the blood than to its oxygen content. A modest fall in CO_2 can suppress breathing while even a major decrease in blood oxygen content fails to stimulate breathing.

A side-effect of the interactions between oxygen binding sites of vertebrate haemoglobin, and the relative independence of its interactions with carbon dioxide, is susceptibility to carbon monoxide poisoning, as described in Box 4.2.

Box 4.2 Carbon monoxide and haemoglobin

As readers of detective stories know, carbon monoxide (CO), which is rare in the atmosphere except near fires, is one of the fastest, quietest and most effective poisons of humans and other vertebrates. CO binds very strongly to the same part of the haemoglobin molecule as oxygen. In the case of human haemoglobin, the affinity for CO is more than 200 times higher than that for oxygen, so even when CO is present at a concentration of only 0.01% (partial pressure only 0.0013 kPa, see Box 4.1), up to half the oxygen-binding sites are occupied by CO molecules after a few breaths.

○ What similarities between CO and O_2 might enable CO to bind to the oxygen-binding sites?

● The molecules are nearly the same size: M_r of O_2 is $2 \times 16 = 32$, and the M_r of CO is $12 + 16 = 28$.

Contrary to intuitive expectations, the inactivation of such a large proportion of the blood pigment is not the immediate cause of death. It has been known since the early twentieth century that people and animals can survive (although much weakened) a 50% loss of the oxygen-carrying capacity of the blood caused by other means: exposure to oxygen-depleted air, prolonged heavy bleeding or severe anaemia (failure to produce enough red blood cells). Disabling a similar fraction of the blood's haemoglobin by CO poisoning is lethal almost at once because when two of the binding sites of each haemoglobin molecule are occupied by CO molecules, cooperation between the monomers binds the oxygen on the remaining sites so tightly that it cannot be released to the tissues. In other words, a small intake of CO does not fatally impair the uptake of oxygen in the lungs, but stops almost completely its delivery to the cells that need it. The brain, a highly aerobic tissue, is affected within seconds, leading to loss of consciousness and, often within less than a minute, disruption of the neural signals that control breathing.

○ Would exposure to CO and the resulting decrease in bound oxygen stimulate breathing?

● No. Breathing is stimulated mainly by the proportion of carbon dioxide in the blood, which is bound to other parts of the haemoglobin molecule.

Animals (and people) poisoned with CO do not struggle to breathe, but die quickly and quietly. However, people can recover completely from being deeply unconscious and unable to breathe if their lungs are artificially ventilated for long enough for the CO to be exhaled, restoring the haemoglobin's capacity to deliver oxygen to the tissues.

○ Would insects be poisoned by carbon monoxide?

● No. Most insects do not have haemoglobin (or other blood pigments).

Carbon monoxide has a much lower affinity for other kinds of blood pigment, making it selectively toxic to vertebrates, especially metabolically active, highly aerobic species such as mammals and birds.

THE BOHR EFFECT

The affinity of haemoglobin for oxygen is affected by several other factors, of which the best understood are pH and the partial pressure of carbon dioxide, which dissolves to form carbonic acid (H_2CO_3) which dissociates into bicarbonate (HCO_3^-) and H^+ ions. Professor Christian Bohr[*] who was professor of physiology at the University of Copenhagen around the end of the nineteenth century, was among the first to study the physiology of fresh blood *in vitro*. He demonstrated that the affinity of blood for oxygen is decreased by an increase in the partial pressure of carbon dioxide, causing the oxygen dissociation curve to shift to the right in graphs such as Figure 4.6. This effect, named the Bohr shift or **Bohr effect**, can be found in most kinds of haemoglobin (and many other blood pigments).

Figure 4.6 Bohr effect: the effect of carbon dioxide on the affinity of oxygen for haemoglobin, which promotes the delivery of oxygen to tissues that need it. The S-curve (red) was measured at pH 7.2 and the R-curve (blue) at pH 7.5. The point marked 'a' corresponds to P_{O_2} of the arterial blood and the point marked 'v' to P_{O_2} of the venous blood.

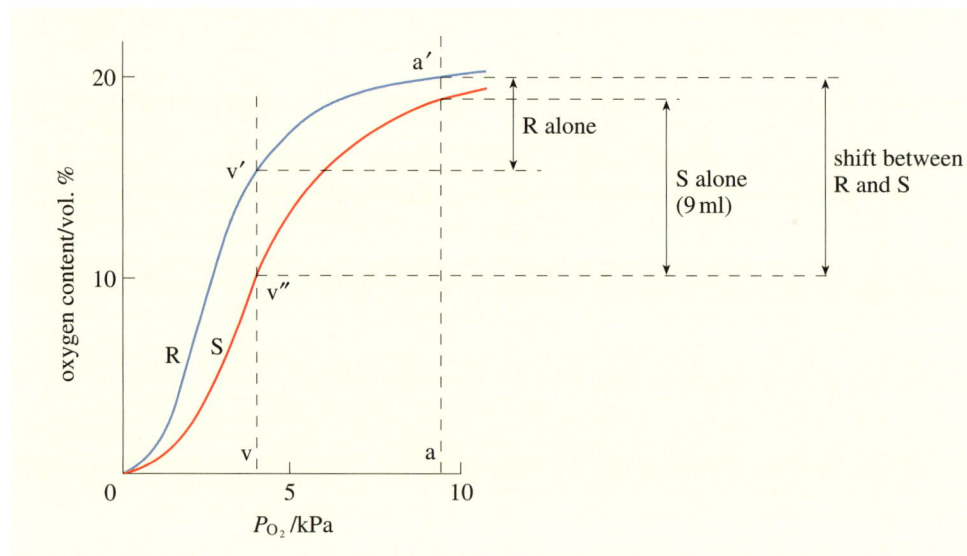

The Bohr effect is thought to arise mainly from the drop in pH that occurs when the increased amounts of carbon dioxide dissolve to form bicarbonate and H^+ ions, but the carbon dioxide itself might have a small direct influence on the affinity of haemoglobin for oxygen.

○ As well as carbon dioxide, what other end-product of energy metabolism *in vivo* could alter the pH of an active tissue?

● Lactic acid is produced by anaerobic metabolism and often builds up during prolonged strenuous activity, causing a fall in pH large enough to induce the Bohr shift.

In actively respiring tissues, levels of carbon dioxide (and in many cases, of lactate as well) are high. The Bohr shift has important implications for the delivery of oxygen to the tissues because it enables haemoglobin to give up a larger proportion of the bound oxygen to tissues that need it than would occur if the acidity did not alter oxygen binding.

[*] The father of the distinguished physicist Niels (1885–1962) and mathematician Harald (1887–1951), and grandfather of the physicist Aage Bohr (1922–).

A pigment that has no Bohr shift follows curve R in Figure 4.6. The maximum P_{O_2} in arterial blood is 9.4 kPa (at the point marked 'a') and the P_{O_2} in the venous blood is only 4.0 kPa (at the point marked 'v'). The points for v are less than those for a because the haemoglobin has deposited some of its oxygen to the tissues as it passed through the capillaries. The corresponding values for the oxygen content of blood (i.e. the amount bound to the haemoglobin) are shown as a′ for arterial blood and v′ for venous.

○ How much oxygen is delivered to the tissues per 100 ml of blood?

● If the oxygen content of arterial blood is about 20% volume (i.e. 20 ml oxygen per 100 ml of blood) and that of venous blood is approximately 15% volume (as on curve R), then each 100 ml of blood passing to the tissues yields about (20 − 15) ml = 5 ml of oxygen.

If the pigment followed curve S exclusively i.e. there were no Bohr shift, the oxygen transfer to the tissues would be approximately (19 − 10) ml = 9 ml (see Figure 4.6).

In fact, the two curves relate to the same pigment and the degree of oxygenation is measured at two *different* partial pressures of carbon dioxide, one of which (curve R) is at a lower P_{CO_2}, typical of arterial blood (higher pH), and the other (curve S) is at a higher P_{CO_2} typical of venous blood (lower pH); the difference between them is attributable to the Bohr shift.

○ What is the oxygen transfer to the tissues, given a Bohr effect of this magnitude? Remember that the P_{CO_2} at the respiratory surface is low (see point a′ on the curve R), and the uptake of carbon dioxide by the capillary blood means that the degree of oxygenation of red blood in the tissues is represented by point v″ on the curve S.

● In these circumstances, the oxygen transfer to the tissues is (20 − 10) ml = 10 ml, the range shown as 'shift between R and S' in Figure 4.6.

This higher value for the oxygen transfer to the tissues (around an extra 10% compared with curve S alone) is a consequence of a shift in the dissociation curve that occurs when the respiratory pigment passes from the respiratory organ to the tissues or vice versa. A change in the pigment's affinity for oxygen in this way causes increased transfer of oxygen to the tissues over and above what it would be if the dissociation curve was unaffected by carbon dioxide and followed either of the curves (R or S) alone.

These observations and experiments on blood *in vitro* do not by themselves prove that the effect has any physiological significance *in vivo*.

○ What additional measurements would be needed to establish that the Bohr effect occurs *in vivo*?

● Accurate measurements of the pH inside erythrocytes (i.e. that actually experienced by the haemoglobin), and of the partial pressure of carbon dioxide (see Box 4.1) and pH at the respiratory surfaces and in the tissues, and how they change during strenuous oxidative metabolism would help to demonstrate whether this effect occurs *in vitro*.

The venous blood of resting people is at pH 7.36, which is slightly lower than that of arterial blood (pH 7.40), and has a partial pressure of carbon dioxide of 6.13 kPa, slightly higher than that of arterial blood (5.33 kPa). These differences are sufficient for an appreciable Bohr shift, and indeed there is significantly less oxygen in venous blood than would be predicted from the oxygen dissociation curve of arterial blood (Figure 4.5). At rest, about half the oxygen taken up in the lungs is delivered to the tissues by each passage of the blood. During exercise, carbon dioxide and lactate accumulate around the capillaries in muscle and other active tissues, so thanks to the Bohr effect, more oxygen is delivered to them, supporting ATP synthesis by oxidative phosphorylation.

These efficient mechanisms for delivering oxygen over quite long distances have made possible the evolution of large bodies that contain metabolically demanding tissues such as the liver and the brain, and numerous dense, powerful muscles for swimming and other forms of locomotion.

○ Does the leghaemoglobin in the root nodules of leguminous plants (family Fabaceae) have similar properties to vertebrate haemoglobin?

● Yes. It binds atmospheric oxygen, thereby maintaining a low partial pressure of oxygen in the root nodule, but a little is released to support aerobic respiration in the bacteria.

○ Is maintaining a steady, fairly low concentration of oxygen in tissues also important in animals?

● Yes. Excessive oxidation produces free radicals which can damage proteins, DNA and lipids, and which if abundant for long periods, are believed to promote ageing (see Halliday and Pond, 2001, and Saffrey, 2001).

In spite of superficial contrasts, the functions as well as the properties of plant and vertebrate haemoglobins are in fact quite similar. Small differences in molecular structure of haemoglobin often confer important differences in physiological properties of the blood. In general, the Bohr effect is weaker in blood from large animals than from physiologically similar small ones because the tissues of the latter use oxygen at a higher rate.

In animals that have several distinct life-history stages in different habits, the haemoglobins may adjust accordingly. For example, many tadpoles live in stagnant water in which only a little oxygen but much carbon dioxide is dissolved. The haemoglobin in their blood has a higher affinity for oxygen but a weaker Bohr effect than that of the more athletic frogs that breathe normal air.

○ What advantage would these combinations of properties be to tadpoles and frogs?

● The high affinity for oxygen of tadpole haemoglobin is necessary for their gills to take up scarce oxygen, and should not be impaired by the acidity caused by much dissolved carbon dioxide. Frogs need haemoglobin that can deliver oxygen efficiently to exercising (i.e. more acid) tissues. Their respiratory surfaces (skin, lungs and mouth) are very rarely exposed to concentrations of carbon dioxide high enough to reduce their capacity to take up oxygen.

○ What genetic mechanisms are necessary for the evolution of these alternative forms of haemoglobin?

● These different haemoglobins must be transcribed from different genes, which are activated at different stages of the life cycle. The genes presumably arose by duplication, and each acquired mutations that adapted its product to different habitats.

However, the intricate structure of haemoglobin, with its separate binding sites for oxygen and carbon dioxide, means that much of the molecule is highly conserved in all classes of vertebrates. The structures and homologies of the genes for haemoglobins have been intensively studied and their probable changes in evolution established, as described in Chapter 5.

SUMMARY OF SECTION 4.3

1 Vertebrate blood is contained entirely within vessels and is pumped around the body under pressure. Mechanical properties of the walls of the larger vessels accommodate changes in pressure generated by the pumping heart and prevent bursting and excess bleeding.

2 Gas exchange and nutrient transfer take place at the capillaries that permeate all cellular tissues.

3 The blood of almost all vertebrates contains the pigment haemoglobin, which is of lower molecular weight than that of most invertebrate haemoglobin, and is confined inside erythrocytes. It binds oxygen and (to a different part of the molecule) carbon dioxide.

4 The numerous cells make vertebrate blood rather more viscous than invertebrate blood. The composition of the blood, including the size and abundance of erythrocytes, differs greatly between species and changes during the life cycle.

5 Haemoglobin binds more strongly to oxygen at higher pH and releases it more readily at lower pH. This Bohr effect, and the shape of the oxygen dissociation curve, arise from changes in the molecular configuration of the pigment — and hence its capacity to bind oxygen — with pH.

6 The Bohr effect determines the efficiency of delivery of oxygen from the respiratory surface to the tissues that utilize it. Metabolic activity increases the acidity in the tissues, which releases oxygen from the haemoglobin.

7 Haemoglobins of different species, and in some cases different stages in the life history of the same species have oxygen dissociation curves and Bohr effects that adapt them to their habits and habitats.

8 Breathing is controlled by neurons that are acutely sensitive to the carbon dioxide concentration of the blood.

4.4 SKELETAL TISSUES

The skeleton is clearly important for large, powerful animals, especially those living on land. As well as supporting the muscles in locomotion, vertebrate skeletal materials protect the elaborate nervous system and the eye, ear and other sense organs (forming the skull (cranium) from which the name craniate is derived), and form the basis of the feeding apparatus (i.e. the jaws and teeth).

Most of the diverse structural tissues of vertebrates (including bone, tooth and cartilage) exploit the mechanical properties of the extracellular protein collagen, and its capacity to be hardened by calcium salts that form around its long, tough fibres. Many kinds of invertebrates have small quantities of collagen, often as a major component of cartilage or the body wall (Chapter 1), but the protein is found in a greater variety of forms, and in much larger quantities, in vertebrates. As well as occurring in chemically different forms, i.e. as products of different genes, collagen is also organized and assembled in a wide range of different ways that confer contrasting mechanical properties on tissues, equipping them to act as tubes, struts, ropes, weapons, shock absorbers and protective coverings.

The pressurized blood system permeates even very dense structural tissues, nourishing living cells buried deep within the large quantities of extracellular material. Thus many vertebrate skeletal tissues, as well as being internal, are living, and so are capable of growth, repair and remodelling to adapt to changes in habits and habitats.

○　How do animals with shells or cuticles grow?

●　Arthropods and nematodes grow by ecdysis (see Sections 1.2.3 and 2.2.2): the old cuticle is shed and a new, larger one hardens. Molluscan shells grow only at the edges of the mantle (see Section 1.3.1).

The timing or mode of growth of vertebrate internal skeletal tissues is not limited in these ways.

4.4.1　BONE

Bone is one of the most important uniquely vertebrate tissues. It consists mainly of an extracellular matrix of collagen fibres of diameter up to 100 nm, and hydrated crystals of an inorganic salt, *hydroxyapatite,* $3[Ca_3(PO_4)_2]Ca(OH)_2$, in the form of small enlongated crystals about 4 nm by 40 to 400 nm. The hydroxyapatite crystals lie within and between the collagen fibres, and together they form sheets (lamellae) a few micrometres in thickness. Bone is thus a composite material, strong in both tension and compression: its high mineral content makes it stiff and strong under compression, but the presence of collagen makes it strong in tension (i.e. when pulled or bent) and stops cracks from running through the brittle crystals of hydroxyapatite.

In most vertebrates (teleost fish being the main exception), there are living cells, called *osteocytes,* embedded in these materials. They secrete the collagen molecules that assemble into fibres and create the appropriate chemical conditions for mineral precipitation. The osteocytes are nourished by numerous blood vessels that permeate the tissue in small channels (called Haversian canals). In most vertebrates, the skeleton consists largely of compact bone, which is up to 90% calcified and is permeated only by cavities containing the osteocytes and the blood vessels. The density of compact bone is up to $2.4\,g\,cm^{-3}$, mainly because of the high density ($3.2\,g\,cm^{-3}$) of the hydroxyapatite mineral.

Terrestrial vertebrates, especially mammals and birds, have several more specialized kinds of bone, which make the skeleton lighter, or stronger, or able to grow, repair and remodel itself more rapidly. The main differences are in the abundance and orientation of the cells and extracellular collagen fibres, and in the degree of mineralization. One of the best known is Haversian bone (shown in Figure 4.7), in which most of the lamellae form concentric cylinders (that appear as rings in sections), called osteons, around central canals containing blood vessels. Osteons usually run approximately parallel to the long axis of weight-bearing bones. Other kinds of bone form struts, which confer strength for minimal increase in weight, and the core of most large bones is filled with soft, less dense tissue called bone marrow. This adaptation is taken to extremes in flying birds (and the extinct flying reptiles, pterodactyls) whose long bones have cavities containing air sacs.

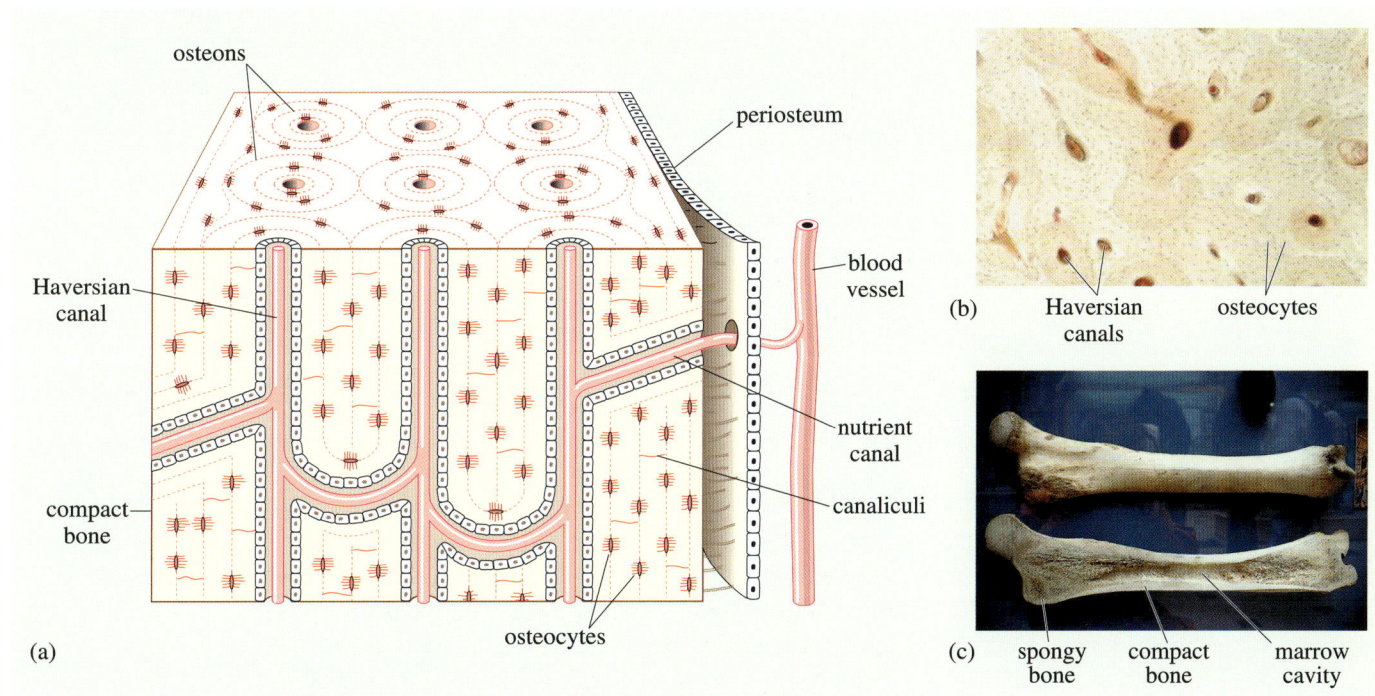

(a)

(b) Haversian canals   osteocytes

(c) spongy bone   compact bone   marrow cavity

Both the cellular and the extracellular components of bone are added to and replaced during growth and, particularly in mammals, during adult life, so bones can be repaired, strengthened, reshaped or even totally eliminated in response to functional demand. Bone-dissolving cells (called osteoclasts) from the marrow attack the functional bone through the vascular channels, thereby making way for bone-forming cells (called osteoblasts) to form new bone.

Bones are linked to each other with ligaments and to muscles with tendons. Both these tissues are composed almost entirely of extracellular proteins, mostly collagen, with very little cellular material. Tendons are diverse in internal structure and mechanical properties, ranging from stiff, sometimes even ossified, to very flexible but almost inextensible or elastic, capable of recoiling completely after being stretched by up tp 10% of their slack length.

Figure 4.7 (a) The organization of osteocytes, blood vessels and collagen fibres in Haversian bone. (b) Cross-section of Haversian bone of a rat. (c) Longitudinal section of a dried mammalian femur. The shafts are composed of dense, compact bone with Haversian canals. In life, the core is filled with soft marrow, which consists of adipose and haemopoietic (blood-forming) tissues, crossed with struts of spongy bone.

Bone cracks surprisingly easily when bent or stretched; tiny cracks would coalesce into major fractures but for the ability of living bone to repair itself as soon as a weakness appears. Osteoblasts (and a surprisingly wide range of other kinds of cells) are sensitive to mechanical forces and form new bone in response to local stress, wear and breakage. Exercise promotes faster repair of breakages, and, if frequent and strenuous, leads to the deposition of additional bone, thus forming larger, denser bones. Withdrawing such forces for as little as a few days leads to net loss of bone and severe weakening of the skeleton.

○ Why do the skeletons of astronauts in space and invalids lying in bed become weak? Which bones are likely to be least affected?

● Body weight is the main force acting on the back and leg bones, and is high when standing or running, and low when lying down. Gravity is minimal in space, so the astronauts' bodies, like other objects, are weightless. Without such forces, the skeleton of the back and legs become weak. The jawbones would be least affected since the main forces acting on them arise from biting and chewing, which do not depend upon gravity.

Even after exercising for several hours a day on machines that impose large forces on all the limb bones, astronauts' skeletons are often barely strong enough for them to be able to stand up when they return to Earth.

This 'micro-maintenance' explains why the vertebrate skeleton seems to be so much stronger and more reliable inside living animals than implements made from dead bone. As a material for making tools and weapons, bone is inferior to wood, metals and most kinds of stone because it splits so easily, particularly if allowed to dry. Deer antlers are among the few types of bone that are naturally exposed to desiccation. Because this material is tough and hard even when dry, it was the preferred choice of Stone Age people for use as picks, awls and chisels.

○ Do molluscan shells share this limitation, i.e. being much weaker when freshly dead than when part of the living animal?

● No. 'Dead' gastropod shells cannot be repaired if shattered as those in contact with the mantle secretory surface are, but they remain robust enough for hermit crabs to adopt them.

In spite of this difference, experiments undertaken to develop a procedure for promoting the formation of new bone reveal some similarities between nacre and bone that suggest they have cellular mechanisms in common.

Bone acts as a reserve of essential metabolites such as calcium ions, which may be present in only small quantities in the diet, and are needed in large quantities for nourishing young and the formation of eggs. Because of the profuse blood supply, this component of bone can be withdrawn very rapidly in response to calcium-releasing hormones. Depleting the mineral content of bone makes it less dense but much more fragile, as happens in many elderly people and in lactating mammals.

○ Is bone an ideal skeletal material for actively swimming aquatic animals?

● No. Bone is very dense, so animals with skeletons composed of it would sink in water.

Some bony fish (including probably many of the most ancient, that are known only as fossils) spend most of the time on or near the bottom. Others offset their dense skeleton with buoyancy devices including air-filled swimbladders that form as a pouch from the anterior region of the gut. Chondrichthyans (sharks, skates, rays and dogfish) never have swimbladders; their skeleton is composed of cartilage, which in many species becomes calcified, but is always significantly less dense than bone. The compromise between skeleton strength and maintaining neutral buoyancy is thought to be the main reason why all very large, actively swimming marine fishes are chondrichthyans. For example, the basking shark (*Cetorhinus maximus*) grows to over 14 m in length and can weigh about 4 tons in air. It feeds on copepods and other small planktonic animals filtered on its enormous gills, and breeds around the coast of south-west England and Ireland, where it is quite common in summer.

TEETH

Teeth are composed mainly of **dentine** which is a specialized form of bone, with an extensive calcified matrix secreted by dentine-forming cells.

○ What is the main contrast in the arrangement of the cellular components of bone (Figure 4.7) and dentine (Figure 4.8a)?

● Osteocytes are entirely enclosed in the calcified extracellular matrix but in teeth, the main part of the cells are condensed in the pulp cavity with only fine processes extending into the dentine.

The fine tubes that contain the processes of the dentine-secreting cells are called dentinal tubules. The pattern they form in the extracellular matrix is readily preserved in dead and fossilized tissue and can be used as a taxonomic character.

The more heavily mineralized extracellular matrix makes teeth more resistant than bone to both impact and abrasion, but most only retain their mechanical properties when kept moist (i.e. in the mouth). Tusks are enlarged canine or incisor teeth that protrude externally. Their dentine forms ivory, in which the dentinal tubules are arranged in a more complex pattern that resists splitting due to drying, and imparts greater strength, especially in bending and twisting.

Tusks have evolved in several different lineages of mammals (elephants, walruses, narwhal whales, hippos, etc.), and each kind of ivory has a slightly different arrangement of tubules that forms a distinctive regular pattern (Figure 4.8b). As people discovered during the Stone Age, ivory is a much better material than bone or most other kinds of teeth for making needles, awls, joints and other precision implements that must be tough, strong and hard-wearing.

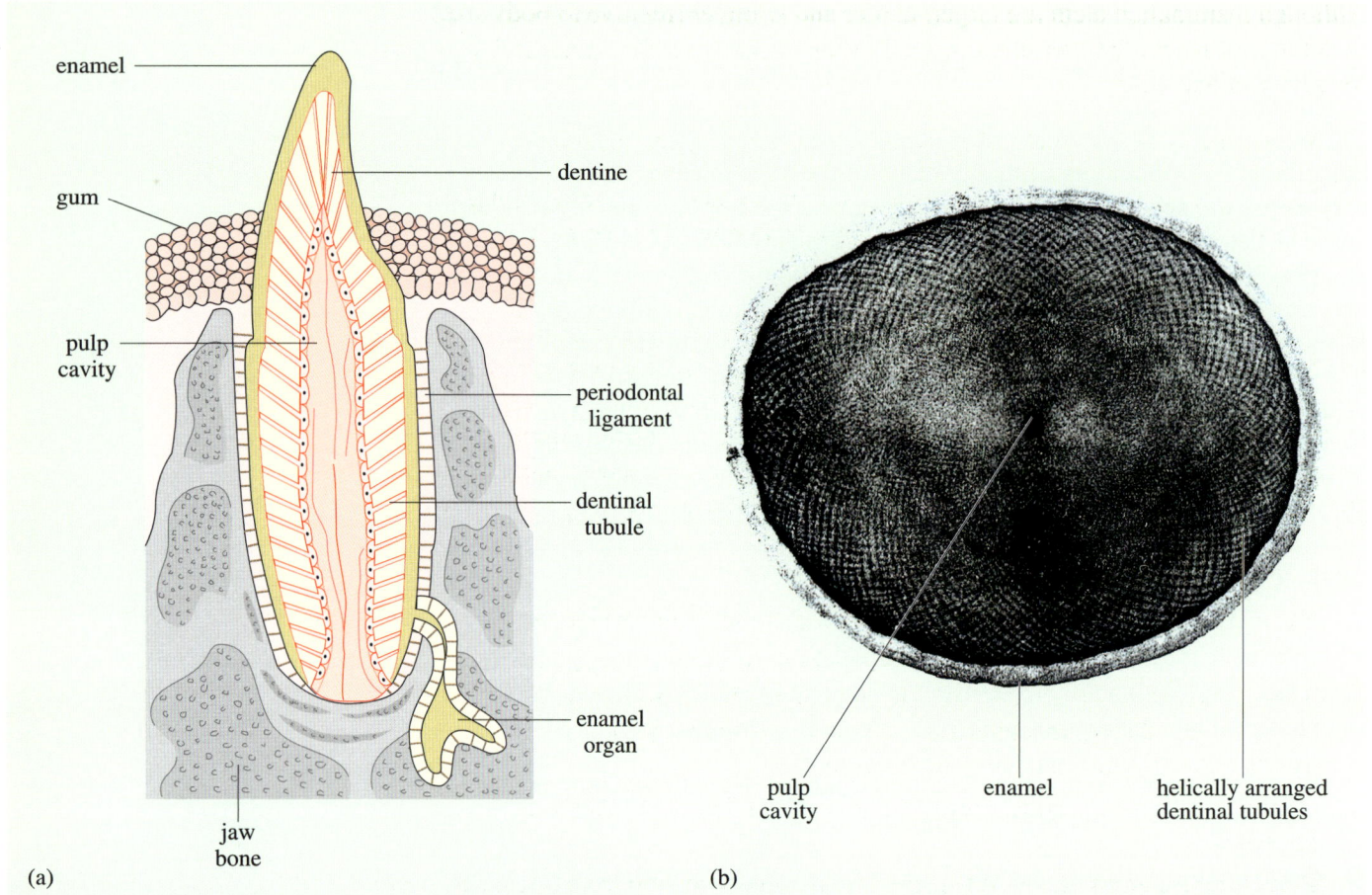

Figure 4.8 (a) Vertical section of mammalian canine or incisor tooth. (b) Cross-section of an elephant tusk (an enlarged incisor tooth), showing the thin outer layer of enamel (which lacks cellular material), and a tiny central pulp cavity. Most of the area is dentine, containing dentinal tubules with a characteristic lattice pattern. In life, each tubule contained a long process of a dentinal cell whose cell body was in the pulp cavity.

Fish, amphibians and reptiles often have teeth on other bones lining the mouth, as well as or instead of on the bones that form the jaws. Some fish also have tooth-like structures on the bones supporting the gills. The numbers of teeth are variable and the dentition is added to as the animal grows. In long-lived species such as crocodiles and alligators, the teeth are replaced in a fairly regular sequence, and those that are broken or torn out by struggling prey are readily replaced by newly formed ones.

Mammalian teeth are much more complex in structure and are set in sockets rather than growing directly out of the bone as in lower vertebrates. They form only on the upper and lower jaws, and fit together to form intricate cutting or grinding surfaces operated by finely controlled muscles. The molar teeth do not emerge until the jaws are almost fully grown, usually some time after weaning. The incisors, canines and premolars are replaced only once early in life while the skull is growing (in some species, including most marsupials (subclass Metatheria), not at all), so broken or decayed teeth are not replaced during adulthood.

○ Do mammals and other toothed tetrapods live as long as those in which the teeth are replaced by a horny beak?

● No. In general, birds and tortoises have longer lives than mammals and toothed reptiles of similar body mass (see Halliday and Pond, 2001).

Although mammalian teeth are larger, harder and stronger (relative to body size) than those of most other vertebrates, tooth wear seems to set the upper limit to longevity in mammals.

○   What aspects of mammals' diet requires such dental specializations?

●   Many mammals eat leaves, seeds and even the wood of large terrestrial plants (see Pond, 2001), that is, tough abrasive foods that would wear even very hard teeth.

Many successful mammalian herbivores, including rodents, lagomorphs (rabbits and hares) and equids (horses and donkeys), have exceptionally long teeth that erupt slowly throughout life. Elephant molars and premolars are huge (a single tooth can weigh several kilograms) and one tooth in each half of the upper and lower jaws can fill most of the chewing surface. They emerge sequentially, with successively larger teeth replacing the worn predecessors as the jaw grows.

SUMMARY OF SECTION 4.4

1   Structural tissues are particularly diverse and complex in vertebrates.

2   Bone is a living tissue unique to the vertebrates; osteocytes are incorporated into an extensive extracellular matrix of collagen fibres and nourished from blood vessels that permeate the entire structure. Small crystals of hydroxyapatite form in the extracellular matrix.

3   Cells associated with bone continuously repair and remodel the hard materials.

4   Bone is dense and sinks in water; free-swimming bony fish have buoyancy devices that enable them to float. The skeleton of chondrichthyan and certain other (especially large) fishes is cartilage.

5   Teeth are derived from bone. The main tissue, dentine, consists of a heavily mineralized matrix with the cell bodies assembled in the pulp cavity.

6   Most vertebrate teeth are replaced several times during life and in response to breakage or wear. Mammalian teeth are more complex in structure and are replaced only once or not at all, though most specialized herbivores have some continuously growing teeth. Ivory is specialized to resist desiccation, abrasion and impact.

4.5   SKIN

The tissues that comprise the skin have much in common with those of the internal skeleton, especially in lower vertebrates. The two major components of skin are a living layer, called the **dermis**, that contains several kinds of cells, small blood vessels and large quantities of extracellular collagen that forms a flexible mesh (Figure 4.9a). Overlying the dermis, especially in terrestrial vertebrates, is an outer **epidermal** layer of keratin, which is tough, insoluble and chemically inert.

○ Where does keratin occur in other kinds of cells and what is its role?

● Keratin is a major component of the cytoskeleton of epithelial cells (see Loughlin, 2001).

The living epidermal cells form a typical epithelium and divide throughout life. As they mature, the epidermal cells stop dividing and synthesize more and more keratin, which accumulates until it kills the cells that made it. The dead keratin-filled cells stick together, forming an impermeable outer layer that protects the skin from desiccation. Newly keratinized cells form underneath the dead surface, replacing it as it is worn away.

The development of the two layers differs greatly between classes. Keratin is minimal in fish but the collagenous layer is thick, and often ossified to form bony scales, as shown in Figure 4.9b.

Figure 4.9 (a) The generalized structure of vertebrate skin. (b) Large bony scales and stout fin rays in the gar-pike, *Lepisosteus* sp., a primitive group of actinopterygian fishes now found only in rivers and estuaries of North and Central America. (c) The scales are greatly reduced in many teleosts, especially fast-swimming predatory fish such as this mackerel. The body skin is soft and flexible, but dermal bone is still prominent in the skull and operculum (bony flap that covers the gills.)

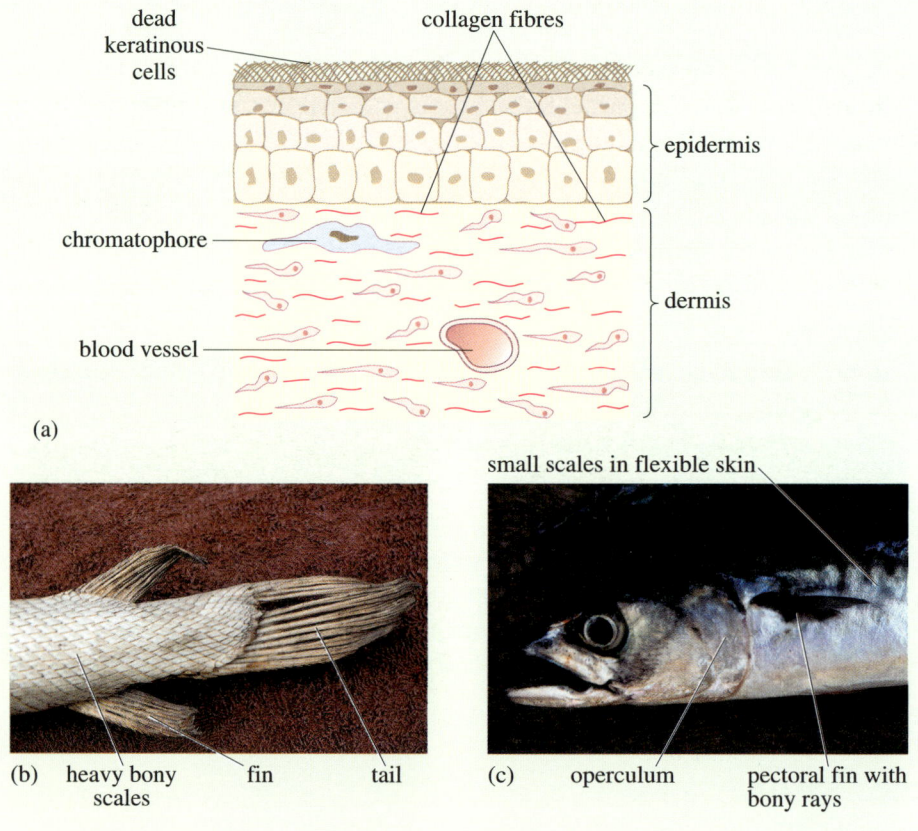

(a) dead keratinous cells — collagen fibres — epidermis — chromatophore — dermis — blood vessel

(b) heavy bony scales — fin — tail

(c) small scales in flexible skin — operculum — pectoral fin with bony rays

○ What are the similarities between the organization of the dermal layer of skin (Figure 4.9) and bone (Figure 4.7)?

● Both consist mainly of a three-dimensional network of collagen fibres and widely-spaced cells nourished by numerous fine blood vessels.

Scales vary greatly in size and shape in different kinds of fishes, from very small for maximum flexibility, to large and heavy for maximum protection, sometimes with an outer layer similar to dentine or enamel. The skin of many extinct amphibians seems to have been similar, but that of all living species serves as a respiratory surface, as well as or instead of the lungs (in adult stages) or gills (in larvae).

○ Which features shown in Figure 4.10 equip the amphibian skin to act as a respiratory surface?

● The reduced keratinous layer allows small blood vessels in the dermal layer to be close to the surface where gases are exchanged. Secretions from the many glands keep the skin moist, prevent desiccation and facilitate respiration.

Keratin is almost absent from tadpole skin and its appearance is one of the many changes that take place at metamorphosis into terrestrial frogs.

Frogs that are aquatic as adults, notably *Xenopus*, and many fish, including agnathans, the most primitive living vertebrates, and some higher teleosts such as eels, secrete mucus in large quantities, but for a different function. By making the skin slippery, mucus deters predators and may also reduce the energetic cost of swimming. The skin of most reptiles and all birds is thickly keratinized, forming a hard scaly outer surface, and glands are localized to a few sites (such as the preen gland near the anus in birds), not widely distributed over the whole surface as in amphibians. The principal exceptions are the sea snakes, whose thin, soft, unkeratinized skin is slimy with mucus (like that of amphibians and eels) and also acts as a respiratory surface.

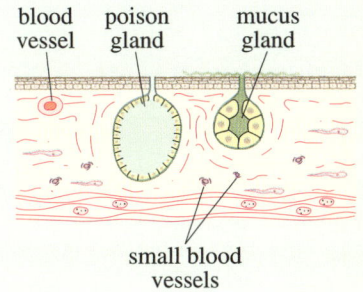

Figure 4.10 Section through the skin of an amphibian.

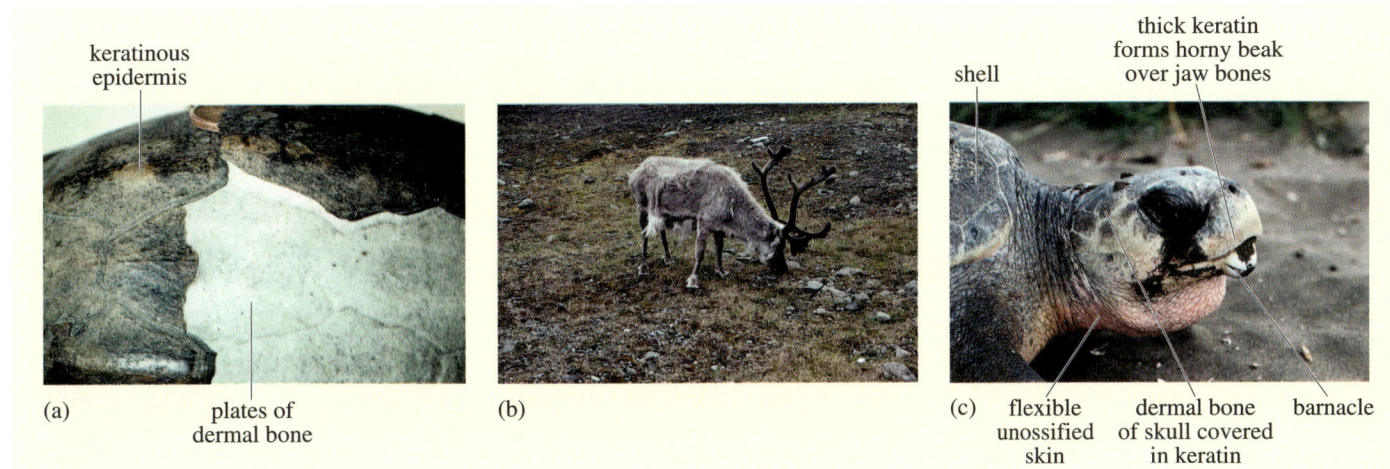

Figure 4.11 Tetrapod skin: (a) The tortoise 'shell' consists of dermal bone (underpinned by components of the ribs) covered with epidermal keratin, usually containing dark pigments *. (b) The epidermal components of the skin, including hair and feathers, are moulted and replaced periodically in almost all tetrapods, often as a means of adjusting body insulation. In deer, including this reindeer (*Rangifer tarandus*), the whole antlers, consisting of a bony core and an outer layer of horny keratin, are also shed and regrow each breeding season. (c) Shedding the epidermis removes ectoparasites, though this barnacle has established itself on the lower jaw of a marine turtle (*Chelonia mydas*).

* This epidermal layer is quite thick in a few species; the tough, translucent keratin mottled with yellow to black melanin from pigment cells is used as 'tortoise shell'.

○ How do these slow-swimming, apparently edible reptiles avoid predation?

● They are highly poisonous, especially to other vertebrates such as fish; their venom contains fast-acting neurotoxins that paralyse muscles (see Pond, 2001).

Dermal bone is also prominent over much of the body of turtles and crocodiles and their relatives. The scales or shell consist of an inner layer of bone covered with keratinous scales (Figure 4.11a). Specialized keratin-forming cells generate scales on reptiles, birds and a few mammals (e.g. the tails of rats and beavers) and the outer layers of beaks (in birds and tortoises), claws, nails, hooves and horns.

Keratinous structures are even more massive and elaborate in birds and mammals, forming feathers or hair (Figure 4.12a) respectively. Whiskers and the spines of hedgehogs (Figure 4.12b) and porcupines are large, specialized hairs, while fused masses of keratin form rhinoceros horns and baleen in the great whales (Figure 4.12c).

○ What other structures are necessary to support these secretions and the growth of hair, feathers, etc?

● Secretions and the formation of hair and its derivatives require much synthesis of proteins and other substances, for which a substantial blood supply is necessary.

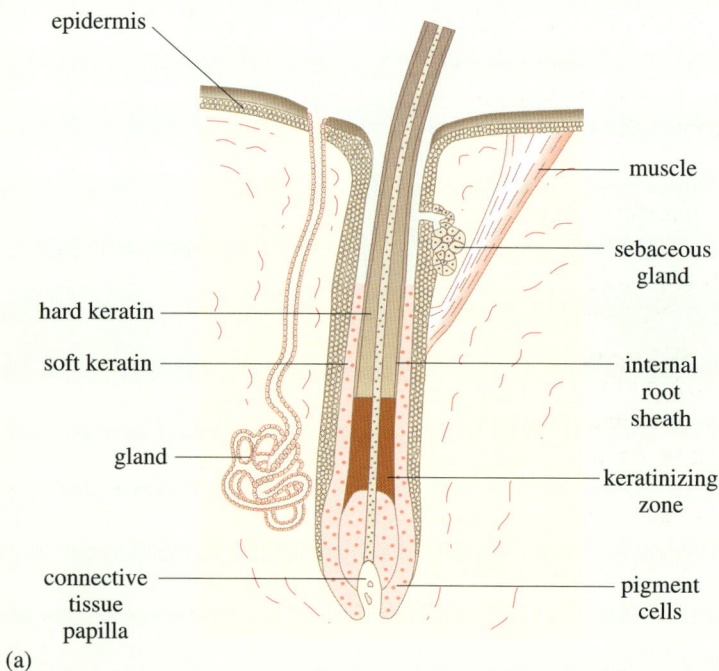

Figure 4.12 Hair and its derivatives: (a) General structure of a mammalian hair. (b) Hedgehog spines are thick, stout hairs that can be bent back or erected by strong muscles. The belly skin has soft, insulating hairs. (c) Baleen is a derivative of hair that replaces teeth in the great whales, in this case a fin whale (*Balaenoptera physalus*). The baleen acts as a net, straining planktonic animals from the gulps of seawater that the whales take into their huge mouths.

Tetrapod skin is always well perfused with blood, and because several kinds of sense organs are also embedded in the skin (see Section 4.5), a rich nerve supply as well. These keratinous structures are exposed to abrasion and other forms of mechanical damage that are almost inevitable for a large animal living on land, but they are continuously replaced from below, thus maintaining their ability to protect the underlying layers of skin that contain glands, blood vessels, nerves and other 'living' components. The outer layer of keratin is shed periodically, both in reptiles, and in mammals and birds, which periodically moult the hair or feathers as well as continually sloughing off the outer layer of skin (Figure 4.11b).

○ How could moulting protect animals from ectoparasites?

● Continuous replacement of the surface makes it much more difficult for ectoparasites to maintain a hold.

However, ectoparasites such as lice, and flea larvae thrive on the dead cells shed from the epidermis and various kinds of crustaceans, especially barnacles, manage to attach themselves securely to the skin of whales and other marine vertebrates (Figure 4.11c). Moulting the old skin or plumage imposes considerable drain on the animals' energy and protein reserves. Many animals eat very little while moulting: birds cannot fly without mature flight feathers, and aquatic mammals and birds are reluctant to swim, apparently because their reduced fur or plumage cannot keep them warm enough.

As well as producing hair, mammalian skin also contains many secretory cells that form sweat, sebaceous and pheromone-producing glands. The production of sweat is one of several mechanisms for dissipating excess heat (but not in fact the most widespread or important) and is unique to mammals. Sebaceous glands secrete lipids that keep the fur in good condition and, especially in otters, beavers, seals and other aquatic species, make it waterproof.

Thin, warm mammalian skin proves a very suitable habitat for microbes and larger organisms of many kinds, especially when it is clad in fur. Almost all of one large group of lice, most fleas and many ticks are highly specialized ectoparasites of furry mammals. Mammals deter infestation by vigorous grooming, and with various skin secretions, including *lysozyme*, an enzymatic protein that kills a wide range of bacteria by breaking glycosidic bonds in their polysaccharide capsules and peptidoglycan walls. Fatty acids and more complex lipids in sweat also make the skin hostile to colonization by bacteria and fungi.

○ Do any secretions from the skin of other vertebrates have a similar role?

● Yes. The antimicrobial proteins, magainins, are found in amphibian skin glands.

Skin secretions of all amphibians (Figure 4.10) seem to deter predators and/or microbes in one way or another. Some make the animal slippery, or so sticky that potential predators end up with a mouthful of debris, but others act as neurotoxins, disrupting the transmission of nerve impulses. The skin of the Amazonian frog *Phyllobates terribilis* has long been rubbed on darts to improve their efficiency: monkeys and other small vertebrates die within minutes from tiny doses of its poison, named batrachotoxin. The active ingredients of skin secretions of other amphibians have recently been studied in detail; some prove to be highly specific for certain receptors in the mammalian brain and peripheral nervous system, and thus may be useful as human drugs.

○ Why are people not harmed by eating animals that have been killed using batrachotoxin?

● Because the meat is usually eaten after cooking, which chemically breaks down many organic molecules and/or disrupts their tertiary structure.

The mechanical properties of the skin of other mammals depend largely upon the abundance and arrangement of the dermal collagen. In most mammals, especially those such as cats which are supple, and able to assume a variety of postures (e.g. when climbing trees and running), the collagen fibres are randomly arrayed, enabling the skin to stretch easily, sometimes to nearly double its resting length (Figure 4.13a). The skin of some large animals is adapted for mechanical protection and has an internal structure similar to that of tendon. The relationship between strain and stress of a sample of rhinoceros skin taken from the upper flank (Figure 4.13a) is more similar to that of typical tendons than skin from a cat. The skin of adult rhinos (Figure 4.13b) is tough enough to resist all but the hardest teeth and claws, enabling them to stand and fight when threatened by large predators, rather than run away.

Figure 4.13 The strains (extension/resting length) produced by imposed stresses (MPa) in collagenous tissues from various mammals (data from Shadwick *et al.*, 1992). (b) The hindquarters of an Indian rhinoceros (*Rhinoceros unicornis*) showing dermal 'armour'.

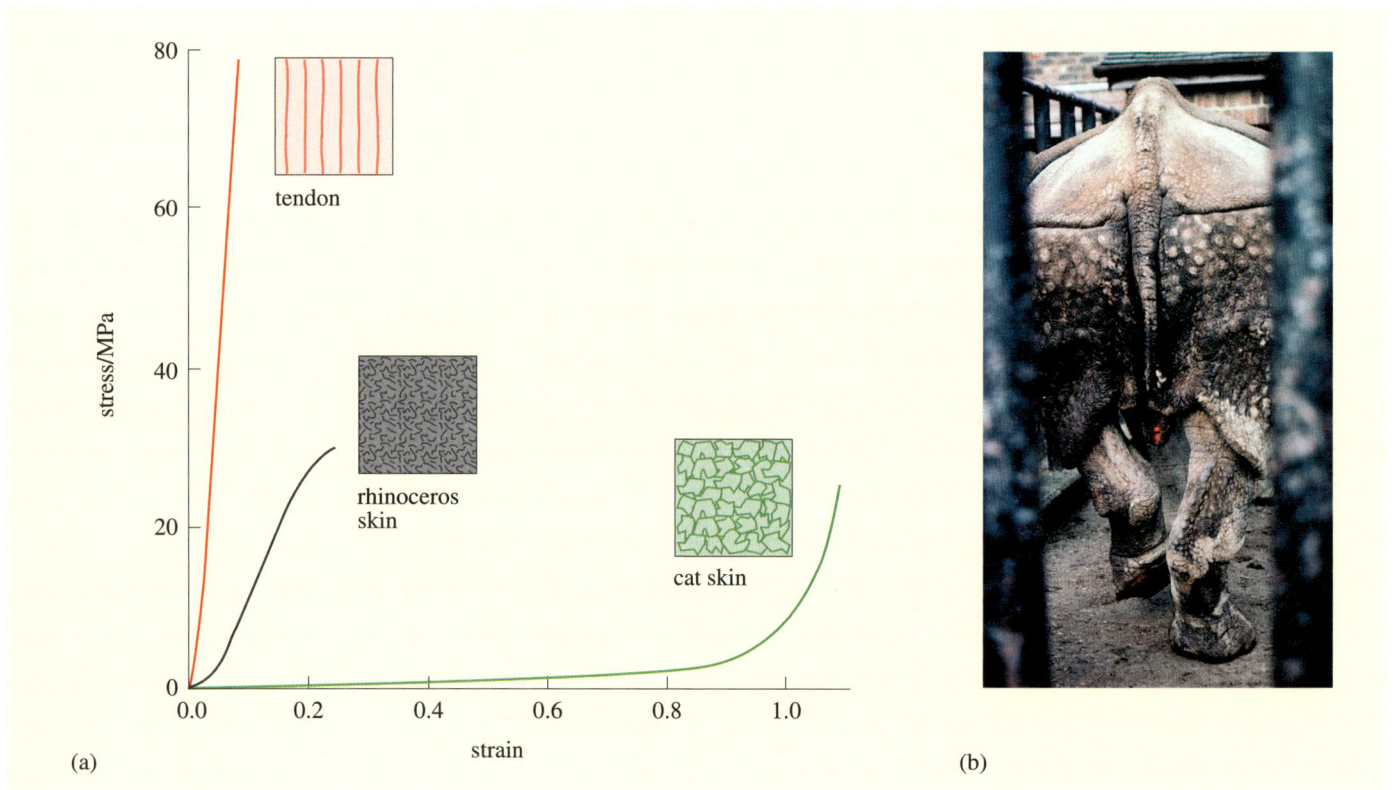

(a)

(b)

The collagenous layer of skin under paw-pads and other places subjected to frequent compression is often made more elastic (i.e. more readily able to mould itself around objects and then revert to its original shape) by the inclusion of adipocytes among the collagen fibres. These tissues, usually called structural adipose tissue or 'fatty connective tissue', protect the bones and joints of the feet from excessive jarring when running and jumping, and enable even heavy animals such as lions, bears and elephants to walk silently over rough ground.

A familiar example is our own heel pad: at each stride, this fatty pad is squashed from a resting thickness of about 1.5 cm (in adults) to about 1 cm. Mechanical measurements similar to those made for Figure 4.13 show that the pads are bouncy, absorbing the energy of impact between the heel and the ground, and recoiling within milliseconds to their original shape when the load is removed. These properties protect the skeleton from the shock of impact at every stride, and the recoil may help to place the bones in the correct position as the body weight is transferred from heel to ball of the foot.

Both the collagenous and keratinous layers of skin also include cells that synthesize pigments, usually black, brown or yellow in mammals but a wide range of vivid colours in most other vertebrate classes. Colours impregnated into the keratinous tissues (see Figure 4.12a) can change completely over days or weeks, as these materials are shed and replaced, as happens when birds moult and replace their feathers, or mammals shed their fur. Pigmented cells embedded in the collagenous layer remain alive (see Figure 4.9a), so can proliferate and/or change shape in response to functional demand. Some fish, especially those native to brightly lit reefs or shallow waters, and certain reptiles, notably chameleons, have several kinds of differently coloured pigment cells under neural control. They can change colour in seconds, producing elaborate patterns, either to blend with the background or for inter-specific and intra-specific communication, signalling aggression or sexual advances.

4.5.1 MAMMARY GLANDS

The skin is the basis for mammary glands (Figure 4.14), the unique tissue that is the principal defining character of mammals. The mammary glands are quiescent except in breeding females. They grow towards the end of pregnancy, and immediately after giving birth, female mammals secrete milk, a nutritious and easily digested emulsion of fats and water containing dissolved **lactose** (a disaccharide composed of glucose and galactose), proteins and various mineral salts.

Both viviparity and parental feeding are known from most classes of vertebrates, including both cartilagenous and bony fishes, but only in mammals are the young fed on a special secretion. Specialized mammary glands and the metabolic apparatus that supports the secretion of large quantities of lipids, proteins, water and other nutrients are well developed in all modern mammals, though the form is substantially different in the three living subclasses (eutherians, metatherians and prototherians).

○ Why does the milk (seen in the lumens of many of the ducts in Figure 4.14) stain the same colour as the cells? What could be the origins of the small 'empty' globules seen in some patches of milk?

● Milk is rich in proteins, which stain in the same way as the intracellular proteins. Milk also contains lipids emulsified in the water and salts. If allowed to cool (as happens in a milk bottle left to stand), the 'cream' separates, forming droplets of almost pure lipid, which are dissolved away by the alcohols and other solvents used in the preparation of such histological slides, leaving blank 'holes'.

Figure 4.14 Section through fixed mammary tissue of a lactating woman stained with haematoxylin and eosin. Numerous small ducts, seen here in cross-section, are lined with secretory epithelia, and supported by collagenous connective tissue. The epithelial cells take up glucose, fatty acids, and amino acids from the blood and synthesize triacylglycerols, lactose and casein, a large flocculent protein, that are secreted together with salts and water into the fine ducts from where they pass to larger collecting ducts around the nipple.

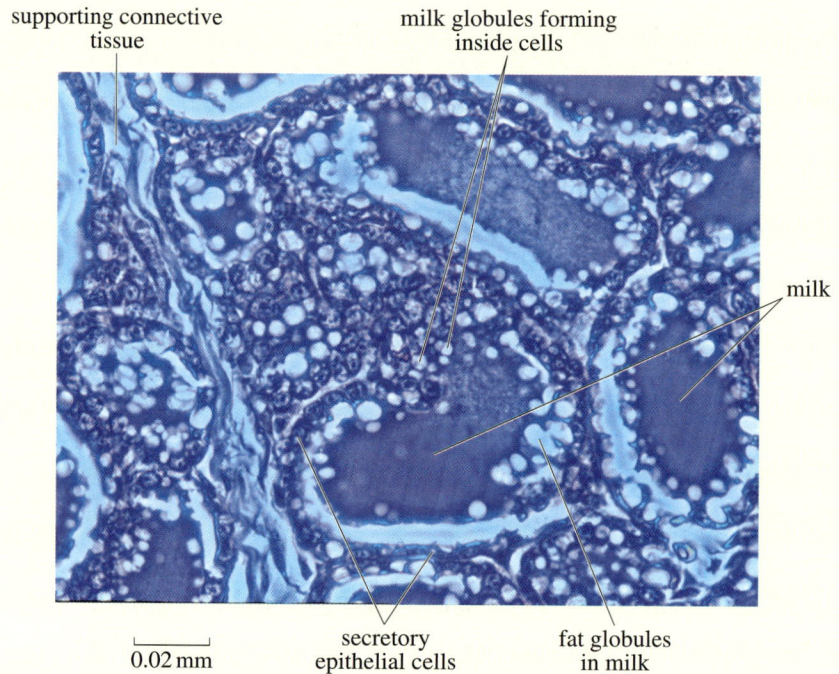

The evolution of lactation involved changes at many different levels: new genes that produce enzymes that synthesize the components of milk, a whole new tissue, and major alterations in the whole-body physiology of both the mother and the offspring. Both molecular and cellular evidence indicate that the mammary glands are derived from greatly enlarged skin glands. All milks contain lysozyme and other proteins associated with defence against bacterial infections, indicating that milk evolved from antimicrobial secretions of the skin.

The mammary glands retain vestiges of their ancestry as producers of antimicrobial secretions: during the first few hours (in rats and other small mammals) or days (in humans and cows), they secrete **colostrum**, a rich mixture of relatively large proteins that confer immunity to microbial pathogens, especially those to which the mother has already acquired immunity. During the first few hours or days after birth, these large molecules pass directly into the neonate's blood because its gut secretes fewer digestive enzymes and is much more permeable than that of older mammals. Most milks contain small quantities of lysozyme and other microbial inhibitors throughout lactation.

○   Why would the transfer of immunity to the neonates take such high priority?

●   Because lactation entails prolonged, close association between the mother and offspring, the ideal conditions for her parasites and pathogens to pass to the vulnerable neonates, which need immune defences as soon as possible.

In some mammals, including humans, certain antibodies cross the placenta and reach the foetus before it is born, but post-natal transfer is still important. Babies, calves and other mammalian neonates that miss the opportunity to ingest colostrum are significantly more susceptible to infectious diseases.

Most of the simple sugars produced by hydrolysis of lactose are incorporated into glycoproteins in the neonate's rapidly growing tissues, not oxidized for ATP production. Regardless of adult diet, the main source of metabolic energy for most mammalian sucklings is fatty acids derived from triacylglycerols, which are very abundant in most milks.[*]

○   Which vertebrate tissue synthesizes large quantities of triacylglycerols from circulating fatty acids?

●   Adipose tissue.

The enzymes that synthesize milk lipids in the mammary glands are very similar to those found in the adipose tissue of mammals and non-lactating vertebrates. The main difference is that the mammary enzymes are less specific than those adipocytes: a wider range of fatty acids, including some with fewer than 10 carbon atoms, find their way into milk triacylglycerols than ever appear in mammalian adipose tissue.

Many mammalian mothers fatten during pregnancy and utilize the lipid reserves to support milk production. The capacity to utilize nutrient reserves in adipose tissue, bone and other body stores for making milk, emancipates mammals from the requirement, which still limits birds, of breeding only where and when foods suitable for the juveniles can be collected in large quantities.

○   From the point of view of the mother's physiology, what is the fundamental difference between feeding the young on milk, and bringing food to the nest (i.e. as birds do)?

●   Milk is a secretion, synthesized from precursors taken up from the blood; such nutrients may come directly from the diet, but milk synthesis can also be fueled from body reserves, which is impossible for most birds.[**]

---

[*] The milk produced by women and most other large primates is exceptionally low in fats and rich in lactose.

[**] In a few kinds of birds, notably pigeons, and perhaps some penguins, both parents produce a deciduous tissue shed from the throat which is fed to the nestlings for a few days after hatching. From the point of view of nutrient storage, this habit is comparable to lactation.

Some mammalian mothers, notably polar bears and many of the great whales, fatten before giving birth and lactate for weeks, supporting one or several fast-growing young without feeding themselves at all. Many others can draw upon body reserves if the food supply dwindles, or does not provide all the nutrients needed to fuel milk production. The evolution of lactation thus entailed major changes in the control of the deposition and mobilization of lipids from adipose tissue, and also of calcium and other minerals from bone.

Large, complex teeth, of which those in the upper jaw exactly match those of the corresponding lower jaw, are among the many mammalian features that can be attributed to the radical alterations in the organization of growth made possible by the provision of milk to the young for some time after birth. Because they can rely upon the mother to provide them with liquid food, young mammals do not need to chew 'real food' until they have grown to as much as two-thirds adult size. Parental feeding thus facilitated the evolution of exactly opposable teeth for efficient chewing. Teeth are not needed at all until weaning, then the 'milk' teeth enable the young to feed itself while the skull completes its growth. With a few exceptions, such as elephants and whales, the jaws are almost full-size by the time the adult set of teeth erupts.

Suckling, by definition, is powered by the facial and respiration musculature, leaving the teeth and jaws to grow unimpeded by the need for efficient function. The muscular lips, cheeks and tongue are not only essential for sucking and chewing, they also contribute to grazing grasses and herbs, and browsing trees and bushes. The folding of the roof of the mouth to form a 'secondary palate', enabling air from the nose to enter at the throat without passing through the mouth, may also have evolved as an adaptation to sucking.

○ What aspect of feeding behaviour in adult mammals depends upon these features of the mouth?

● The secondary palate enables mammals to breathe with large quantities of food in the mouth, thus enabling them to chew for long periods.

Lactation is thus fundamental to many of the distinctive features of mammals, and has enabled them to colonize a wide variety of habitats. The genetic processes by which the uniquely mammalian enzymes that synthesize the components of milk may have evolved are described in Section 5.4.4.

SUMMARY OF SECTION 4.5

1 Skin consists of an inner living layer (the dermis) rich in extracellular collagen. Terrestrial vertebrates also have an outer epidermal layer of dead, keratinized cells that provide mechanical protection and limit permeability. Skin contains glands that secrete mucus, sweat, pheromones and various antimicrobial and anti-predator substances. Either layer may be pigmented.

2 In all modern amphibians (but not most extinct species) the dermal layer is extensive and contains several large glands, but the epidermal layer is reduced. The soft, moist skin serves as a respiratory surface.

3 Feathers and hair are derived from the epidermis and consist almost entirely of keratin, plus variable quantities of pigments.

4   The relative abundance and arrangement of the different components of skin contribute to its immense variety of form in different vertebrates.

5   Mammary glands secrete milk that nourishes the young and these glands are unique to mammals. They are derived from skin glands and the proteins in milk and the enzymes involved in the synthesis of milk components are similar to other vertebrate proteins that have different functions.

6   The capacity for milk production emancipates mammals from the need to breed where special food is available for the young. Lactation also enables tooth formation to be delayed. Many features of mammalian growth processes and breeding strategies are directly related to lactation.

4.6   SENSORY SYSTEMS

All organisms depend on information about their surroundings in order to find food and mates and escape from predators. They must find their way about and also assess important aspects of the environment — temperature, light, the presence of toxins etc. In animals, this information is obtained through *sensory receptors* which are specialized structures that respond to particular features of the environment. They code or translate this information into neural signals that are transmitted to the central nervous system (CNS) where they are interpreted in a way that provides the animal with a coherent and relevant picture of the external world.

Sensory receptors and the organs they form are classified on the basis of the type of stimulus to which they react. There are four basic types of stimuli:

*   chemical,
*   electromagnetic,
*   electrical,
*   mechanical.

4.6.1   CHEMORECEPTORS

Sensory receptors that respond to chemicals are called *chemoreceptors*. In humans, the sense of taste refers to materials in contact with the mouth; smell (olfaction) refers to gaseous substances that reach the nose via the air or water. The distinction between these two senses is misleading because the taste of food is partially due to the aroma that stimulates our sense of smell. In aquatic vertebrates, the distinction between taste and smell is even less useful. Some fishes, for example, have chemoreceptors distributed across the surface of their bodies. Instead of making arbitrary distinctions between taste and smell, chemoreceptors are usually classified according to their location on the body.

The sense of smell is mediated by chemoreceptors usually located in the nasal passages. There are three components to the olfactory circuitry: the *olfactory epithelium* containing the sensory cells, the olfactory bulb and the olfactory tract in the brain. The olfactory epithelium (Figure 4.15) supports the sensory cells. Each cell has a tuft of sensory cilia at the apical end and an axon leading to the olfactory bulb at the basal end. The olfactory bulb in turn sends long axons, collectively called the olfactory tract, to the rest of the brain.

Figure 4.15 A part of the epithelial lining of a vertebrate olfactory organ. Sensory cells are shown in blue and have cilia at the apical ends and axons that project into the olfactory bulb at the basal end.

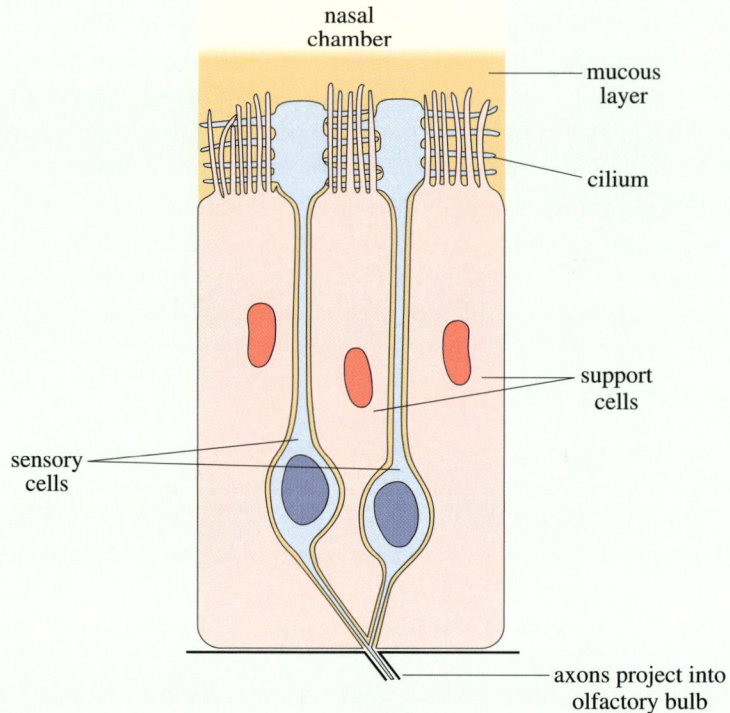

Fishes have a well developed sense of smell; the olfactory receptors are usually located in paired, blind-ending nasal sacs and are independent of the respiratory system. Water, carrying chemicals, flows in and out of these sacs as the fish swims. However, in some groups of fishes, one-way flow of water through the nasal sac is possible because a partial septum divides the sac into incurrent and excurrent apertures. In some groups, these openings become separated and the excurrent opening becomes displaced to the margin of the mouth or opens directly into the mouth (Figure 4.16).

In terrestrial air-breathing vertebrates (amphibians, reptiles, birds and mammals), the olfactory organ is an integral part of the respiratory system. Nostrils provide access to each nasal passage. Most of the air entering the nostrils flows to the lungs but a portion is diverted to a side pocket containing the olfactory epithelium (Figure 4.17). The animal can control the fraction of air diverted to the olfactory system. For example, the action of sniffing increases the air flow to the olfactory epithelium. Many amphibians, reptiles and mammals possess an additional chemosensory organ called the **vomeronasal (Jacobson's) organ**. The organ consists of a single-opening tube lying between the nasal cavity and the roof of the mouth. It acts as a contact receptor which is stimulated by non-volatile chemicals such as protein in urine and secretions of conspecifics, i.e. members of the same species. In salamanders, a pair of grooves which connect the front of the mouth to the nostrils, transport aqueous material preferentially to the vomeronasal organ in the nasal cavities. The fluid moves through these grooves by capillary action and chemicals carried in the fluid may be pheromones or

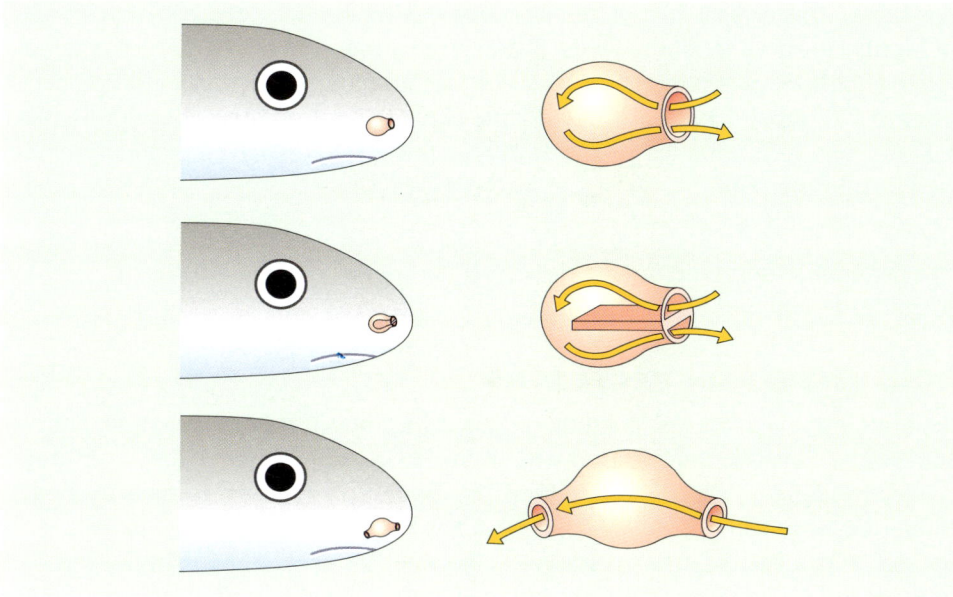

Figure 4.16 Stages in the evolution of one-way flow across the nasal epithelium of fishes. Arrows indicate the flow of water through the nasal sac.

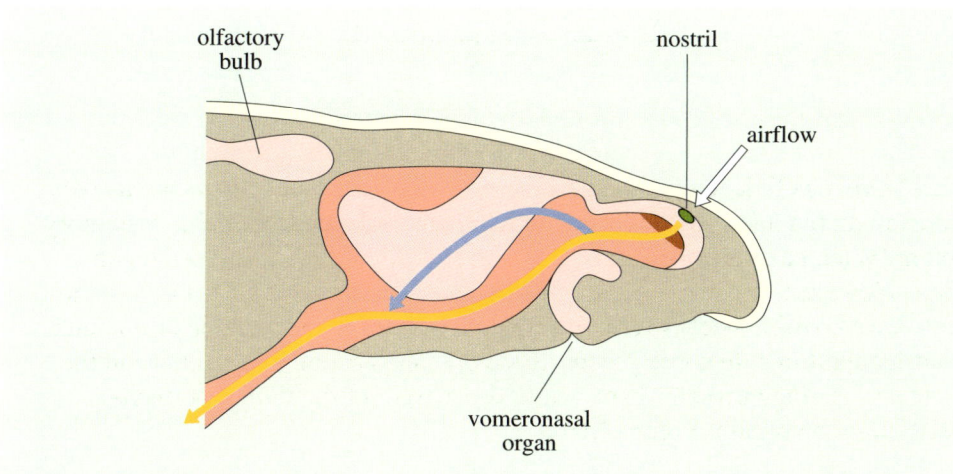

olfactory bulb

nostril

airflow

vomeronasal organ

Figure 4.17 The lizard *Lacerta viridis* illustrates the typical terrestrial vertebrate pattern of primary air flow from nostrils to lungs (yellow line) and partial air flow over the olfactory epithelium (blue arrow). Also shown is the vomeronasal organ in the roof of the mouth, which in lizards and snakes is stimulated by chemicals picked up by the forked tongue.

indicate food. Salamanders continuously touch their nose to the ground — a behaviour called nose-tapping. By doing so, they gather chemical cues important in identifying their home territory. In mammals, much of the intra-specific chemical communication involved in mate attraction, courtship, copulation, aggression and parental care is mediated via the vomeronasal organ. Many mammals lick or touch their nose to urine or secretions to stimulate the organ. After contacting such sources they perform a behaviour called *flehmen* in which the head is raised and the upper lip retracted.

Taste, like olfaction, involves the detection of chemical stimuli by chemoreceptors. The chemoreceptors in this case are the *taste buds*. Taste buds are made up of specialized sensory epithelial cells and associated nerves and are located in the mouth and pharynx of amphibians, reptiles and birds. In mammals, they are distributed throughout the tongue. Many bony fishes have taste buds on their heads, fins or specialized oral structures called barbels (e.g. catfish), which are highly sensitive to amino acids. For example, those on the barbels of channel catfish (*Ictalurus punctatus*) can detect certain amino acids at concentration of between 10^{-9} mol l^{-1} and 10^{-11} mol l^{-1}, the equivalent of no more than 23 mg (less than one-hundredth of a teaspoonful) of the amino acid dissolved in an Olympic-sized swimming pool.

Although there are many similarities between the sense of taste and smell, they have different embryonic origins, innervations, sensitivities and types of molecules to which they respond.

4.6.2 RADIATION RECEPTORS

Radiation travels in waves. Gamma radiation has the shortest wavelength, and radio waves have the longest, and with intermediate wavelengths, they constitute the *electromagnetic spectrum* (Figure 4.18). Radiation carries information about its intensity, wavelength and direction and has several properties that make it a superior transmitter of information over distance. It travels in straight lines at great speed (300 000 km s^{-1}) in air and over great distances with little attenuation (absorption). Changes in the environment can be perceived almost instantaneously, except for the time used in transducing the signal and transmitting it to the central nervous system.

No organism can use all the information available throughout the electromagnetic spectrum since each can perceive only a limited range of wavelengths. Most vertebrates can perceive only a narrow band of electromagnetic radiation of wavelength between about 380 nm and 760 nm, called visible light, because we can see it (Figure 4.18).

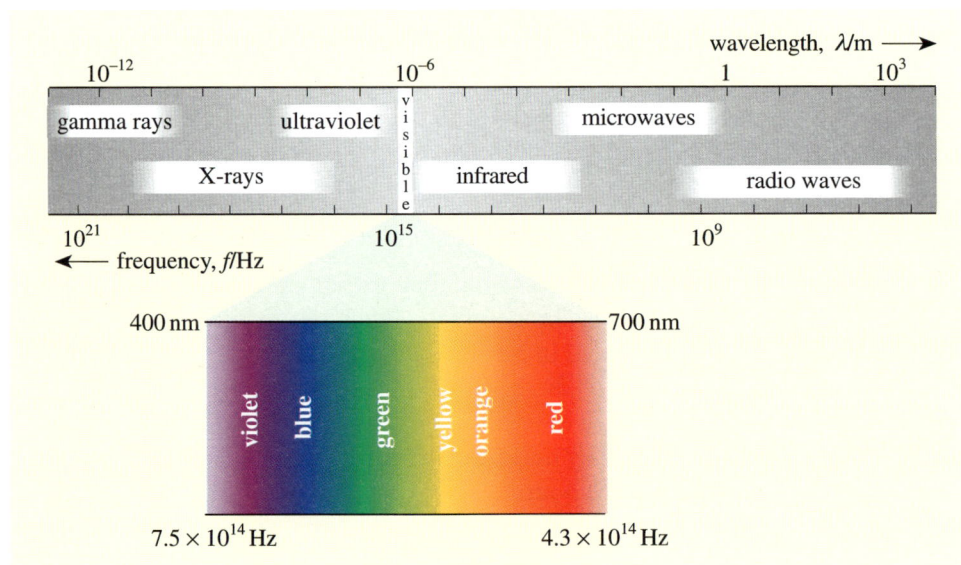

Figure 4.18 Electromagnetic spectrum showing the visible wavelengths and frequencies.

PHOTORECEPTORS

Vertebrates have evolved several sense organs that gather electromagnetic radiation; the most obvious and best understood photoreceptor is the eye. Since the beginnings of biological evolution over 5 billion years ago, sunlight has fuelled life and defined biological time on earth. Light and the light/dark cycle were probably the most important selective forces ever to have acted on biological organisms, and they remain so, and one of the most remarkable consequences of light on earth has been the evolution of eyes. Vision is considered the distance sense *par excellence* of tetrapods and is equally important to many fishes.

Vertebrates have image-forming eyes built on similar principles to that of a single-lens camera, with adjustable lens and aperture, and a light-sensitive retina.

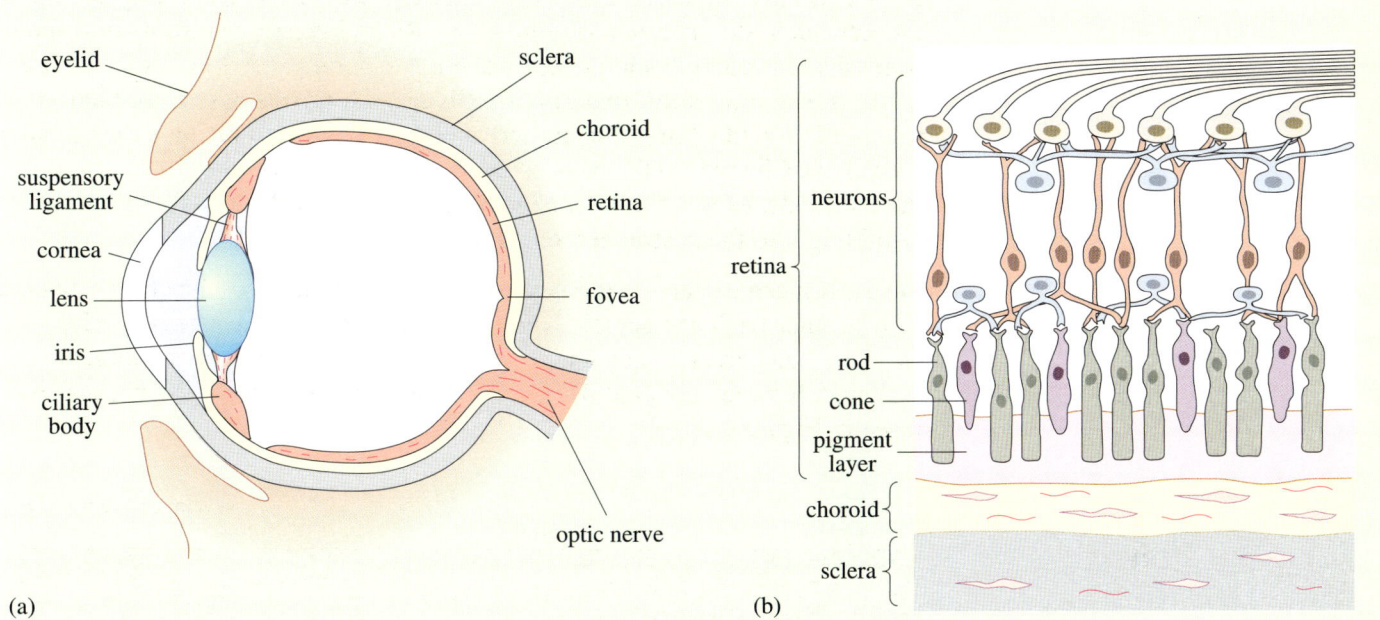

The outer layer of the vertebrate eye (Figure 4.19a), the *sclera* (white of the eye), is tough connective tissue derived from the dermal layer of skin, to which the ocular muscles attach. Contractions in these muscles rotate the eyeball in its orbit to direct the gaze towards an object of interest. At the front of the eye, the sclera becomes a transparent *cornea*, with collagen fibres packed in a regular, crystalline array through which light can pass with minimal absorption or refraction. In terrestrial vertebrates, other folds of skin form the eyelids.

Figure 4.19 Vertebrate eye: (a) General structure of a mammalian eye seen in vertical section. (b) The fine structure of the retina of a higher primate, i.e. a mammal with good colour vision.

○ What form does the keratinous layer take in these skin-derived structures of the eye?

● The cornea and the sclera are collagenous, derived from the dermal layers and have no keratinous layer. The eyelids have a keratinous layer that forms eyelashes in mammals.

The lack of a dead keratinous layer makes the eye more vulnerable to infections from airborne microbes and to mechanical damage than other superficial tissues. Tear glands in the eyelids, which are derived from skin glands, secrete a protective fluid, which cleanses and moistens the exposed surfaces. Tears quickly become copious when the cornea or sclera is irritated by infection, injury or noxious substances.

○ Which component of skin secretions (including tears) kills microbes?

● Lysozyme is an antimicrobial protein secreted from skin glands.

Lysozyme was the first protein for which the tertiary structure was worked out using X-ray crystallography; in the days before it could be easily isolated from egg white (and long before any protein could be produced in genetically engineered bacteria), the principal source of raw material was human tears, collected from volunteers by squirting lemon juice (a mild irritant) into the eye.

The middle layer of the vertebrate eye consists of the *choroid,* the *ciliary body* and the *iris*. The choroid provides nutritional support to the ocular tissues and is pigmented. In some nocturnal vertebrates, it includes a special reflective material which reflects the light back through the eye. The reflected light has a second chance of being absorbed by the photoreceptor cells in the retina, thereby improving the sensitivity and acuity of the image under conditions of dim light. This reflection produces the characteristic eye shine of cats and other nocturnal mammals (and a few birds, including owls) seen at night in car headlights or other strong light. Because the retina is stimulated twice, animals with reflective choroid are even more easily dazzled by bright artificial lights than diurnal animals such as ourselves.

To focus an image on the *retina*, light rays must be bent from their normal parallel lines of travel. The capacity of a transparent material for bending light is called its refractive index, and is very different between air and aqueous materials including the cornea. So when light passes from the air into the cornea, it is bent sharply, and since the cornea is curved, the image is focused on the retina. In fact, the cornea does most of the focusing in terrestrial vertebrates (Figure 4.20a). In aquatic vertebrates, the cornea contributes very little to focusing because the refractive indices of the cornea and water are very similar. In fishes (Figure 4.20b) and in secondarily aquatic mammals such as whales (Figure 4.20c), the *lens* is almost spherical in shape and has a refractive index well above that of water, and is the main agent of focusing. To bring near and distant objects into focus on the retina, small ciliary muscles move the lens forward and backwards respectively along the optical axis. Focusing is therefore achieved by changing the *position* of the lens (like moving the lens on a camera).
In terrestrial vertebrates, with the exception of snakes, the lens is used for fine-focusing and focuses by changes in *shape* (Figure 4.20a). The ciliary body is a tiny circle of smooth muscle around the anterior of the eyeball that is attached to the flexible *lens* through a circular suspensory ligament. In birds and mammals, adjustments to the focus of the eye are achieved by contraction of the ciliary muscles. When they contract, the suspensory ligaments slacken and the lens

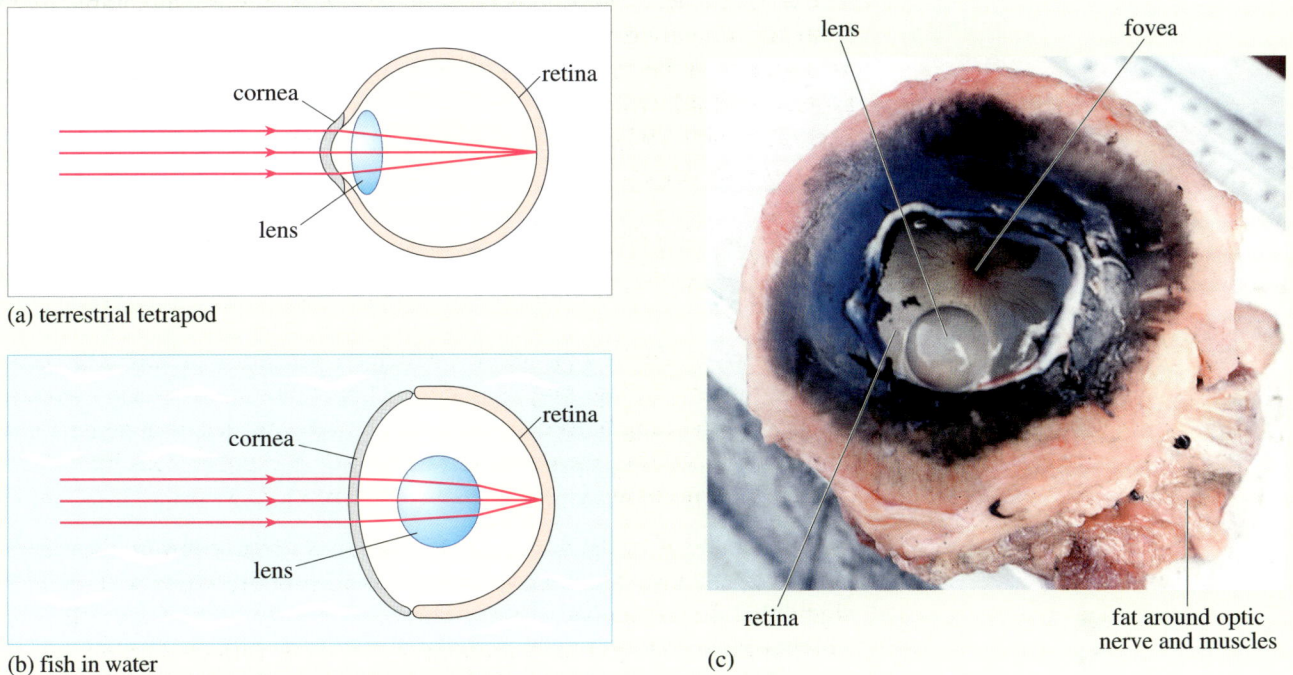

Figure 4.20   Focusing the image on the eye: (a) In terrestrial vertebrates the cornea does most of the focusing. Fine adjustments to focus are achieved by changes in the shape of the lens. (b) The lens of fish is almost spherical and does most of the focusing, because the light is not bent by the cornea. Focus is adjusted by shifting the position of the lens. (c) A partially dissected eye of a fin whale (*Balaenoptera physalus*) seen from the front. The displaced lens is clear and almost spherical.

becomes rounded, enabling the animal to focus at a close range. When the ciliary muscle relaxes, the suspensory ligament tightens and the lens becomes flatter and is able to focus at a greater distance.

The iris is a thin, usually pigmented continuation of the choroid, the middle layer of the eye, that forms the *pupil* — the opening that admits light to the back of the eye (see Figure 4.19a). Tiny smooth muscles within the iris act like a diaphragm to reduce or enlarge the size of the pupil and thereby regulate the amount of light that falls on the sensory cells.

The innermost layer of the eye is the photosensitive **retina** that originates during development as an outgrowth from the brain. The retina itself is composed of three cell layers (Figure 4.19b). The deepest layer of cells within the retina contains the specialized photoreceptor cells of which there are two kinds: **rods** and **cones**. The response of a rod to light is 2 to 5 times slower than that of a cone but a rod is 100 times more sensitive, so rods function best in dim light. Their sensitivity is greatest near the middle of the range of visible wavelength (see Table 4.3), and the images they form are perceived as black, white and shades of grey. Cones respond to bright light, contribute to sharp daytime vision and detect colours. The photoreceptor cells synapse with various kinds of neurons called horizontal and bipolar cells.

Table 4.3 Visual pigments in the human retina. The absorbance of the various pigments was measured by shining a standard light through the region of the retinal cells that contain the photopigment and recording the reduction in intensity of certain wavelengths.

Cell type and pigment	Colour	Wavelength of peak absorbance/nm	Approx. range of absorbance/nm
Rhodopsin in rods	blue/green	498	400 to 600
Short wavelength cones	blue	420	390 to 500
Middle wavelength cones	green	534	400 to 600
Long wavelength cones	red	564	400 to 700

The retinas of all vertebrates possess rods but not all vertebrates have cones. The proportion of cones to rods differs greatly between species. Cones are concentrated near the fovea (the point of maximum acuity on the retina, Figure 4.20c) whereas rods predominate at the periphery. All the photoreceptors in the retinas of animals living in dark, poorly illuminated environments are adapted for spatial pattern analysis and brightness contrast. Such animals are known as monochromats because they see the world in shades of grey. Cones are few or absent, and the retina is composed almost entirely of rods. Vertebrates known to have poor or no colour vision include many of the nocturnal mammals, such as bats and deep-sea fishes. Diurnal tetrapods and fishes that are active under bright conditions have both rods and cones, but cones are usually less numerous.

○ What other evidence indicates that these groups have good colour vision?

● Many fish living in shallow well-lit waters, and many diurnal reptiles and amphibians are brightly coloured (see Section 4.5). The markings play a major role in sexual and social behaviour.

Humans have 100 million rods and 3 million cones in each retina. The eyes of most other mammals, especially nocturnal species, have few or no cones and colour vision is limited or absent.

The cellular structure of rods and cones is similar and both contain light-capturing photopigment molecules which consist of a chromophore, 11-*cis* retinal, which is a lipid closely related to vitamin A. It is bound to a transmembrane protein called **opsin** which is encoded by one of four distinct opsin genes. The best known photopigment is *rhodopsin* which is found in human rod cells (Figure 4.21).

○ In what ways does the relationship between rhodopsin and the membrane (Figure 4.21) resemble that of bacteriorhodopsin?

● Both proteins are large and cross the membrane seven times, with domains on both sides of the membrane as well as embedded in it.

The vertebrate visual pigment is chemically similar to the photopigment found in archaean prokaryotes. There is more on this topic in Section 5.4.3.

In humans, vision begins when a photon of light is absorbed by a molecule of visual pigment. The fatty acid portion of 11-*cis* retinal is isomerized to all-*trans* retinal, which reversibly alters the tertiary structure of opsin and activates transducin, a guanine nucleotide-binding protein, or G protein. The active subunit of transducin activates the effector cGMP phosphodiesterase, which then closes ion channels, thereby hyperpolarizing the illuminated photoreceptors. The electrical signals emitted by the visual pigments reach the visual cortex of the brain. These basic mechanisms are similar in most vertebrates despite the various visual systems that have evolved in environments that are exposed to different intensities and colours of light.

The wavelength to which vertebrate photoreceptor cells are most sensitive depends upon:

1   The variant of retinal it contains. The most common form of retinal found in rhodopsin is called $retinal_1$. A second form, $retinal_2$, when combined with opsin forms a pigment known as porphyropsin. The substitution of $retinal_2$ shifts the absorption peak of the pigment up about 20 nm to 25 nm.

2   The amino acid sequence of the opsin protein. Identical chromophores bound to opsins with slightly different amino acid sequences absorb light of different wavelengths, producing photoreceptors that respond selectively to different colours, as indicated in Table 4.3.

3   The presence in the photoreceptor cells of coloured droplets of oil which broaden the range of wavelength detected, while reducing sensitivity to light. Oil droplets in the photoreceptor cells improve the colour vision of certain birds, amphibians, lizards, snakes and turtles.

In humans, the rods contain **rhodopsin** (Figure 4.21) and each of the cones contains one of three different pigments. With these four visual pigments, humans and many other vertebrates can perceive and discriminate colours in light of wavelengths ranging from less than 400 nm to at least 700 nm.

○   Which of the pigments listed in Table 4.3 is most abundant in the human eye?

●   Rhodopsin. There are 33 times more rods than all three types of cones put together so the pigment in rods must be the most abundant.

Animals that have two cone pigments with different spectral sensitivities are known as *dichromats*. Although they can detect most colours, they cannot perceive red. Dichromatic colour vision is characteristic of many mammals and some marine fish. Animals with good colour vision typically possess three cone pigments and are called *trichromats*. Humans, apes and Old World monkeys, freshwater fish, and most diurnal amphibians, reptiles and birds are trichromats. However, the range of colours that can be seen, the categorization of hues and their brightness differ between species because the peak absorbances of the photoreceptors are different.

Many birds, freshwater fish and turtles that live in shallow, brightly lit waters, have four or occasionally five different types of cone pigments that enable them to perceive light of wavelengths that extend into the ultraviolet range (see Box 4.3) and the infrared (Box 4.4). The different pigments are often very unevenly distributed in the eye. Typically some pigments are located in the upward-looking part of the eye while others are located in the lateral, forward or downward-looking parts. Guppies, for example, use primarily green receptors to look upwards during foraging, and red receptors to view conspecifics from the side.

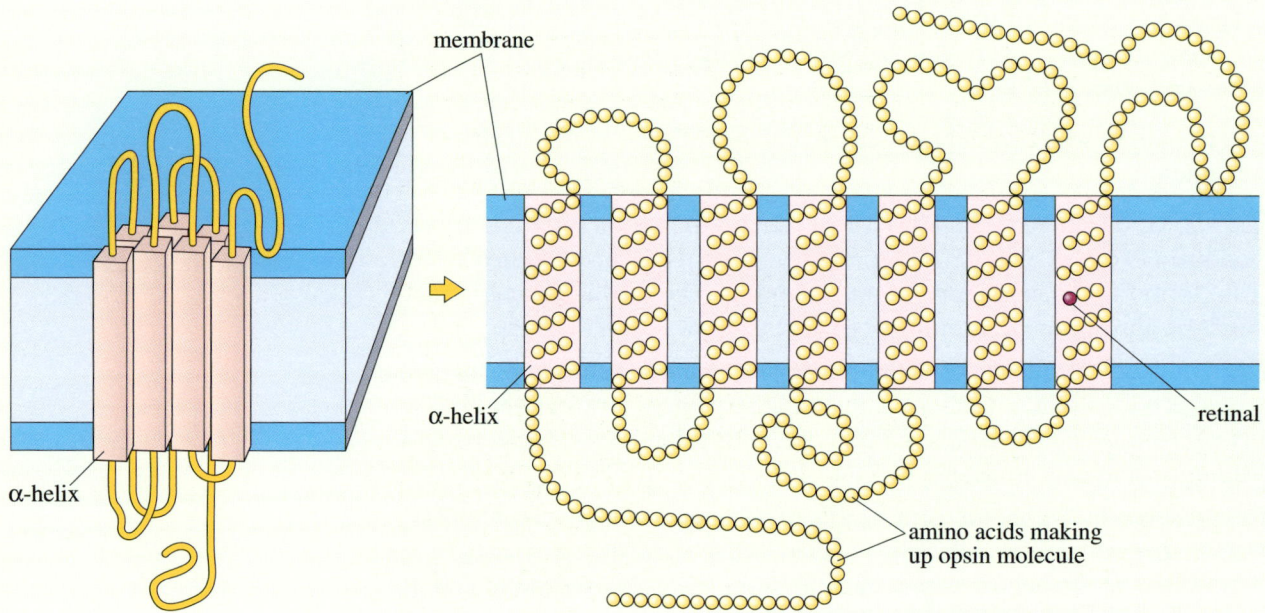

Figure 4.21 The tertiary structure of a rhodopsin molecule in a rod membrane. The chromophore, retinal, which absorbs light, is shown in purple.

The contrasts between water and air present fundamentally different challenges for vision. Water affects light in several different ways. The depth at which an animal lives affects the amount and type of light available. Light intensity diminishes selectively with depth. The first wavelengths to be absorbed are ultraviolet, and then infrared, red, orange, yellow, green and finally blue. Virtually no sunlight penetrates below 100 m even in clear water, but fishes that live at below these depths nevertheless have large eyes.

○ What could deep-sea fish be seeing?

● Cephalopods, ctenophores (and many other kinds of invertebrates) and some fish are bioluminescent so are visible even without sunlight.

BOX 4.3   ULTRAVIOLET VISION

Some vertebrates including hummingbirds, pigeons, turtles, tiger salamanders, several species of fish (roach, Japanese dace, goldfish, rainbow trout) and some mammals such as the rat, house mouse, gerbil and gopher can see ultraviolet (UV, wavelengths around 370 nm, see Figure 4.18) as well as visible colours. Sensitivity to UV light declines during the lifespan of some fish, probably due to changes in the habitat in which they are found. In the brown trout (*Salmo trutta*), the young live in shallow water and feed on plankton, but two-year-old fish move into deeper water where UV light does not penetrate.

A recent study on five closely related species of Puerto Rican anoline lizards shows the relationship between the perception of UV and habitat. The adult males have large, often multicoloured throat dewlaps (Figure 4.22) that are used in courtship and in displays of rivalry between territory holders.

Figure 4.22   An adult male lizard (*Anolis* spp.) displaying his yellow patterned dewlap while courting a female or threatening rivals.

By measuring the spectral reflectance of the throat dewlaps, researchers found that two species showed a high degree of UV reflectance, two species exhibited a low UV reflectance and one species was intermediate in its level of UV reflectance. The three species that exhibited average to high levels of reflectance live in microhabitats that are often exposed to direct sunlight, while the two species with low UV reflectance inhabit the understory of dense forest, where little or no UV light is found. Thus a clear relationship exists between light conditions of the habitat and the properties of the colours of the males' dewlaps. Visual signals used in social and sexual behaviour should be as bright and attention-catching as possible under the conditions where they are usually displayed.

BOX 4.4   INFRARED RECEPTORS

Infrared radiation, which lies just to the right of the visible band of light on the electromagnetic spectrum (Figure 4.18), emanates from the surface of any object with a temperature above absolute zero, that is, any object warmer than −273 °C. All natural objects in the biosphere are above this extremely low temperature and so give off infrared radiation both during the day and at night. Most terrestrial animals can perceive infrared radiation as heat over much of the body surface. A few animals have specialized sense organs that are extremely sensitive to this kind of radiation.

The best known specialized infrared receptors are the paired facial pits or pit organs on the heads of some snakes (Figure 4.23). In advanced snakes such as rattlesnakes, which eat only endothermic prey (mostly rodents and other small mammals), the pit organs are located, one on each side, between the nostril and the eye. The direction of the predatory strike is guided by infrared radiation from its prey: a blindfolded snake strikes at a dead rat pulled near it if the body is warmer than the environment, or even at a shining light bulb wrapped in a cloth, but not at dead prey at the same temperature as the surroundings. These experiments show that neither vision nor smell is required to elicit strike behaviour.

The sensitivity of facial pits has been recorded in the nerves leading from the organ. Sound, vibration or light of moderate intensity (with the infrared part of the spectrum filtered out) has no effect on the activity of the nerve. However, if objects of a temperature different from the surroundings are brought into the area around the head, there is a measurable change in nerve activity, regardless of the temperature of the intervening air. The pit is covered by a transparent membrane and it has been suggested that

a rise in temperature in one chamber behind the membrane could cause an expansion of gas and consequent deformation of the membrane which is in turn sensed by a suitable receptor. However, this explanation is improbable, as a small cut in the membrane that opens the chamber to the outside air causes no loss of sensitivity.

Two other possibilities have been considered: the effect is photochemical, which means that infrared is absorbed by a specific compound, analogous to the light-absorbing compounds in the eye, or the pit organ is sensitive to a slight temperature rise that is caused when infrared radiation reaches it. Evidence suggests that the latter explanation is correct — the mode of reception of the pit organ is entirely thermal. However, its mode of operation is still not known. Pit vipers are extremely sensitive to infrared radiation. A pit membrane that is warmed by as little as 0.003 °C excites the sensory receptors, sensitivity sufficient to detect a mouse 30 cm away.

Infrared receptors are also found in primitive, non-venomous snakes such as pythons which kill by constriction. Such snakes are less sensitive to infrared than pit vipers and they can detect a mouse at distances of about 7 cm to 15 cm. Infrared receptors are also found on the faces of vampire bats that feed on the blood of other mammals. The receptors apparently help these bats detect warm blood vessels beneath their host's thick fur.

Organs that sense infrared radiation do not occur in aquatic animals, partly because of the low penetration of infrared radiation in water and also because direct contact of the body surface with a medium of high thermal conductivity and thermal capacity (e.g. water) would make it impossible to perceive the small amounts of heat involved.

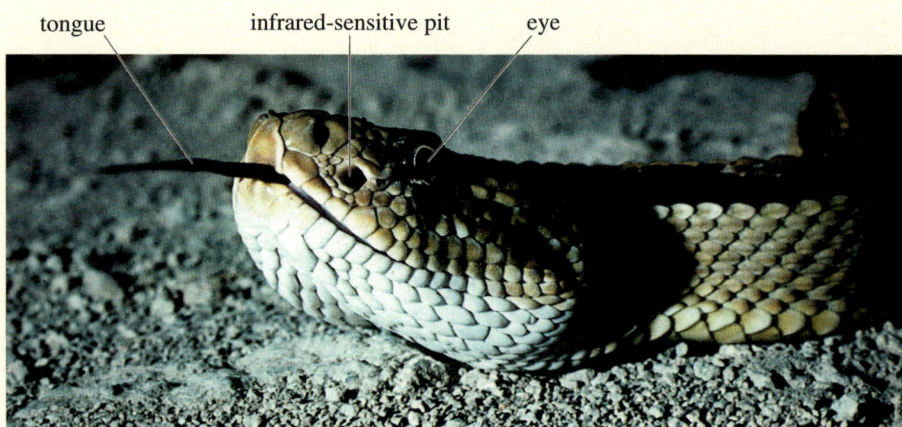

tongue   infrared-sensitive pit   eye

Figure 4.23 The head of a rattlesnake (*Crotalus* spp.) showing one infrared-sensitive pit organ.

4.6.3 ELECTRORECEPTORS

Electroreceptors are sensory receptors that detect weak electric fields. They probably evolved in primitive vertebrates and have been lost and re-evolved a number of times in various groups. Among living fishes, they are found in all elasmobranchs, some teleosts, including catfishes and sturgeons, and lungfish. The possession of an electric sense provides several advantages to aquatic animals because the environment is rich in electric fields produced by both living and inanimate sources. Such fields can provide information about prey or mates, local electrogenic landmarks and the animal's orientation with respect to currents induced by the Earth's magnetic field. In addition to having an electric sense, several groups of fish have electric organs that consist of modified muscle, and are capable of producing an electric discharge which may serve as signals for conspecifics or may be used to stun prey or avoid predation (see Box 4.5).

Electroreceptors located in the skin can be broadly classified into two categories. Tuberous receptors are found only in electric fish and lie buried under the skin in an invagination beneath a loose layer of epithelial cells. They respond specifically to high-frequency discharge rates (several hundred hertz) characteristic of electric fish. For each species, the receptor cells are most sensitive to the discharge rate of that species' electric organ. Ampullary receptors are found in both electric and non-electric fish. They are broadly tuned to low-frequency fields and have a wide range of sensitivities. They open externally through minute pores in the skin.

Electroreceptors can be remarkably sensitive, responding to voltage gradients as low as 10^{-8} V cm^{-1}. A similar gradient would be the equivalent of an ordinary 1.5 V torch battery placed with one end in London and the other end roughly 1500 km away in Warsaw; the voltage gradient would be 1 V per 1000 km, or 1 V per 10^8 cm, the magnitude of the minute fields that can be sensed by fish electroreceptors.

BOX 4.5   ELECTRIC FISHES

Some electric fishes can generate electricity as well as sense electric fields. The South American electric eel (*Electrophorus*), for example, can deliver discharges of between 500 V and 600 V, which are powerful enough to kill other fish and possibly much larger animals. The electric organ of the electric ray (*Torpedo*) and the electric catfish (*Malapterurus*) can generate a sudden jolt of voltage to stun prey or thwart a predator. In most electric fishes, however, the electric organ produces a mild electric field around the fish, for example *Sternarchoramphus*. These discharges are useless in defence or offence, but they can be used to obtain information about the environment because electrically conductive or non-conductive objects have different effects on the flow of current produced by the fish.

Living animals such as other fishes have a relatively high salt content of body fluids, which makes them conductive and causes the lines of the electric field to converge. Materials such as rocks and wood are non-conductive and the current lines diverge. Electroreceptors that are sensitive to these distortions are common in fresh water fishes that live in murky waters or that hunt at night and have poorly developed eyes. Under these circumstances, an electric sense has many advantages, enabling fish to navigate and to detect prey. Patterns of electric organ discharge are also used in intraspecific communication to recognize the sex and species of other fishes and also detect behaviours such as threat, submission and courtship.

Many fishes that do not actively generate their own electrical field nevertheless possess electroreceptors that are sensitive to the weak electric activity that results from the ordinary muscle function of other animals. Most chondrichthyans, sturgeons, catfishes and others have abundant electroreceptors across their heads, especially concentrated around the mouth. Their role in detecting minute electrical signals from living animals has been investigated in the dogfish (*Scyliorhinus*), a small predatory shark that occurs in shallow coastal waters.

Figure 4.24 Feeding responses of a small shark (*Scyliorhinus canicula*): (a) When a hungry shark passes in the vicinity of a flounder completely buried in the sand, it detects the flounder and immediately attacks it. (b) To exclude olfactory cues, the flounder is covered with an agar chamber perfused with water that exits at some distance (arrow). The shark nevertheless attacks the prey at the correct location. (c) When pieces of fish are placed in the agar chamber, the shark searches for food where the perfused water exits. (d) If the agar chamber is covered with a non-conducting plastic film, the signal is attenuated and the shark passes without noticing anything. (e) An artificial electric field of the same magnitude as generated by the breathing movements of the flounder excites the shark, which immediately attacks this 'prey'.

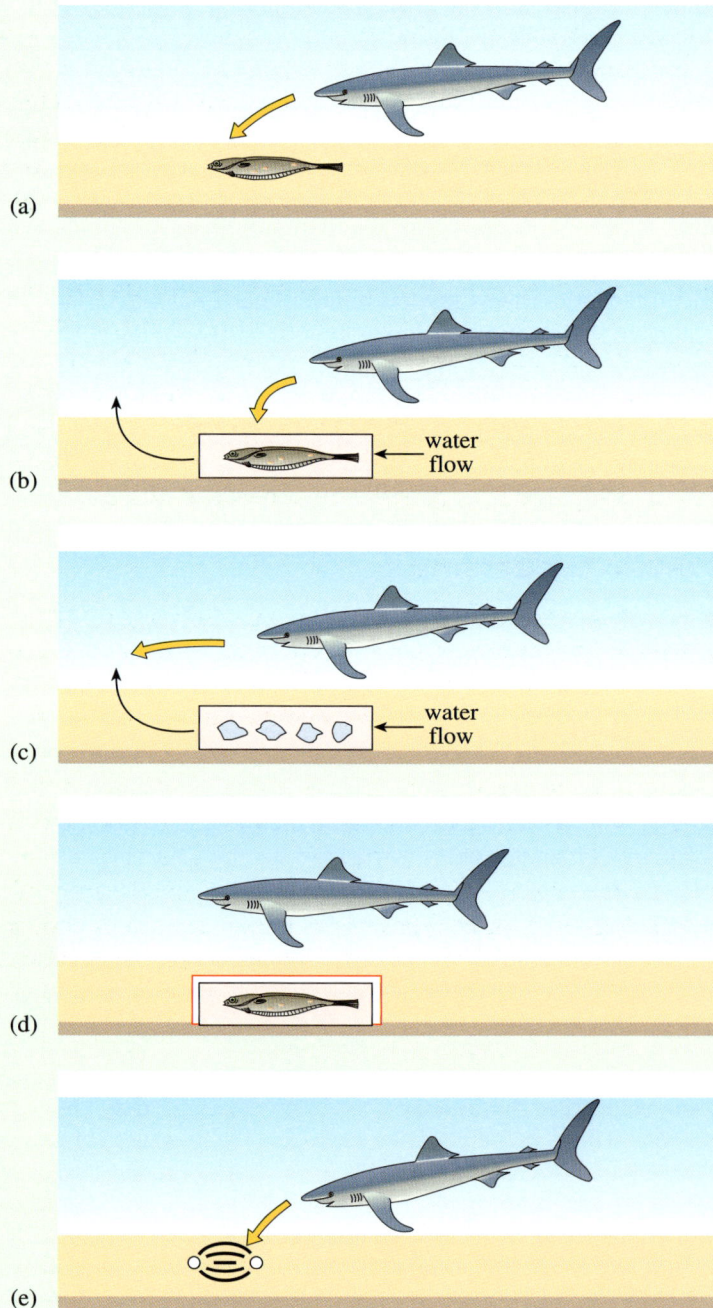

Experiments show that sharks and dogfish can detect a fish that is resting and completely covered by sand from the minute electrical potentials produced by the muscles that power its breathing movements (Figure 4.24).

Several other kinds of aquatic vertebrates have an electric sense. The lateral line (see Section 4.6.4) of the axolotl (*Ambystoma*) (an aquatic salamander which feeds on small invertebrates, tadpoles and fish) has two types of sensory neurons. The mechanosensitive neurons are extremely sensitive to movements in the water, and the electrosensitive units, which react to minute voltage gradients of the same order of magnitude as those produced by small animals, do not respond to any but the roughest mechanical stimulation. The electric sense of the duck-billed platypus is also derived from mechanoreceptors but not those of the lateral line. The platypus seeks food in muddy streams and usually dives with its eyes, ears and nose shut. It has been known for a long time that its bill is packed with extremely sensitive mechanoreceptors. More recently it has been shown that electroreceptors are also present, and the platypus can follow weak electric fields emitted by its prey.

Fire ants (*Solenopsis* spp.) and perhaps other arthropods may also be able to detect electric currents. Attraction to electric fields with certain properties is thought to explain why these ants, which have an unusual and very powerful sting out of proportion to their small size, frequently invade computers, xerox machines, microwave ovens and other electrical equipment. At least two exotic species were accidentally introduced into the southern USA during the last century and are becoming a serious nuisance in homes and offices.

4.6.4 MECHANORECEPTORS

Mechanoreceptors are sensory cells that respond to displacement, velocity or acceleration. One basic mechanoreceptor is the **hair cell** * which transforms mechanical stimuli into electrical signals. Each hair cell has two types of processes (which resemble hairs when seen under the microscope) on their apical surfaces (Figure 4.25): one group of relatively short stereocilia and one longer kinocilium. A **neuromast organ** (Figure 4.25) is a small collection of hair cells, supporting cells and sensory nerve fibres and is the most common type of vertebrate mechanoreceptor, occurring separately on the skin or as a component of a more complex sense organ. The neuromast, or some modification of it, is the fundamental component of all three types of mechanoreceptor systems: the **organ lateral line system** that detects water currents, the **vestibular apparatus** that senses changes in equilibrium (i.e. balance), and the auditory system that responds to sound.

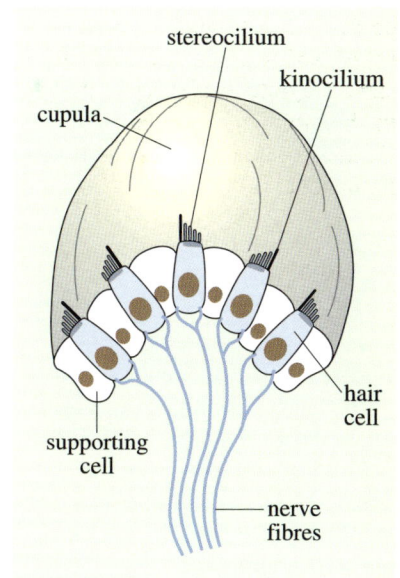

Figure 4.25 A generalized vertebrate neuromast organ showing hair cells.

* Hair cells are so called because of their fine processes. The 'hairs' are cilia and do not consist of keratin like true mammalian hair.

THE LATERAL LINE SYSTEM

All kinds of fish and aquatic amphibians possess a system of lateral line canals, concentrated on the head and extending along the sides of the body and tail (Figure 4.26), but lateral lines are unknown in higher tetrapods, including secondarily aquatic reptiles (e.g. sea snakes, turtles and crocodiles), birds and mammals (e.g. seals, whales). The canals can be recessed in a groove or sunken and covered by surface skin that has pores through which currents of water flow over the neuromast organs (Figure 4.26a). The neuromasts respond directly to water currents providing the animal with information about its direction of movement and about disturbances in the water. Hair cells in the canals are orientated with their most sensitive axis parallel to the canal. About half are orientated in one direction and the rest in the opposite direction. In the absence of water flow, the neuromast generates a continuous series of electrical pulses. Water flowing in one direction generates an increase in discharge rate while water flowing in the opposite direction causes the discharge rate to fall below its resting rate (Figure 4.27).

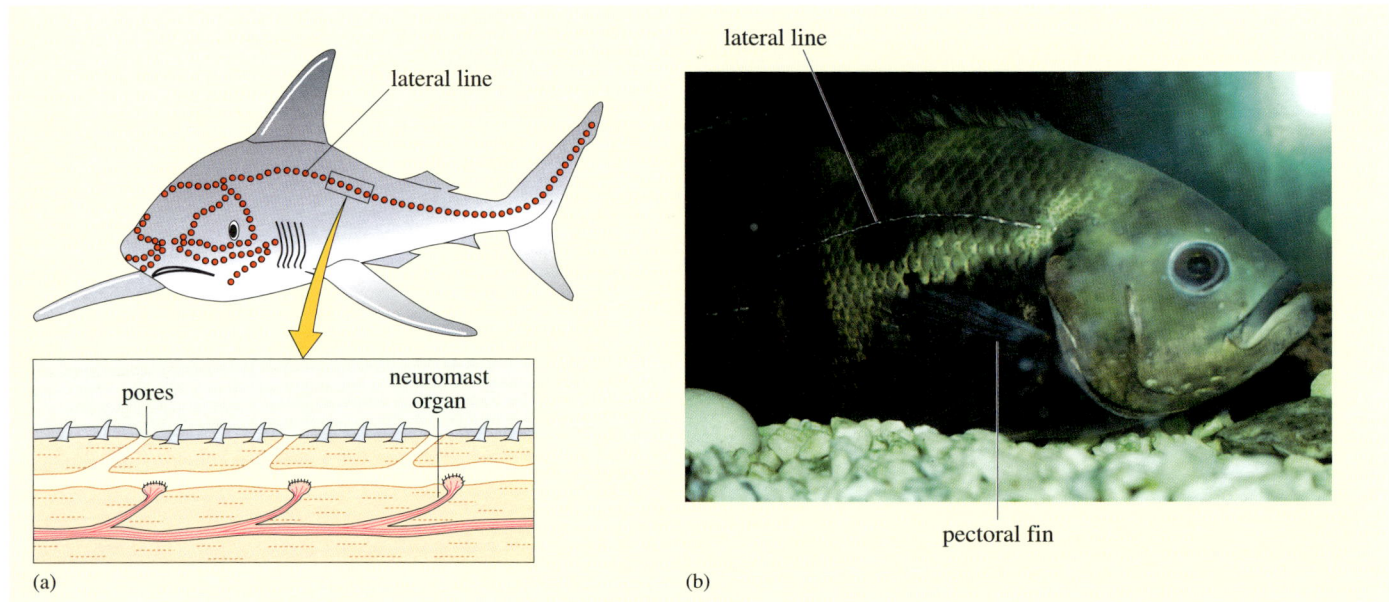

Figure 4.26 (a) The arrangement of the lateral line organs in a shark (class Chondrichthyes) showing the neuromast organs in a sunken canal. (b) The lateral line on a cichlid fish (*Astronotus ocellatus*, class Osteichthyes).

Using this system, even cave fish, which are blind, can navigate around obstacles in their environment and find their prey. Some fishes use their lateral line canals as a kind of 'distance touch' so that they are able to detect compression in the water in front of them as they approach a stationary object ahead. Captive fishes whose lateral line organs are inactivated collide with aquarium walls and other solid objects. In a school of fish, the movements of individuals are extremely well coordinated and to a human observer appear perfectly synchronized. Schooling fish often swim at a constant pace and maintain characteristic individual distances. The school as a whole executes complicated manoeuvres that require individuals to respond exceedingly fast to speed and direction changes of their neighbours. The mechanoreceptors of the lateral line are the main sense organs controlling schooling behaviour. If a fish is blinded, it is still able to match speed and direction changes of its neighbours but destruction of the lateral line nerves makes a fish unable to school normally.

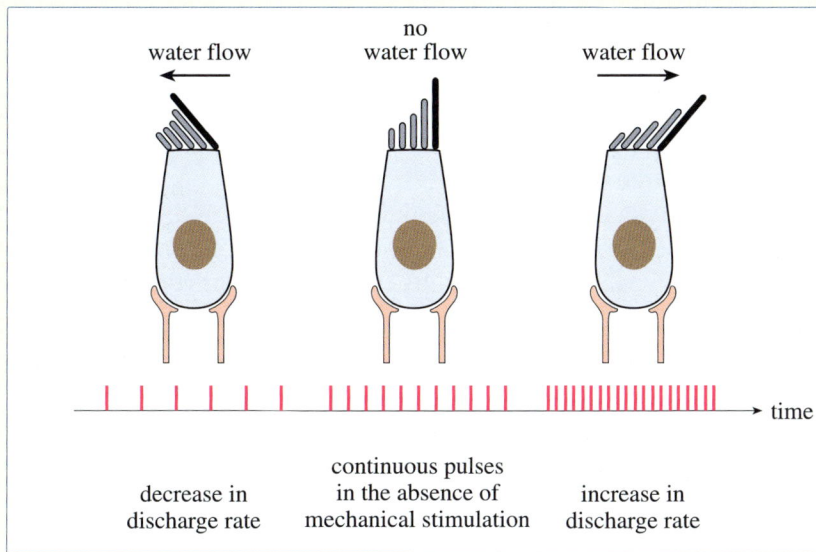

Figure 4.27   The discharge rate of action potentials produced in neurons that arise from the hair cells or neuromast organs in the lateral line in response to water flow.

The lateral line organs of aquatic amphibians are simpler than those of teleost fishes and do not form complex canals. The lateral line system is well developed in the larval or tadpole stages of frogs but degenerates at the time of metamorphosis and does not return. One exception is the lateral line of the aquatic frog *Xenopus*. In *Xenopus laevis* neuromasts are sensitive to water currents flowing at only 0.025 mm s^{-1}.

4.6.5   EARS AND HEARING

Hearing is a development of the ancient mechanisms of vibratory perception in the lateral line organs that arose early in the evolution of vertebrates. There are significant differences in ear structure between groups of fishes but receptor cell structure and the basic function of the ear and auditory system are similar among all vertebrates. Hearing ability provides vertebrates with means of tracking prey and avoiding enemies, thereby enhancing survival. Later developments enabled animals, particularly birds and mammals, to use their ability as a means of communication.

Although we naturally think of hearing as being the normal function of ears, in ancestral vertebrates hearing was apparently unimportant and perhaps absent: equilibrium (orientation in relation to gravity) was the primary sensory attribute of the auditory organ.

THE INNER EAR AS AN ORGAN OF EQUILIBRIUM

Equilibrium is a basic function of the vestibular apparatus of the inner ear that is similar in all vertebrates. It arises from structures that form part of the lateral line system of fishes and consists of three semicircular canals and at least two connecting compartments: the *sacculus* and the *utriculus* shown in Figure 4.28.

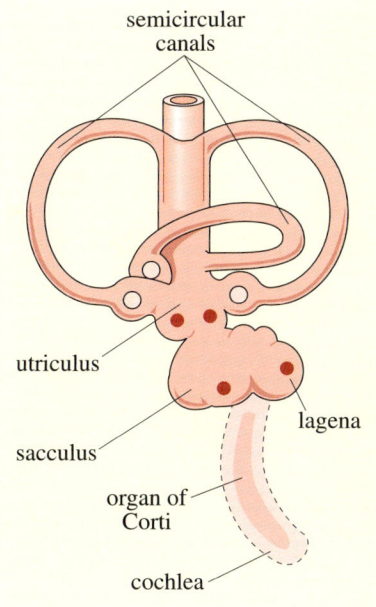

Figure 4.28   The generalized plan of the semicircular canals and connecting compartments in the vertebrate inner ear. In reptiles, birds and mammals, the lagena is expanded to form the cochlea. In mammals, the cochlea is coiled like a snail shell.

The sensory receptors within the semicircular canals are called *cristae* and those within the sacculus and utriculus are the *maculae*. Both the maculae and the cristae can be regarded as expanded neuromast organs, having hair cells embedded in an overlying gelatinous cupula. In the maculae, the gelatinous cupula becomes a thickened structure in which are deposited crystals of calcium carbonate forming 'earstones' or *otoliths* (sometimes called otoconia). The utricular maculae and to a lesser degree, those of the sacculus and lagena register the position of the head and linear acceleration by the tilt of the otolith and the bending of its sensory hairs. They do not however, furnish any information about turning movements, which are detected by the semicircular canals.

The semicircular canals are orientated roughly in the three planes of space and respond to angular acceleration produced when the head is rotated or turned. Displacement of liquid in one or more of the canals displaces the cupulae, bending the sensory hairs and generating neural signals.

THE INNER EAR AS AN AUDITORY ORGAN

In reptiles, birds and mammals, the lagena is a specialized region of sound reception which is involved in hearing and develops as an enlargement of the sacculus.

In water, there is very little difference between the acoustic impedance of living tissues and the water around it, which means that a fish absorbs incident sound energy so well that its entire body vibrates in sympathy with the sound waves. Therefore, there is no differential movement of body parts that could stimulate the sensory cells in the ear. In sharks, rays and bony fish, the otoliths, which are deposited on the cilia of the hair cells of the sacculus (see Figure 4.28), are three times denser than the surrounding tissues. The greater mass of the otoliths makes their motion lag behind that of the surrounding tissues. The bodies of the hair cells rest on a base of tissue, while the stereocilia are in contact with the otoliths. Sound therefore produces relative motion between the bodies of the hair cells and their stereocilia, which stimulates the hair cells and generates action potentials in the auditory nerve. However, because of the mass of the otoliths, such a system can only detect low frequencies (200 Hz or lower).

Among tetrapods, the parts of the inner ear devoted to equilibrium remain relatively unchanged. However, the lagena region attains size and importance as the site of the *basilar papilla* on which the hair cells rest. In addition to the basilar papilla, amphibians have a second auditory organ that is involved in hearing known as the *amphibian papilla*. The amphibian papilla is sensitive to high-frequency sounds (1 to 5 kHz) and the basilar papilla is sensitive to sounds below 1200 Hz.

Both birds and mammals have greatly refined their hearing ability by the expansion of the lagena into a long, coiled **cochlea** and a similar expansion of the basilar papilla in the basilar membrane that runs the length of the cochlea duct. The hair cells are contained within a structure known as the organ of Corti which rests on the basilar membrane.

BOX 4.6   THE PRINCIPLES OF SOUND PRODUCTION AND PROPAGATION

What is sound? Sound is generated when normally uniformly distributed air molecules are pushed together and apart by a vibrating object. Imagine a tuning fork. One way to make a tuning fork emit sound is to 'twang' the prongs by squeezing them together and then releasing them. The movement of the prongs causes movement of the air molecules close to their surface, pushing together and pulling apart the molecules close by. When the air molecules are pushed together, the local air pressure increases and when they are pulled apart it decreases. These changes in pressure are not restricted to nearby air. The movement of air molecules around the prongs causes air molecules further away to move and so on. Thus air molecules are pushed together at some points and away at others, transmitting (or propagating) the sound wave through the air. Figure 4.29 illustrates what happens in the air surrounding a tuning fork at a particular instant after 'twanging' the prongs.

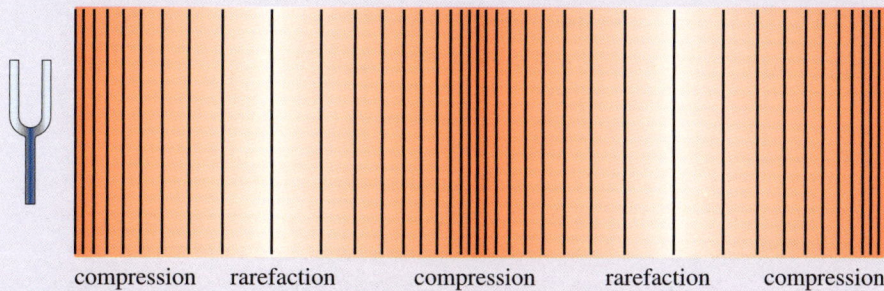

compression   rarefaction   compression   rarefaction   compression

Figure 4.29   A tuning fork sets up vibrations in the air around it. Regions of higher pressure (compressions) are indicated by lines that are close together while regions of lower pressure (rarefactions) are indicated by lines spaced fairly widely apart.

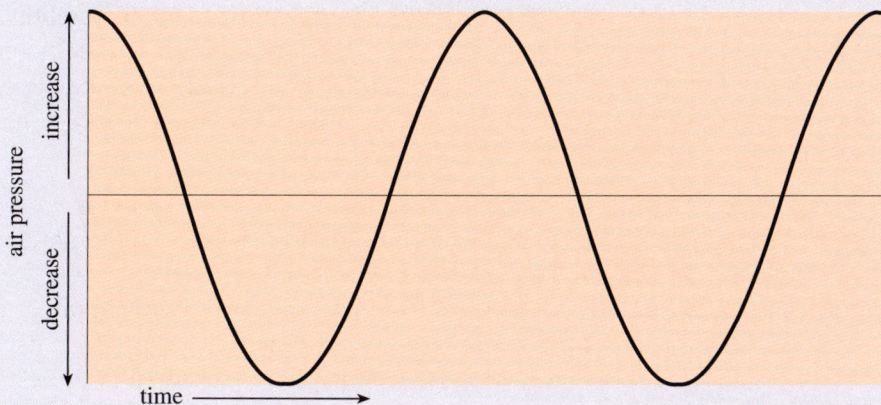

Figure 4.30   The changes in air pressure with time recorded at a single point. A signal produced by a tuning fork and represented in this way is known as a **sinusoid**. The horizontal line is at the value of air pressure for undisturbed air.

In this way, a sound wave can be transmitted a considerable distance. If the increases and decreases in air pressure were measured at some fixed point, they would show a pattern similar to that shown in Figure 4.30. Since this illustration

213

shows pressure changes, it represents an *acoustic* signal. Sinusoids can be considered a basic component of more complex signals and there are an infinite number of them. One crucial way in which sinusoids differ from one another is the time taken before they repeat, which is shown in Figure 4.31.

You can see the lower sinusoid takes longer to repeat than the upper. The shape that repeats is called a cycle and the time it takes to repeat is called the **period**. The period can be expressed in any units of time, such as seconds or milliseconds. The same information can be specified by the **frequency** with which the cycle repeats, given by the number of cycles (or periods) that occur in one second. The unit of frequency is the hertz (abbreviated to Hz), which is just another name for 'cycles per second'.

○ Which of the two sinusoids in Figure 4.31 has the higher frequency?

● The top one since it has a shorter period, i.e. repeats itself more often in one second.

The frequency of a sinusoid is the primary determinant of its pitch so that the top trace would be perceived as having a higher pitch than the lower trace. Frequency is only one of the characteristics that specify particular sinusoids. There are other ways in which they can differ.

Both traces in Figure 4.32 have the same frequency, yet they are not identical because the waveforms start at different points in their cycles. The top one starts at zero

and decreases while the bottom starts and zero and increases. This property is known as a **phase** and is expressed in degrees. There are 360° in one full cycle so the two sinusoids in Figure 4.32 are 180° out of phase. Sinusoids of the same phase are said to be *in phase*.

○ When two waveforms differ in phase by 360° are they in phase?

● Yes, they are one complete cycle apart and therefore in phase again.

Sinusoids vary not only in frequency and phase but also in how big they are, their amplitude. Amplitude is the magnitude of the excursions on the vertical axis, as shown in Figure 4.33. Although these two sinusoids are the same frequency and phase, the lower one has greater amplitude than the upper. Differences in amplitude of auditory signals arise from differences in the magnitude of the displacement of a vibrating object. So, the amplitude of a signal produced by a tuning fork depends on the force with which its prongs are 'twanged'. Signals that differ in amplitude differ subjectively in their loudness.

Any sinusoid can therefore be uniquely specified by its frequency, amplitude and phase. Frequency refers to how fast the sinusoid moves up and down, amplitude to how big the up-and-down movements are, and phase to when the sinusoid starts. Sounds made up of a sinusoid of a single frequency are known as *tones*. Most sounds we hear consist of a mixture of several sinusoids of different frequencies that add together to form what is known as a *complex* sound.

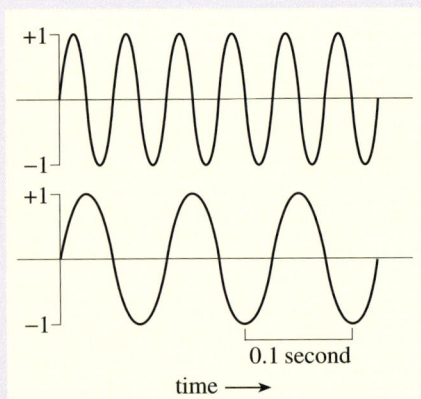

Figure 4.31 Two sinusoids that differ only in frequency.

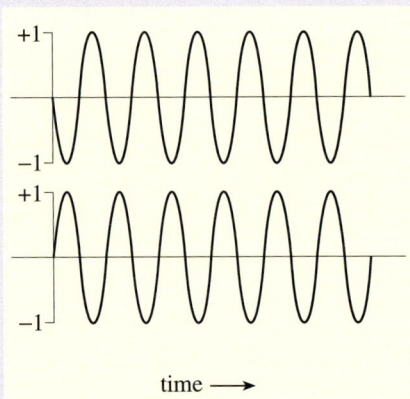

Figure 4.32 Two sinusoids that differ only in phase.

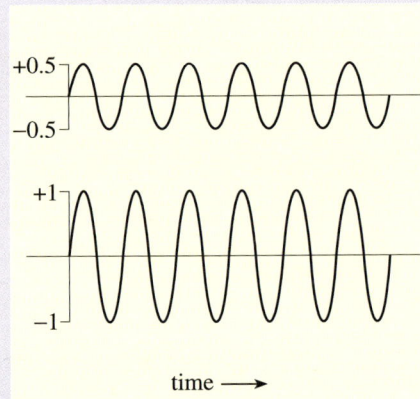

Figure 4.33 Two sinusoids that differ in only amplitude.

THE MIDDLE EAR

When a sound wave reaches the ear, the changes in air pressure (Box 4.6) cause the eardrum to vibrate. In general, the function of the middle ear is to transmit these vibrations of the eardrum (tympanum) to the inner ear, where the sensory receptors (hair cells) are found.

Fish do not have a tympanum. Several groups of teleosts use the swimbladder as a pressure detector to transfer sounds from the surrounding water to the inner ear. In some species, the swimbladder just touches the auditory portion of the skull and no special coupling is present. The proximity of the swimbladder increases the sensitivity of the ear slightly and extends the frequency range that the fish can hear up to about 1 kHz. In certain fish including herrings, the swimbladder has finger-like extensions that intrude into the ear on each side which extends the range of frequencies detected to about 4 kHz. However, the best hearing is found in the Ostariophysi, the dominant group of freshwater teleosts, which includes goldfish, minnows, guppies and most of the small fish kept in freshwater aquariums. Their swimbladder is connected to the inner ear through several tiny bones called *Weberian ossicles* (Figure 4.34) that carry vibrations to sound detectors in the sacculus and lagena enabling the fishes to perceive sounds of frequencies up to several kHz.

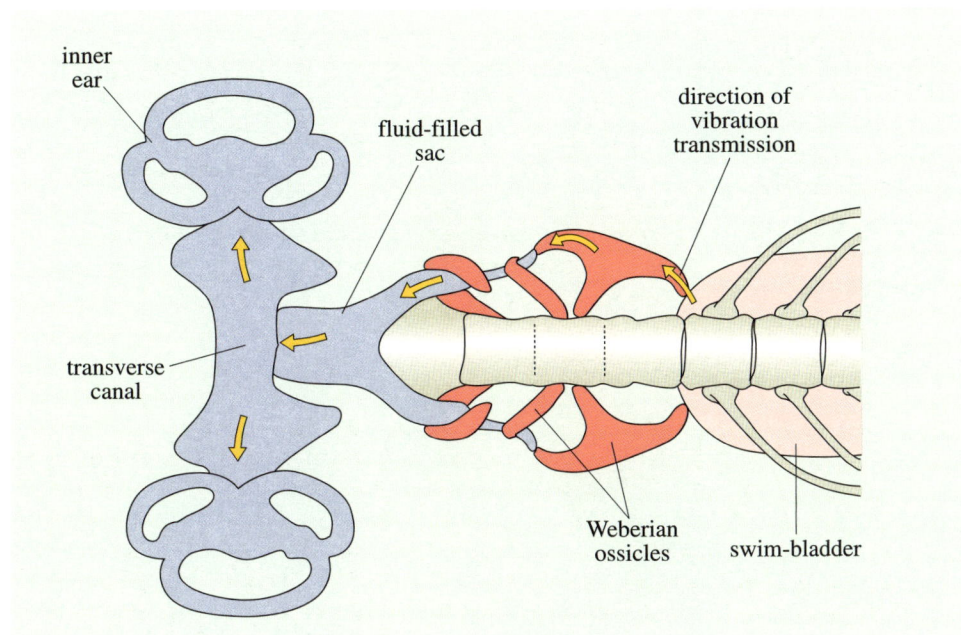

Figure 4.34 The transfer of sound to the inner ear of ostariophysan fishes by the Weberian ossicles. These small, stiff bones are modified processes of the first four vertebrae. Arrows show the direction of transmission of sound from the swimbladder, through the Weberian ossicles, to the fluid-filled transverse canal and to the inner ear.

○ Why would improvements in hearing be an advantage to freshwater fish?

● Fish living in lakes, ponds and rivers experience a very wide range of predators: birds, mammals, reptiles and sometimes large amphibians as well as other fish. The presence of vegetation and sediment often obscures visibility, so hearing potential predators is very important.

During the transition from water to land, vertebrates (unlike fish) had to deal with the problem of sound transmission from air to the fluid-filled cavity of the inner ear. You will appreciate this problem if you have ever tried talking to someone who is underwater. Sound waves that hit the air/water interface are largely reflected and very little energy is transmitted, which would happen if sound passed directly from the outside (air) to the inner ear (fluid). To deal with this problem, tetrapods evolved from one to three small bones in the middle ear, called the middle ear ossicles. In mammals, the three ossicles act as a lever system to transmit vibrations to the oval window which is the point at which sound enters the inner ear from the middle ear. Their function is to transform sound waves in air to sound waves in fluid and they amplify sound.

Amphibians, reptiles and birds have only one ossicle, that spans the distance from the eardrum (tympanic membrane) to the inner ear. This bone is called the **stapes** (stirrup) in mammals and their ancestors, and the columella in birds and reptiles. It is derived from the **hyomandibular** bone that in fish functions as a brace between the jaw and the braincase (Figure 4.35a). In modern and extinct fish, and in Palaeozoic amphibians such *Acanthostega*, muscles attached to this bone ventilate the lungs with air, and the bone itself functions in jaw suspension.

Figure 4.35 Diagrammatic cross-section of the head showing the bones associated with hearing in various vertebrates: (a) The hyomandibular bone in fish. (b) The stapes in amphibians. Homologous are shown in the same colour.

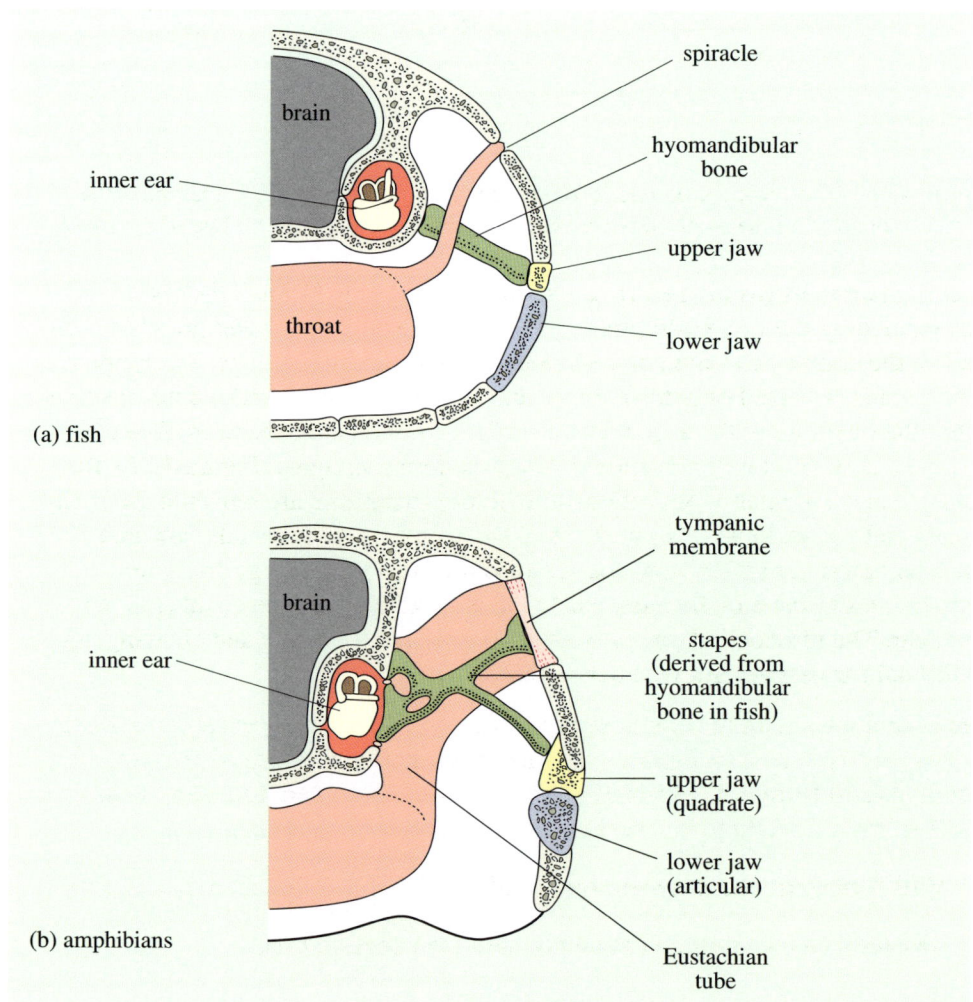

Early amphibians were the first vertebrates to encounter the problem of hearing airborne sounds. Detailed studies of fossilized skeletons indicate that *Acanthostega*'s stapes was a modification of the hyomandibular that was co-opted into use as a sound transmitter (Figure 4.35b). Additional evidence that the stapes is homologous to the hyomandibular comes from the observation that in mammalian fetuses, the stapes develops from the second gill arch (see Section 4.2). In fish, this same embryonic structure becomes the hyomandibular bone.

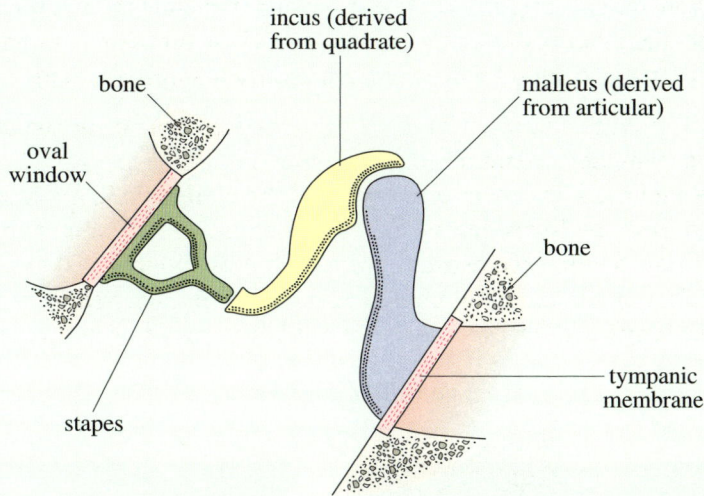

Figure 4.36 The three ear ossicles (colour coded as in Figure 4.35) in mammals that transmit sound from the external tympanum membrane to the oval window of the inner ear.

In early mammals (as in modern amphibians, reptiles and birds), the jaw hinge was formed by the **quadrate** (upper jaw) and the **articular** (lower jaw) bones as well as the dentary and squamosal bones. In later mammals, a new jaw hinge was formed, that involved only the dentary and squamosal bones and did not involve either the quadrate or the articular bones. These bones were therefore free to assume a new function or disappear. In all modern mammals (Figure 4.36), the quadrate and articular bones articulate with the stapes and are renamed the **incus** (anvil) and the **malleus** (hammer). They are now detached from the jaw and function in the transmission of sound, transmitting sound energy from the tympanic membrane in the outer ear to the **oval window** of the cochlea in the inner ear. The remodelled jaw proved more efficient for biting and chewing, and additional ear ossicles allowed better hearing.

The homologies with two jawbones found in reptiles, amphibians and birds can be seen during fetal development: the malleus, incus and stapes of modern mammals still form in the same positions in which they were found in adults of fossil animals thought to be ancestral to mammals.

○ What other structures appear transiently during development in mammals?

● The pharyngeal clefts and the notochord (see Section 4.2).

THE EXTERNAL EAR

What most people call the ear in mammals is correctly termed the **pinna**, an external cartilagenous flap that surrounds the external orifice and helps to focus the sound on the middle ear. Muscles attached to its base allow the pinna to be turned towards the source of a sound. Most mammals, except whales and some seals, have a visible pinna and it can be relatively very large in species with acute hearing, especially bats and some rodents and rabbits (Figure 4.37). Birds and lower tetrapods never have a pinna, though the ears of birds such as parrots, which have very sensitive hearing, are covered with specialized feathers. In some advanced reptiles such as lizards and crocodilians, and in mammals and birds, the external ear also consists of an *external auditory meatus* that leads from the external opening of the ear to the eardrum. In reptiles it is a short, indented tube whereas in mammals and birds it is elongated. The eardrum is therefore recessed at the bottom of the external ear cavity. In many frogs and tortoises however, the eardrum is clearly visible (Figure 4.38a) and lies flush with the body surface. Snake ears do not have a tympanic membrane, but, despite the popular opinion that they are deaf, laboratory studies prove that snakes can hear. Recordings of electrical activity from areas of the brain to which auditory nerves travel confirm that the inner ear of snakes is sensitive to seismic vibrations and airborne sounds, although at only a limited range of frequencies.

Figure 4.37 The pinna in various mammals: (a) The pinnae of bats are relatively enormous and their role in collecting and focusing sound is supplemented by additional structures on the face, which give bats their characteristic grotesque appearance. (b) Most mammals, especially small prey species, can orient the pinnae towards a source of sound, including sounds coming from behind. In this jack rabbit (*Lepus californicus*, actually a kind of hare living in deserts of the south-western United States), the very large pinnae are also richly perfused with blood and dissipate excess body heat. (c) Thermoregulation is also a function of the relatively large pinnae of African elephants (*Loxodonta africana*), which flap their ears when overheated, and as a signal of threat to conspecifics. (d) The pinna is greatly reduced in all marine mammals, and is absent in cetaceans including the white-beaked dolphin (*Lagenorhynchus albirostris*).

○ Can birds (with a single ear ossicle and no pinna) hear high-frequency sounds and discriminate between fast-changing notes as well as mammals (with three ear ossicles) can?

● Yes. Sound is extremely important in the social, sexual and parental behaviour of birds and includes many high-pitched notes. Many birds 'twitter', i.e. communicate using fast-changing notes.

Figure 4.38   The head of (a) a large tortoise (*Geoemyda grandis*) and (b) a lizard (*Sceloporus*) showing the eardrum. In tortoises, the eardrum lies flush with the body surface whereas in lizards, it is recessed at the bottom of a short external ear cavity.

Hearing is also well developed in mammals, especially in nocturnal groups such as bats, rodents and primitive primates and in all the marine mammals, especially whales and dolphins (cetaceans) and seals, walruses and sea-lions (pinnipeds). In spite of these fundamental differences in the structure of the ear, birds and mammals have evolved almost equally good hearing and use sound in hunting and in communication in similar ways.

4.6.6 AUDITORY SPECIALISTS

The capabilities of the nervous system are stretched to the limit in the interactions between a predator and its prey. A hunting animal detects and locates its prey using information given out inadvertently by the prey, often using remarkably sophisticated sensory systems. Bats and owls hunt at night when visually-guided predators are at a disadvantage, and they track prey mainly by sound using highly specialized auditory systems.

HEARING IN THE BARN OWL

The nine species of barn owls are different enough from other owls to be classified in their own family, the Tytonidae, of which the common barn owl, *Tyto alba* (Figure 4.39a) is the most abundant species, found throughout the temperate and tropical areas of the world. *T. alba* often lives close to human settlements and farm land, nesting in barns or belfries as well as in hollow trees and holes in banks or rocks. Like most other owls, they remain paired for long periods (sometimes for life) and return to nest in the same place year after year. They hunt in open areas at night and 95% of their prey consists of small mammals such as field mice. Amphibians and other birds make up the other 5%.

Barn owls locate their prey using the faint rustling noises made by small animals as they move through the undergrowth. Like predators that hunt on the ground, barn owls must be able to locate their prey quickly and precisely in the horizontal plane. But they have an added task: since an owl hunts from the air, it must also be able to locate the source of a sound in the vertical plane by determining the angle of elevation above the animal it is hunting. The barn owl has solved these problems very effectively and is able to locate sounds to within a range of one to two degrees (about the width of a little finger at arm's length) in the *azimuth* (the horizontal dimension) and the *elevation* (vertical dimension). Humans are about as accurate as a barn owl in azimuth but are three times worse in elevation. Monkeys and cats are about four times worse in locating sounds in the horizontal dimension.

The most distinctive feature of the barn owl is its face, which is round and relatively large, and covered in stiff feathers arrayed in tightly packed rows (Figure 4.39a and b). This feathered structure is called the *facial ruff* and forms a surface that is a very efficient reflector of high-frequency sounds. Two troughs run through the ruff from the forehead to the lower jaw, each about 2 cm wide and 9 cm long. The troughs are formed by the feathers and serve a similar purpose to the pinna of the human ear: to collect high-frequency sounds from a large volume of space and channel them into the ear canals. The ear openings themselves are hidden by the preaural flaps, flaps of skin that project to the side next to the eyes.

The entire facial structure is hidden under a layer of fine feathers that are acoustically transparent, i.e. sound passes through them. These features of the face make the barn owl extremely sensitive to very faint sounds (e.g. a mouse chewing) and they enhance its ability to determine the direction from which the sound is coming. To locate the source of a sound, the owl must determine the direction of propagation of sound waves from information detected by its ears, situated on either side of its head. One kind of information available is the difference in the time of arrival of sound at the two ears. A sound coming from one side of the head reaches the ear on that side of the head first, while a sound emanating from directly in front of the bird reaches both ears at the same time. Using these interaural time differences, the owl is able to locate accurately the source of a sound in the horizontal plane.

However, in order to locate a sound source in two dimensions, information about the elevation of the source is also required. In the owl's case, this information is supplied by differences in the intensity (or loudness) of the sound at the two ears. The use of differences in intensity of the sound between the ears to provide elevational cues is possible because of a unique characteristic of a barn owl's ears: they are asymmetrical. The ear openings and the protective preaural flaps are vertically displaced, the left ear flap being above the midpoint of the eye and the right ear flap below. There is also a slight asymmetry in the facial ruff with the right trough oriented upward and the left one downward (Figure 4.39b).

○   How could this asymmetry in the ears provide the owl with elevational cues?

●   The right ear would be more sensitive to sounds from above the horizontal plane and the left ear more sensitive to sounds from below. As the sound moves up, it becomes louder in the right ear and softer in the left. As the source moves down, the sound becomes louder in the left ear and softer in the right.

(a)

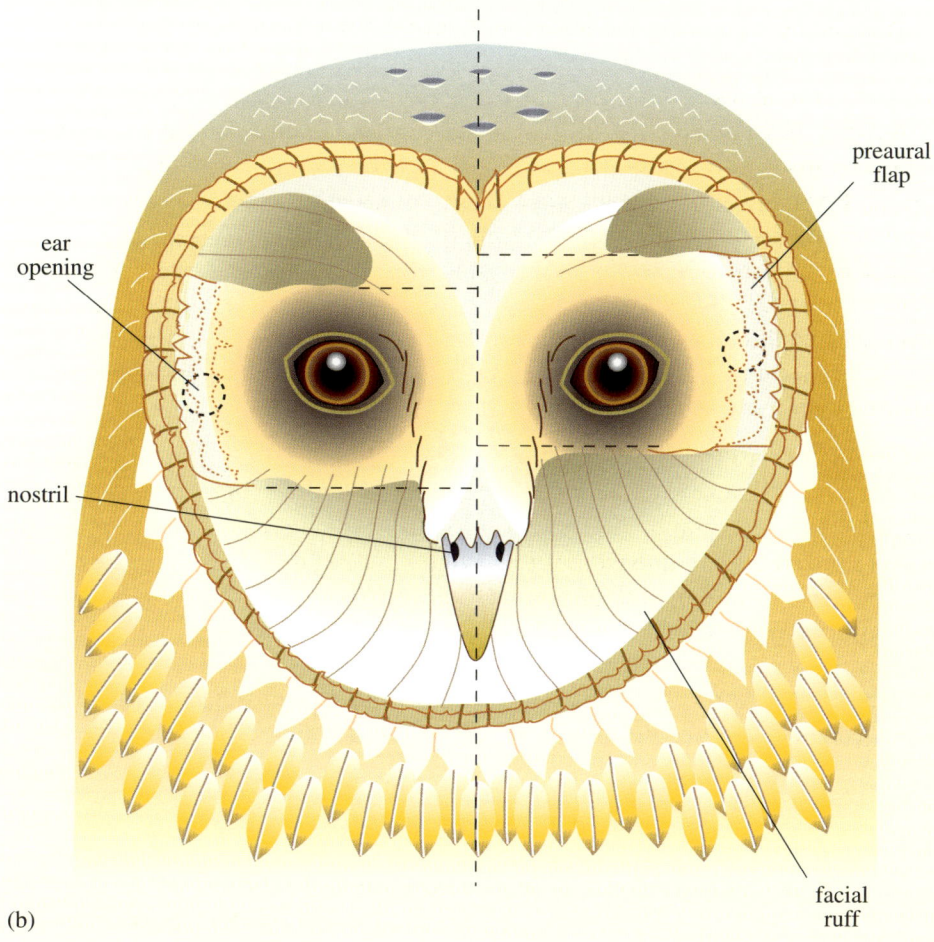

(b)

Figure 4.39   (a) Barn owls. (b) Facial structure and specialized feathers in a barn owl.

The ruff feathers are very efficient at directing high-frequency sounds into the ear canals. In fact, the amount of amplification and directional sensitivity imparted by the ruff feathers varies greatly with the frequency of the sound due to the nature of the sound waves themselves. When sound waves encounter an object, they can bend around the object or be reflected back, depending on the wavelength of the sound and the size of the object. If the wavelength is long (low frequency, see Box 4.6) compared to the size of the object, the waves tend to propagate around the object; if they are short (high frequency) they tend to be reflected. In the case of the ruff, only frequencies above 3000 Hz are reflected efficiently into the ear canals. Directional sensitivity of the ears below 3000 Hz is therefore poor since these frequencies are not reflected efficiently by the ruff. With higher frequency sound waves, each ear is very sensitive to the direction of the sound; a small change in sound direction gives rise to a large change in perceived intensity so that the information about vertical location is very precise.

Evidence that the facial ruff provides the owl with information about the vertical angle of a sound source has been obtained by experimentally removing the facial ruff.

○ What would be the effect of removing the ruff feathers?

● With the ruff feathers removed, high-frequency sounds would not be effectively reflected into the ears and so differences in the intensity of the sound at the two ears would be lessened. The owl is unable to localize sound in elevation and always faces horizontally regardless of the true elevation of the source, but it can still locate in azimuth.

Although some difference in intensity between the ears may be retained because of the asymmetry of the ear openings and preaural flaps, it is not sufficient to enable the owl to identify the elevation of the sound. The importance of differences in sound intensity at the two ears as a means of localizing the elevation of a sound source has also been demonstrated by plugging one ear.

○ What effect would this procedure have on any disparity in intensity between the ears?

● It would be increased.

The loudness of the sound in the plugged ear would be dampened relative to the unplugged ear and so the difference between the ears in the intensity of the sound reaching them would be greater.

○ What effect would plugging one ear have on the owl's ability to locate sounds?

● The owl would be able to locate the sound source in the horizontal plane using differences in time of arrival of the sound at the two ears (which is not affected by plugging one ear). However, it would not accurately locate the sound source in the vertical plane because intensity differences between the ears would be artificially high.

In order to determine whether owls use *specific combinations* of disparities in intensity and timing in locating sound sources, researchers conducted experiments where they inserted tiny earphones into the ear canals. The earphones were connected to separate sources of sound, so the intensity and timing of the sounds reaching each of the owl's ears could be varied independently. The owls responded to sounds from the earphones as they would to sounds arising from outside the head. When the sound in one ear preceded that in the other ear, the head turned in the direction of the leading ear. More specifically, if volume was held constant but sound reached one ear slightly before the other, the owl turned its head in the horizontal plane and the longer the second sound was delayed, the further the head turned. Similarly, if intensity was varied but timing held constant, the owl moved its head up and down. Finally, when sounds were delivered to each ear that differed both in timing and intensity, the owl moved its head in both the horizontal and vertical plane. Combinations of timing and intensity differences that mimicked those from a speaker placed at certain sites caused the animal to turn its head exactly to those same sites.

○   What does this result indicate?

●   The barn owl's brain combines sensory information about timing and intensity of sound at the two ears to determine the horizontal and vertical coordinates of a sound source.

To determine how the brain carries out this binaural (information from both ears) fusion of sounds, a microelectrode was inserted into a single neuron in the brain of an anaesthetized animal. Holding the electrode in place, a speaker was moved across the surface of an imaginary globe around the owl's head. Using this technique, researchers found that certain neurons reacted only to sounds emanating from restricted areas in space. For example, if the owl was facing forward, one neuron might respond only if the sound emanated from a position extending 20° to the left of the owl's line of sight and 15° above or below it. The size and shape of the specific region to which the neuron responds is called its **receptive field**. These space-specific neurons were found in the auditory area of the midbrain, and each responds to a stimulus only when both time and intensity difference fall within the range to which it is tuned. It is not excited either by the correct time difference alone or the correct intensity difference alone. Receptive fields are therefore formed by 'tuning' the neurons to specific combinations of time and intensity differences.

This example illustrates how superficial and internal anatomical characters combine with elaborate neural mechanisms to achieve acute sensitivity and fine discrimination between sounds. Owls have several other impressive adaptations to their habits as nocturnal predators, including soft, fine wing feathers that move silently through the air enabling the predator to pounce on unsuspecting prey, and adaptations of the eyes.

○   What features characterize eyes adapted to see in very dim light?

●   The eyes are very large and the 'eye shine' from the reflective choroid layer is bright.

To fit large eyes into a small head, owl eyes are tubular, so they cannot be turned from side to side or up and down to follow a moving object in the same ways as those of most birds and mammals, including ourselves. Instead, the whole head moves, if necessary twisting the neck through 180°. This adaptation to nocturnal hunting explains why owls appear to stare at observers with their large shining yellow eyes.

SUMMARY OF SECTION 4.6

1   Chemoreceptors that mediate taste and smell may be widely distributed over the body surface in aquatic vertebrates or localized to regions of the nose and mouth.

2   Eyes consist of photoreceptors that contain a light-absorbing pigment linked to membrane proteins, forming a retina. Visual pigments determine the sensitivity to different wavelengths and differ in abundance and composition in animals living in different habitats.

3   More complex eyes have various accessory tissues including a lens that focuses the light on the receptors and may itself be adjustable.

4   Stereoscopic vision requires overlap of the visual fields of the two eyes and confers a sense of depth perception, which is important for animals that hunt by sight.

5   Advanced snakes that prey on birds and mammals have specialized infrared receptors that enable them to detect and locate faint sources of heat.

6   Many fishes can sense small electric fields produced by other animals' muscles using electroreceptors. Some have muscles specialized to generate large and/or rapidly alternating electric currents.

7   Mechanoreceptors contain hair cells that transform mechanical stimuli into electrical signals. They respond to displacement, velocity or acceleration of air or water. Fish and aquatic amphibians have both lateral lines and ears but the former are lost in higher tetrapods. The ear serves in balance as well as in hearing.

8   The ears of birds and mammals differ greatly in structure, but functionally they are remarkably similar. The sensitivity of hearing is enhanced by one or more middle ear ossicles which transmit sound from the outer ear to the inner ear where the receptors (hair cells) are located.

9   Adaptations of the skull, face and feathers enable barn owls to hear faint sounds and orientate towards them. The eyes are adapted to function in very dim light.

4.7   CONCLUSIONS

This chapter has described a few of the tissues and organs that underpin the large size and great diversity of vertebrates. As explained in Chapter 5, many of the basic molecular components of vertebrate tissues are also known in much simpler animals. But vertebrates have a wider range of recognizably different tissues than invertebrates. Their structural tissues (skeleton, teeth etc.) are structurally elaborate, and resist wear and injury by being both hard and tough, but also by being continuously repairable. The mechanical toughness of the vertebrate skin is due mainly to its extracellular materials but it also includes many metabolically active cells that produce a wide range of secretions, including mucus and milk. These tissues and the complex sense organs containing many highly specialized cells could not function without continuous supplies of oxygen and nutrients carried in the blood and vascular system.

Vertebrate blood contains far more cells (and more kinds of cells) than that of invertebrates. With more cells, the blood carries more oxygen per unit volume and repels pathogens more quickly, but it is much more viscous. The cardiovascular system pumps this thicker blood under pressure, so even the most remote tissues are perfused even in very large species. As well as integrating the large amount of sensory information, the nervous system coordinates adjustments to the rate and direction of blood flow. Both the blood and the vascular system have adapted to function in a wide variety of difficult environments, enabling vertebrates to colonize a wide range of habitats, from high mountains to the deep sea.

REFERENCES

Clements, M. and Saffrey, J. (2001) Communication between cells, in *The Core of Life, Vol. II*, J. Saffrey (ed.), The Open University, Milton Keynes, pp. 241–304.

Davey, B. and Gillman, M. (2001) Defence, in *Generating Diversity*, M. Gillman (ed.), The Open University, Milton Keynes, pp. 151–200.

Halliday, T. and Pond, C. (2001) Longevity, in *Generating Diversity*, M. Gillman (ed.), The Open University, Milton Keynes, pp. 201–236.

Loughlin J. (2001) Cell movement, in *The Core of Life, Vol. II*, J. Saffrey (ed.), The Open University, Milton Keynes, pp. 305–343.

Pond, C. (2001) Dealing with food, in *Generating Diversity*, M. Gillman (ed.), The Open University, Milton Keynes, pp. 37–88.

Saffrey, J. (2001) Cells and tissues, in *The Core of Life, Vol. I*, J. Saffrey (ed.), The Open University, Milton Keynes, pp. 1–51.

Shadwick, R. E., Russell, A. P. and Lauff, R. F. (1992) The structure and mechanical design of rhinoceros dermal armor, *Philosophical Transactions of the Royal Society of London series B — Biological Sciences* **337**, pp. 419–428.

Walker, C. and Swithenby, M. (2001) Proteins: structure and catalytic function, in *The Core of Life, Vol. I*, J. Saffrey (ed.), The Open University, Milton Keynes, pp. 53–106.

Wells, R. M. G. and Baldwin, J. (1994) Oxygen transport in marine green turtles (*Chelonia mydas*) hatchlings: blood viscosity and control of hemoglobin oxygen-affinity, *Journal of Experimental Biology* **188**, pp. 103–114.

FURTHER READING

Carr, C. E. (1990) Neuroethology of electric fish, *BioScience,* **40**, pp. 259–267. [A simple popular article.]

Pough, F. H., Janis, C. M. and Heiser, J. B. (1996) *Vertebrate Life* (5th edn), Prentice Hall, New Jersey. [The best of several modern textbooks of vertebrate zoology, covering a wide range of topics. Sixth edition expected in 2002.]

Kardong, K. V. (1996) *Vertebrates: Comparative anatomy, function, evolution* (2nd edn), McGraw Hill. [Particularly strong for tissues, developmental biology and sensory systems.]

King, A. J. (1999) Making sense of hearing, *The Biologist,* **46**, pp. 77–81. [A general article on human hearing.]

Konishi, M. (1993) Listening with two ears, *Scientific American*, **268**, pp. 66–73. [A simpler popular article.]

Knudsen, E. I. (1981) The hearing of the barn owl, *Scientific American,* **245**, pp. 113–125. [A simpler popular article.]

Schmidt–Nielsen, K. (1997) *Animal Physiology* (5th edn), Cambridge University Press, Cambridge. [A useful textbook of comparative physiology, with emphasis on vertebrates, packed with good ideas and useful, accurate facts.]

Simmons, P. J. and Young, D. (1999) *Nerve Cells and Animal Behaviour* (2nd edn) Cambridge University Press, Cambridge. [A short, inexpensive semi-popular book that includes a section about the sensory mechanisms that enable owls to hunt in the dark.]

THE MOLECULAR BIOLOGY OF ANIMAL DIVERSITY

5.1   INTRODUCTION

Twenty years ago, molecular and cell biology and so-called comparative or taxonomic biology were largely separate endeavours whose practitioners had little in common, and in many institutions, rarely met. Molecular biology was conducted in the laboratory, mostly using long-established laboratory species that could be bred in huge numbers, and was primarily concerned with genetic and physiological mechanisms. The 'model organisms' that molecular biologists studied intensively represent organisms that, for one reason or another are ideal for studying basic biological processes (Pond, 2001). An early laboratory organism in molecular biology was the bacterium *Escherichia coli*, followed by the unicellular eukaryotes *Saccharomyces cerevisiae* and *Schizosaccharomyces pombe* (both yeasts, but very different organisms) and animals, including the soil worm *Caenorhabditis elegans*, the fruit-fly *Drosophila melanogaster* and the mouse *Mus musculus*. We shall see in this chapter how the study of the molecular biology of particular model organisms has revolutionized evolutionary and comparative biology.

The other kinds of biology involved field work, often in remote places, studying habits and adaptations in a huge diversity of organisms. Towards the end of the 20th century, molecular biology was extended to a wide range of wild organisms. Many of its techniques became cheaper and quicker and would work on much smaller biological samples.

○   What technique has revolutionized molecular biology by allowing study of very small biological samples?

●   The polymerase chain reaction (abbreviated to PCR).

PCR offers the possibility of recovering specific DNA segments from vanishingly small DNA samples, in sufficient quantities for identification and study (Saffrey and Metcalfe, 2001).

At the same time, methods for the transportation, maintenance and breeding of 'exotic' animals, plants and microbes improved, bringing more species into the laboratory than ever before. People became more concerned about habitat destruction and the increasing rarity of many wild species, and the problems of breeding them in captivity. All these factors promoted the application of gene and protein analysis to organisms that had previously been impossible to study in this way.

This chapter outlines some of the spectacular discoveries about the origins of animal diversity made using molecular techniques, and shows how they can be integrated with traditional taxonomy and functional biology.

We discuss several cases where molecular genetics has shed light, often yielding startling conclusions, on developmental processes and evolutionary relationships between apparently disparate tissues and groups of organisms. We examine the processes by which novel gene functions may arise, and how they lead to changes in body plan and adult structure and properties.

These techniques also enable biologists to study more aspects of evolutionary processes experimentally. The process of natural selection has long been amenable to experimental study (a classic example being the peppered moth *Biston betularia*: see Stevens, 2001), but now the origins of the genetic and functional diversity upon which selection acts can also be directly analysed by experiment.

5.2 THE UNIFORMITY OF BIOLOGICAL PROCESSES AND THE DIVERSITY OF GENES

The process of evolution by common descent leads us to expect that developmental and other biological processes, even in widely disparate organisms, proceed by homologous means, driven by homologous genes and networks of interactions.

5.2.1 HOW GENES CHANGE IN EVOLUTION

Because homologous genes share a degree of sequence similarity due to their common ancestry, the term 'homologous' is often used more loosely, to mean 'similar'. If the sequence of a specific human gene is compared to its mouse counterpart, we can see that, though the two genes clearly encode very similar proteins, they are not identical. That is, there are sequence differences between them. How do genes change in sequence during the evolutionary history of a species? DNA replication is remarkably accurate but an average error rate can be estimated, typically one error per 10^7 bases copied. (See Figure 5.1.) These changes, when they occur in the germ-line, are heritable and provide the raw variation upon which natural selection acts, leading to evolutionary change.

Many such sequence changes do not in fact lead to a structural change in the encoded protein and consequently do not alter fitness. Examples include many of the possible changes in the third nucleotide of a codon, and those that occur within non-coding regions of the genome such as introns.

○ Why do sequence changes altering the third base of a codon often cause no change in the amino acid sequence of the protein produced by the gene?

● There are 64 possible combinations of three nucleotides ($4 \times 4 \times 4 = 64$). The genetic code uses 61 of these codons to code for 20 amino acids (recall that three codons act as stop signals, and do not correspond to amino acids). Some amino acids are encoded by several codon sequences, which often differ only in the third base, so in these cases changes in the third base do not alter the amino acid incorporated into the protein.

So homologous genes from two species may differ in DNA sequence but there may be no difference in the amino acid sequence of the protein they encode. Because they do not alter the phenotype, the frequencies of these altered genes cannot be changed by natural selection. The differences arising from such non-selectable changes in DNA can be measured in related living species such as primates, and indicate the time since lineages diverged from each other. Comparison of the DNA sequences of two distantly related species, such as humans and mice, which are primates and rodents respectively, is consistent with their most recent common ancestor lying much further back in time. In other words, the length of time since the human and mouse lineages diverged is greater than that since the human and other primate lineages diverged. Since the rate of sequence change is roughly constant, the ancestry of these lineages is reflected in the amount of sequence divergence between homologous genes, measured in living species.

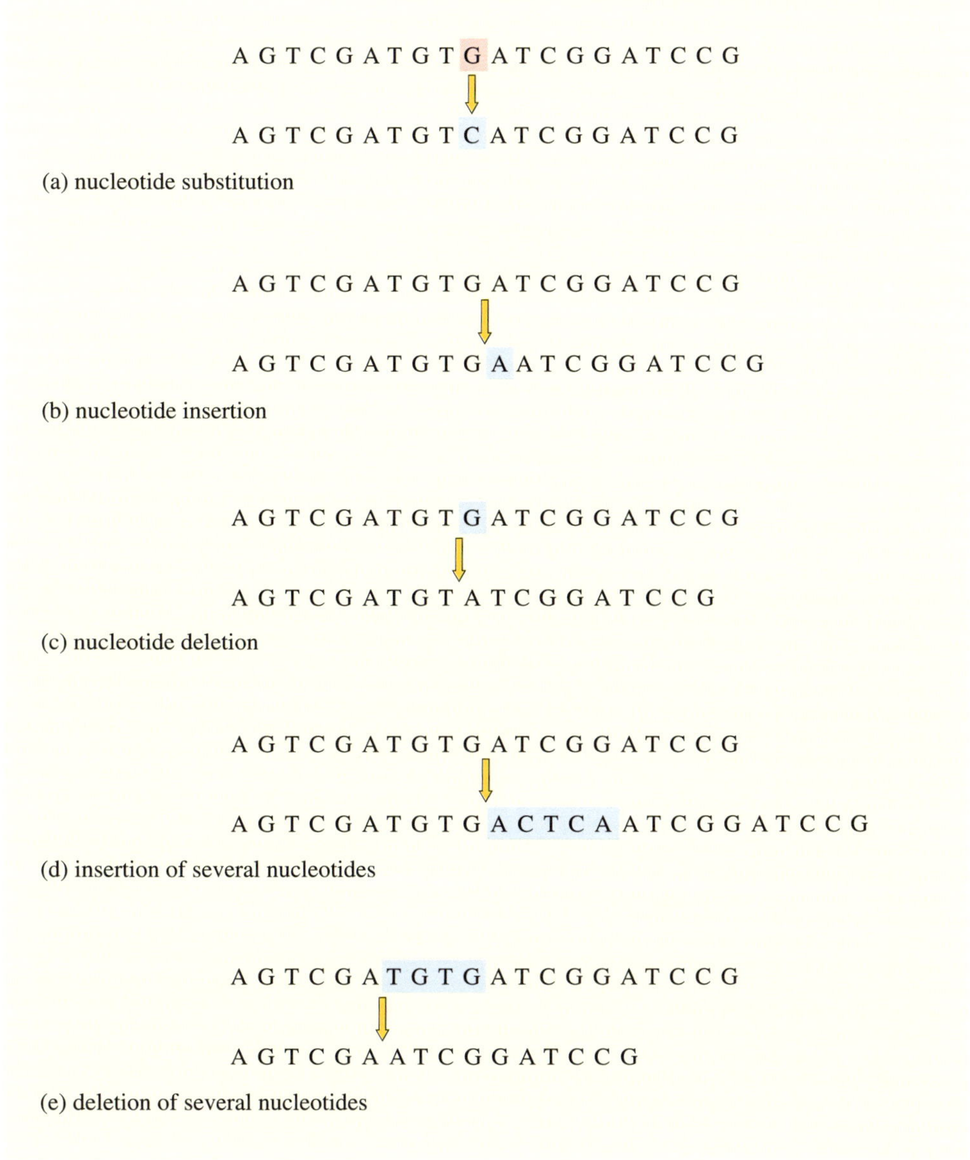

Figure 5.1 Some of the changes that may occur to DNA sequences: (a) substitution of a single nucleotide; (b) insertion of a single nucleotide; (c) deletion of a single nucleotide; (d) insertion of a sequence of several nucleotides; (e) deletion of a sequence of several nucleotides. Note that inserted and deleted sequences may be very much larger than indicated here.

A G T C G A T G T G A T C G G A T C C G

↓

A G T C G A T G T C A T C G G A T C C G

(a) nucleotide substitution

A G T C G A T G T G A T C G G A T C C G

↓

A G T C G A T G T G A A T C G G A T C C G

(b) nucleotide insertion

A G T C G A T G T G A T C G G A T C C G

↓

A G T C G A T G T A T C G G A T C C G

(c) nucleotide deletion

A G T C G A T G T G A T C G G A T C C G

↓

A G T C G A T G T G A C T C A A T C G G A T C C G

(d) insertion of several nucleotides

A G T C G A T G T G A T C G G A T C C G

↓

A G T C G A A T C G G A T C C G

(e) deletion of several nucleotides

So we would expect more closely related species to have more similar gene sequences. For example, one would expect the sequences of the human and mouse (both vertebrates) *Pax-6* genes to be more similar to each other than to the fruit-fly homologue. (*Pax-6* encodes a transcription factor, and we discuss its developmental role in Section 5.5.) We can use such observations of sequence similarity to construct phylogenetic trees reflecting the interrelationships between species. However, sequence variation does not involve solely nucleotide substitution — the insertion or deletion of nucleotides also occurs, though there may be constraints due to the necessity of maintaining an accurate reading frame.

○ What sizes of insertion or deletion would be easily incorporated into the coding region of a gene?

● In order to maintain the correct reading frame, it is likely that deletions or insertions involve multiples of three bases — remember that the genetic code is read in units of three bases (codons).

Replication error is not the only means by which gene sequence has changed through evolutionary history.

○ In Table 5.1, you can see that the far-right column has been left blank. Calculate the average number of bases pairs (bp) per gene by dividing the genome size by the number of genes, and then fill in the blank spaces. You will not be able to calculate the values for mouse and newt.

Table 5.1   Genome sizes of several species.

Species	Genome size (base pairs)	Estimated number of genes	Average number of base pairs per gene
Homo sapiens (human) *	3×10^9	30 000–40 000	
Mus musculus (mouse)	3×10^9	†	†
Triturus spp. (newt)	20×10^9	†	†
Drosophila melanogaster (fruit-fly) *	170×10^6	13 500	
Caenorhabditis elegans (nematode worm) *	97×10^6	19 900	
Plasmodium falciparum (protoctist) *	30×10^6	6500	
Saccharomyces cerevisiae *	12×10^6	6000	
Escherichia coli *	4×10^6	4405	

* indicates that the complete genome sequence has been determined. † indicates that no good estimate for gene number, and thus base pairs per gene, is available.

● The values are: human, 75 000–100 000 bp; fruit-fly, 12 592 bp; *C. elegans*, 4874 bp; *P. falciparum*, 4615 bp; *S. cerevisiae*, 2000 bp; *E. coli*, 908 bp.

○ Why does *E. coli* have much shorter genes than the other species listed in Table 5.1?

- It is a prokaryote, and therefore its genes do not have introns. All the other species listed are eukaryotes.

○ Can you conclude that bigger genomes always have more genes?

- No. While there is a *general* correspondence between genome size and total number of genes, a comparison between *C. elegans* and *D. melanogaster* shows that the correlation is not close.

Table 5.1, which presents an overview of genome statistics, shows that different species have very different numbers of genes. Clearly, species do not evolve just by selection acting on fitness differences attributable to changes in gene sequence — the overall number of genes has changed during evolution. As a rule of thumb, the more complex an organism, the more genes it has. (However, it is clear from the table that the overall genome size is only loosely correlated with the number of genes present.) How do these 'extra' genes arise?

Genes may be duplicated by a variety of mechanisms (Saffrey and Metcalfe, 2001). Once a gene encoding a particular product is duplicated, it is self-evident that **genetic redundancy** is immediately created. In other words, although sequence changes to one copy of this gene may have a severe impact on the function of its product, such changes with have no effect on the fitness of the organism while the second copy of the gene is still intact and operating normally. The organism suffers no ill-effects should one of the newly duplicated genes mutate in a way that prevents it from producing its original product. Mutation and natural selection can thus establish divergence of the two genes during subsequent generations: each gene can follow its own evolutionary path. In many cases, one of the duplicated genes simply acquires one or more mutations that prevent production of its product, or abolish all function from its product. Such a permanently inactivated gene is a pseudogene. A gene family is a collection of genes (some may be functional, some may be pseudogenes) that have diverged in this way from a single ancestral sequence.

○ A hypothetical gene encoding a vital enzyme has been duplicated. What would be the consequences of (a) mutation of the third codon to a stop codon in one of the genes; (b) a mutation that alters the tertiary structure of the enzyme encoded by one of the genes in a way that changes its substrate specificity?

- (a) The mutated gene probably no longer encodes a functional protein. The genetic redundancy generated by the initial gene duplication event has been lost, and one of the two genes has become a pseudogene. (b) The enzyme encoded by the mutated gene has, in effect, gained a novel biological function. In this case, the genetic redundancy has been lost because the two genes have become functionally different.

One of the duplicates may be recruited for a different function and evolve separately from its homologue. By means of gene duplication, large families of structurally related genes appear during evolution.

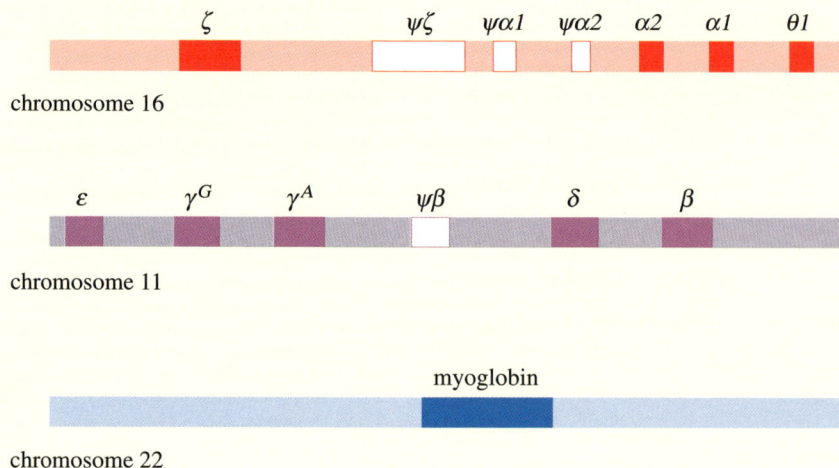

Figure 5.2 The gene family of the human globin genes, showing the physical arrangement of the genes within the genome. There are two clusters of globin genes, located on different chromosomes: one cluster lies on chromosome 11, and the other on chromosome 16. For each cluster, functional genes are shown as dark-coloured boxes, and non-functional pseudogenes by white boxes. The gene encoding the related muscle protein myoglobin is found as a single gene on chromosome 22, and is shown as a dark blue box. The possible evolutionary relationship between these sequences is illustrated in Figure 5.3.

Figure 5.2 shows such a gene family, that of human globin genes. The functional significance of small differences in the molecular structure and physiological properties of haemoglobin in vertebrates were discussed in Section 4.3.2. The α (alpha) family forms a group on chromosome 16, while the β (beta) family forms a group on chromosome 11. Both families have non-functional pseudogenes, which are denoted by ψ (the Greek letter 'psi') within their names. The related gene myoglobin lies on chromosome 22.

Figure 5.3 shows how the globin genes are related to each other by sequence similarity, with the added dimension of time. The rate at which sequences change through evolution remains relatively constant, and can be used as a '**molecular clock**' by which approximate dates of divergence from common ancestors can be deduced. The molecular clock for such studies can be calibrated from the evolutionary history of homologous genes isolated from several species and by comparing the resulting phylogenetic tree with the fossil record. In Figure 5.3, the approximate dates at which each gene duplication event is thought to have occurred are shown at each branch-point in the diagram, and are correlated with major events in animal evolution. This time-scale is approximate, and deduced from known rates of selectively neutral changes in nucleotide sequence and the fossil record.

The probable sequence of events in the evolution of the human globin gene family is as follows. First, the ancestral gene was duplicated. One copy underwent no further duplication, and became the myoglobin gene. Myoglobin has similar structure and properties to haemoglobin, but is found in muscles, not in blood. It is the main cause of the red coloration of muscle tissue, where it promotes oxygen transfer from the blood to the contractile proteins, and stores oxygen for use when supplies from the blood are insufficient.

The other copy of the globin gene underwent further duplications later in evolution, yielding several genes of the α- and β-globin families, which were favoured by natural selection and so were retained. The first duplication event of haemoglobin genes was ancient: 600–800 Ma ago, i.e. before the first known chordate fossils.

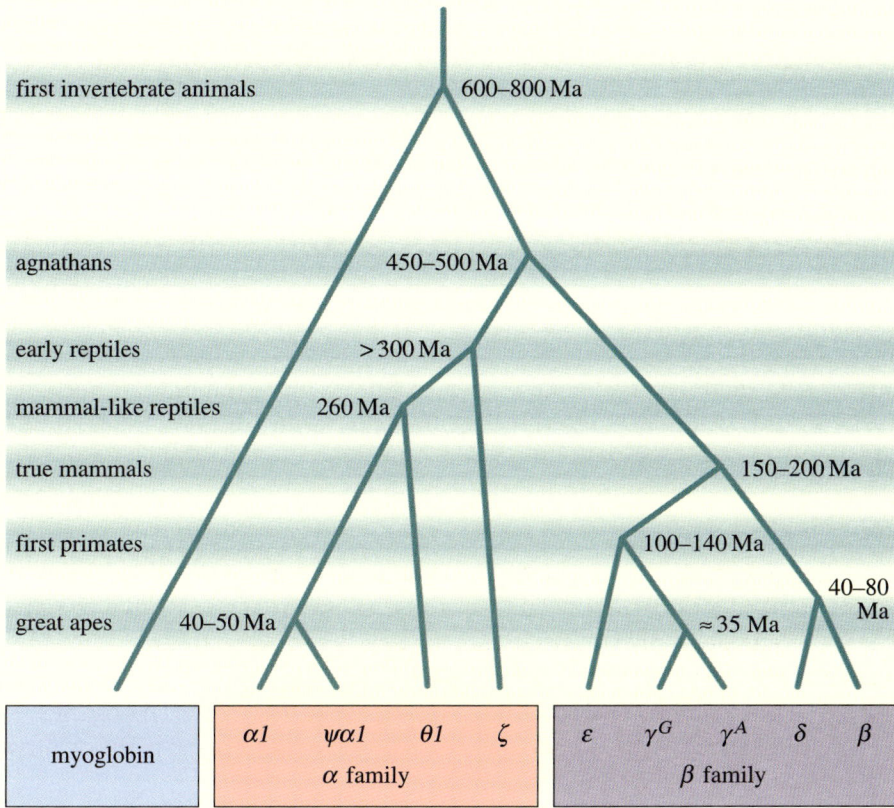

Figure 5.3 The ancestry of the human globin genes deduced from sequence comparisons. These genes include most of those shown in Figure 5.2. The time-scale on the diagram runs from top to bottom, i.e. more ancient gene duplication events lie towards the top and more recent events to the bottom. The appearances in the fossil record of animal taxa are shown as horizontal, blue bars, but do not correspond specifically with the gene duplication/divergence events.

○ In addition to the sequence differences by which myoglobin and haemoglobin can be distinguished, what change in the control of gene expression occurred subsequent to the divergence between the two lineages of genes?

● The expression patterns of the two lineages of genes changed. Myoglobin is expressed in muscle, whereas haemoglobin is expressed in red blood cells.

Similar, probably even more complex, changes in the globin genes produced the specialized blood pigments found in living species of other classes of vertebrates (see Section 4.3.3).

5.2.2 CHROMOSOME CHANGES IN EVOLUTION

Duplication events (such as those discussed above) are not confined to single genes. When chromosomes of closely related species are compared, chromosome rearrangements are frequently observed. Chromosome rearrangements include various types of change to the chromosomes, such as deletions (where a section of chromosome has been lost), duplications (where a section of chromosome has been duplicated) and inversions (where a section of chromosome has been flipped in orientation relative to the 'normal' chromosome order) (Stevens, 2001).

In the case of certain insects (such as *Drosophila* and several other species of Diptera, such as mosquitoes), in which many tissues have giant polytene chromosomes, it is possible to identify even small rearrangements with considerable accuracy. In several cases where chromosome rearrangements have been examined in closely related species, an association of particular inversions with particular species can be seen. Most of these analyses have been carried out in insects with giant polytene chromosomes, although a similar situation has been observed in European populations of mice. It is thought that chromosome rearrangements also play an important role in reproductive isolation during the process of speciation, inferred by the study of rearrangement distribution within the species of a lineage.

Duplication of sections of genomes has been inferred by general sequence analysis. In fact, whole **genome duplication** and even triplication are relatively frequent events in the evolution of a number of plant species. Notably, modern wheat varieties are known to be hexaploid (i.e. the genome has triplicated from the initial diploid state to give six copies of the haploid genome) derivatives of the ancestral forms. While more common in plants, cases of apparent genome duplication in animals are known. For example, a whole genome duplication appears to have occurred during the evolution of bony fish. An evolutionarily more recent example is seen in the genome of the African clawed frog *Xenopus laevis*, which is double the size of that of other living species in the genus, implying that a genome doubling event has occurred at some time after the appearance of the genus *Xenopus*.

It is clear, then, that total gene numbers can increase during evolution by duplication of single genes, groups of genes or whole genomes.

SUMMARY OF SECTION 5.2

1   The genomes of several laboratory organisms have been characterized in great detail, and in some cases the complete genome sequence has been determined.

2   The size of an organism's genome is only loosely correlated with the organism's size or complexity.

3   Measurements from extant species show that through the course of evolution, the number of genes in the genomes within a given lineage has changed.

4   Changes in gene number can arise by gene duplication, the duplication of sections of chromosome and duplication of the entire genome.

5   Duplicated genes cause evolutionarily transient genetic redundancy, but the redundant genes often acquire further mutations, resulting in the evolution of novel proteins.

5.3   WHAT GENOME PROJECTS TELL US

The Human Genome Project is well known to the public. At the time of writing (summer 2001), the determination of the complete sequence is nearing completion, with the first draft of 85% coverage having been released. Perhaps less newsworthy, but very important from a scientific perspective, are the projects in which smaller genomes have been sequenced, including those of several well characterized laboratory organisms.

The complete DNA sequence of organisms of diverse phyla are now becoming available to researchers. Organisms for which complete genome sequences exist include the following:

*   Several species of prokaryotes, including important experimental organisms such as *E. coli*, and some major pathogens;
*   Protoctists, particularly the medically important parasites such as the malaria parasite *Plasmodium falciparum*;
*   Nematoda, such as the valuable laboratory organism *Caenorhabditis elegans* — there are many nematode parasites of humans, for which *C. elegans* is a useful model;
*   Arthropoda, for example the fruit-fly *Drosophila melanogaster*;
*   Chordata — the first draft of the human (*Homo sapiens*) genome was published in 2001;
*   Plantae — the genome of the small flowering plant *Arabidopsis thaliana* has been sequenced, and at the time of writing, several plants, such as rice and maize, are being subjected to whole-genome sequencing;
*   Fungi — the genomes of the yeasts *Schizosaccharomyces pombe* and *Saccharomyces cerevisiae* have been sequenced.

The list of organisms for which large quantities of sequence data are becoming available is continually growing. This vast pool of data provides many benefits to modern biological research into biomedical questions, developmental biology and evolutionary biology, amongst others.

How do the genomes of the model organisms compare? As you can see from Table 5.1, there is no linear correspondence between genome size and number of genes. The number of genes currently remains an estimate, even in cases where the entire genome sequence has been determined.

5.3.1   MECHANISMS OF GENOME EVOLUTION

We have already seen how genes for new kinds of protein can evolve by gene duplication followed by divergence of the sequences and the function of the proteins they encode. The globin genes shown in Figures 5.2 and 5.3 are an example of such a gene family that probably arose from such processes. However, duplication events can involve very much larger sections of a genome than single genes, as deduced from genome sequences, such as those of bacteria, in which whole-genome duplication appears to have been frequent.

Duplication of large sections of the genome can often be deduced from changes to the chromosome complement, when two closely related species are compared. Molecular analysis of groups of genes involved in the specification of segment identity during development (Section 5.6) reveals that the number of these gene clusters has increased during the evolution of vertebrates by repeated duplication of entire clusters.

5.3.2 MOLECULAR PHYLOGENY

Molecular phylogeny is the interpretation of the differences in sequence of genes encoding homologous proteins, which are greater in species that are more divergent. Comparative analysis of sequences of homologous genes isolated from many species can provide confirmation of evolutionary relationships between species, or indeed challenge established views of their taxonomic affinity (Ridge, 2001).

These analyses are extremely complex, and best handled by computers using programs based on specific algorithms to take into account sequence variations such as base pair changes, deletions and insertions. For many DNA sequences, only certain changes are favoured by selection. In the case of protein-coding sequences, not only must reading frames be maintained, but any changed gene sequence must still encode a functional protein.

5.3.3 THE RATES OF GENOMIC CHANGE

In Section 5.2 we discussed the mechanisms by which new genes can arise during evolution. The explosion in molecular biological research has resulted in a huge amount of DNA sequence data which can be used to answer questions about the rate at which these events happen, and about the subsequent evolution of duplicated genes. As the technology for DNA sequencing has improved, and of course with the many genome projects currently completed and under way, these data have increased greatly in quantity.

Computer analyses of sequence data can provide estimates of the frequency of gene duplications, and their ultimate fate. It appears that gene duplication occurs at a rate of 0.01 event per gene per million years. A duplicated gene may have one of three fates: it may be inactivated by the accumulation of disabling mutations, it may acquire a novel function, or the product of either copy of the gene may be altered by mutation. The majority of duplicated genes lose the capacity to express a functional protein as a consequence of accumulated mutations — in other words, they become pseudogenes, so are lost to the organism.

The recently published draft sequence of the human genome sheds some light on the origins of genes. There are numerous sections of the human genome that appear to have arisen by processes of duplication. Furthermore, several dozen genes appear to be derived from bacteria, by horizontal transfer. There is no reason to suppose that the human genome is unique in these characteristics, and it is therefore to be expected that similar processes have been occurring throughout the evolution of animals.

In fact, amongst the exciting observations and deductions from the first analysis of the draft sequence is that there are considerably fewer genes than had been expected by many workers in the field. In Table 5.1, this value is given as 30 000–40 000 genes. The uncertainty is due to the fact that interpretation of the sequence is still ongoing. However, the likelihood is that there are around 38 500 genes in the human genome. If you compare this figure with that of the nematode *C. elegans* (also given in the table), you can see that the human genome contains only about twice as many genes as *C. elegans*. Furthermore, a high proportion of the genes identified in one organism have their counterparts (homologues) in the other. How we reconcile the radically different body plan, complexity, lifestyle and many other characteristics with a surprisingly similar set of genes are the topics of the following sections.

SUMMARY OF SECTION 5.3

1 Large-scale genome sequencing projects provide a wealth of data that are relevant in examining biological processes.

2 Sequence data can be used to investigate the relatedness of different species.

3 Genome projects shed light on biological processes that give rise to novel genes, including duplication of sections of the genome and acquisition of gene sequences from other species.

4 The wealth of sequence data available from these projects allows us to estimate the evolutionary lifespan of duplicated gene sequences, and the frequency with which they arise.

5.4 BIOLOGICAL DIVERSITY AND THE EVOLUTION OF NOVEL PROTEINS

So far you have read about the similarities between organisms at the level of DNA sequence. However, this chapter is concerned with the differences as well as similarities between organisms, and how we can explain the diversity of animals at a molecular level. Here we consider a number of topics, selected to illustrate processes that can give rise to new genes, which specify novel morphological or biochemical activities that generate functions and properties on which natural selection can act.

We begin by comparing two concepts: (a) *gene duplication* as the origin of novel genes, an example being collagen (Section 5.4.1), and (b) *gene sharing* — the use of one gene product for two distinct purposes — using the example of the crystallin proteins that form the lens and cornea of the eye (Section 5.4.2). In Section 5.4.3 you will discover how gene duplication opened up a route to the evolution of genes encoding photoreceptors and hence colour vision. Then, in Section 5.4.4, we describe a molecular mechanism underlying one of the most significant changes to occur in vertebrate evolution — the appearance of lactation, the defining characteristic of mammals.

Later, in Sections 5.5 and 5.6, we consider the role of gene duplication and divergence in the evolution of structure, with specific reference to several biological systems. The elucidation of the molecules and processes by which eye development is controlled has exciting implications for the evolution of eyes, which challenges the established view of the origins of eyes in different groups of animals. Eyes are almost ubiquitous amongst animals, although as indicated in Sections 1.4.2, 2.4 and 4.6.2, they take a variety of forms. The evolution of eyes is a recurring theme in this chapter.

○ How does the gross structure of insect eyes differ from vertebrate eyes?

● Typical insect eyes are compound — that is, they are composed of a large number of individual units, or ommatidia, each with a simple lens (Section 2.4). Vertebrate eyes are more like a camera, with the image focused by the lens upon a single light-sensitive retina (Section 4.6.2).

Interestingly, nearly all cephalopod molluscs, such as the squid (Section 1.4.1), have eyes of similar structure to those of vertebrates, and their lenses share many of the features that we discuss here. The development of branched airways in insects and vertebrates, while involving very different tissues and systems, appears to share a molecular basis when the genetic controls of developmental pathways are compared, as described in Section 5.5.4.

5.4.1 COLLAGENS

Collagens are a specific class of extracellular proteins characteristic of animals, and are found in all animals from sponges (Section 1.2.1) to vertebrates (Sections 4.2, 4.3.1, 4.4, 4.5 and 4.6.2). Collagens are particularly abundant and diverse in vertebrates. The processes by which extracellular collagens are assembled into the complex quaternary structures important for their function as mechanical support are summarized in Figure 5.4.

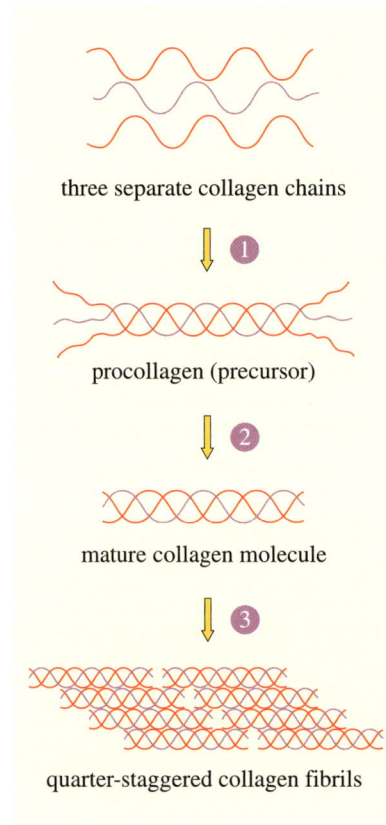

three separate collagen chains

procollagen (precursor)

mature collagen molecule

quarter-staggered collagen fibrils

Figure 5.4 Processing of collagen molecules, and their assembly into fibrils. Step 1: three collagen chains are assembled, in the endoplasmic reticulum and Golgi, to form the procollagen precursor, which is a triple-helical structure for most of its length. Step 2: C-terminal and N-terminal sections are removed by specific proteases, to produce a mature collagen molecule. Step 3: the resulting triple helices are assembled into quarter-staggered fibrils.

Fibrillar collagens have a simple primary structure in which the majority of each collagen molecule is composed of repeating units of three amino acid residues. Within this pattern, however, there is some variability: each unit has the sequence glycine-X-Y, where X is frequently proline, and Y is frequently hydroxyproline, as illustrated in Figure 5.5.

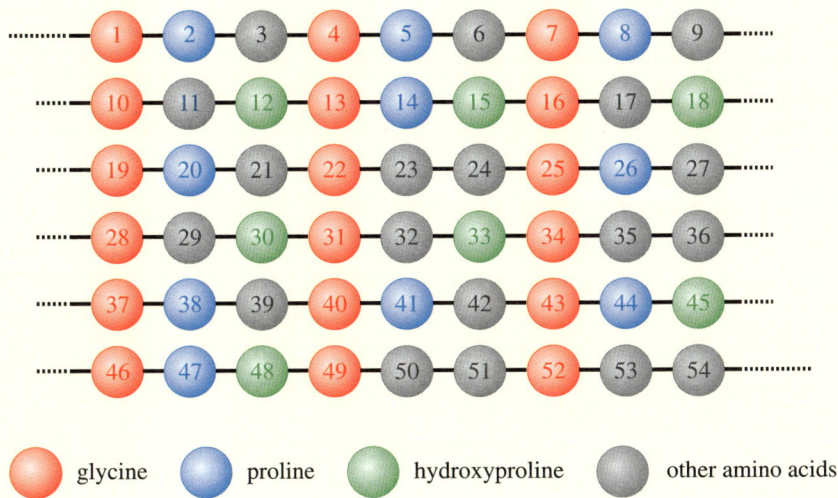

Figure 5.5   Typical primary structure of a sequence of 54 amino acid residues from the central α-helical domain of a fibrillar collagen. The repeated structure of three amino acids, of which the first is always glycine, the second usually proline and the third usually hydroxyproline, is emphasized by the colour-coding of the amino acid residues.

○   How many nucleotides are required to encode each repeated unit of three amino acid residues?

●   Nine nucleotides. Three amino acid residues require three codons, each of three nucleotides.

The structure of the central portion of collagen can therefore be written as [Gly-X-Y]$_n$, where n represents the total number of repeated units. Interestingly, hydroxyproline is not incorporated into the collagen as it is synthesized at the ribosome. Rather, it is formed afterwards by hydroxylation of proline residues, a highly specific post-translational modification which is catalysed by the enzyme prolyl hydroxylase. This reaction occurs on each polypeptide chain before it assumes its helical form. Proline residues that lie to the N-terminal side of a glycine are targeted for this modification. The simple [Gly-X-Y]$_n$ repeating sequence, shown in Figure 5.5, determines the structure of the mature protein: the specific amino acids of the repeated units enable groups of three collagen molecules to associate in a triple-helical arrangement.

The genomic organization of several collagen genes has been determined, and the repetitive central α-helical domain of the procollagen molecule is found to be encoded by a series of exons of 45 or 54 bp.

○   What is the relationship between the size of these exons and the protein segment they encode?

●   Both the 45 and the 54 bp exons are multiples of 9 (= 3 codons of 3 base pairs each), so they encode 5 or 6 repeats of three amino acid residues respectively.

The correct assembly of collagen polypeptides into fibrils is crucial to normal developmental processes. The consequences of defects in this process can be seen as genetic disease in humans, affecting tissues in which the genes for each class of mutated collagens are expressed.

Vertebrates have many forms of collagens, which produce a wide range of extracellular materials, tendons, cartilage and skin. It is known from the study of human disease and from other sources that these forms are produced by distinct, but similar genes which together form a large gene family.

Table 5.2 lists five major types of human collagens, and indicates the diversity found in such an apparently simple molecule. During the evolution of animals, the repertoire of collagen types has obviously blossomed. Much duplication of collagen genes has occurred during animal evolution, and many of these genes have evolved new and complementary functions.

Table 5.2   Types of collagens in humans (and many other vertebrates).

Type	Distinctive chemical or structural features	Tissue distribution
I	low hydroxylysine* content	skin, tendon, bone, dentine, connective tissue
II	high content of hydroxylysine*, heavily glycosylated, fibre thinner than type I	cartilage, notochord
III	high content of hydroxyproline, low content of hydroxylysine*, little glycosylation; inter-chain disulfide linkage between chains at carboxyl-end cysteines	skin, uterus, blood vessels
IV	very high hydroxylysine* content, almost fully glycosylated, rich in hydroxyproline, low alanine content	kidney tubules, lens capsule, basement membrane of epithelial and endothelial cells
V	high content of hydroxylysine*, heavily glycosylated, low alanine content	basement membrane of smooth and striated muscle cells

* Hydroxylysine is common among the 'other amino acids' in Figure 5.5 and (like hydroxyproline) is formed post-translationally, by hydroxylation of lysine residues. The hydroxylysine hydroxyl group forms a covalent bond to sugars, which is why a high hydroxylysine content is associated with heavy glycosylation.

5.4.2 CRYSTALLINS AND LENSES

A transparent lens is essential for all but the simplest eyes. Here, we are mostly concerned with the biochemical basis of transparency and the capacity to refract (bend) light.

Vertebrate lenses are derived from modified epithelial cells that contain soluble proteins, known as **crystallins**, in high concentrations. The transparency of crystallins arises from specific packing of the proteins within the cell. The proteins spontaneously (or sometimes aided by chaperone proteins) assemble into a crystal array that allows light of visible wavelengths to pass through with minimal scattering. Ion pumps in the cell membranes maintain the internal concentrations of ions that keep the protein molecules at the appropriate spacing.

○ Why do animal lenses become milky in appearance several hours after the animal has died? (You may have observed this phenomenon in fish.)

● Maintaining the concentrations of ions necessary for the correct (crystalline) packing of crystallins depends upon the ion pumps in the cell membranes. After death, the ion pumps can no longer be supplied with energy, so ion concentration gradients cannot be generated, and the crystalline packing of the crystallins becomes disordered, resulting in increasing amounts of light scattering in the lens, causing the lens to appear cloudy.

A number of crystallin genes from a variety of species have been cloned and sequenced, and their study reveals a lot about their evolutionary history.

Crystallins in vertebrates fall into two groups on the basis of their sequence (Table 5.3). The first group contains both the α (alpha)- and the βγ (beta-gamma)-crystallins, which are specialized lens proteins. The genes for α-crystallins are structurally related to those for heat-shock proteins, originally characterized as proteins expressed when an organism is subjected to a heat shock (Walker and Swithenby, 2001). Heat-shock proteins, a group of chaperones, can be expressed ubiquitously in all kinds of cells, and appear to have a function in maintaining correct protein conformation. Presumably during the evolution of vertebrates, gene duplication generated extra copies of the genes for some of the heat-shock proteins which, relieved of the selective pressure to protect against protein denaturation caused by heat and other forms of shock, evolved the properties of lens crystallins. The genes encoding βγ-crystallins appear to be related to those encoding certain proteins that form the coats of bacterial spores. It is unclear how and when the gene moved from bacteria to vertebrates, and how it came to be expressed specifically in the eye. The preliminary analysis of the human genome sequence identifies scores of genes whose sequences closely resemble those of bacteria, from which they may have originated in the distant past.

The genes for a second group of crystallins, the non-βγ- and non-α-crystallins, present a rather different story. These proteins are not specific to the eye, and in fact appear as a wide variety of enzymes used elsewhere in the organism for various biochemical processes. For example, the δ (delta)-crystallin found in reptiles and most birds is identical in sequence to a liver enzyme, arginosuccinate lyase. It only assembles into a transparent array in cells where it is present at a very high concentration (and with few other proteins present) and in the appropriate ionic environment. In some cases, a crystallin may be encoded by the same gene as the functional enzyme; in others, the genes are closely related members of the same gene family. Table 5.3 lists several known cases of enzymes that are also known to function as a crystallin. The case where one gene encodes a single polypeptide that is utilized in these two very different biological processes — in this case, catalysis and lens function — is referred to as **gene sharing**, and can be seen as a specific evolutionary strategy. The acquisition of gene sharing might provide selective pressure for gene duplication and subsequent divergence of the duplicated sequences.

A further twist to the crystallin tale is that several very different enzymes can form crystallins in the lenses of living vertebrates (Table 5.3). In other words, there is no correlation between the enzyme's particular catalytic function and its role as a crystallin. Squid eyes are not only anatomically convergent with fish eyes (Section 1.4.1), they are also biochemically convergent. Crystallins in the lens of these molluscs consist entirely of glutathione S-transferase (an enzyme that helps in detoxification of reactive oxygen species in both plants and animals). The optical properties of squid crystallin are similar to those of vertebrate crystallins, despite the fact that the two transparent proteins are derived from different ancestral proteins and hence involve different genes.

Table 5.3 Proteins recruited as crystallins. Adapted from Wistow (1993).

Crystallin group	Species in which it occurs	Taxonomic status	Related or identical protein (and its metabolic role)
Ubiquitous crystallins			
α (alpha)	all vertebrates so far studied		identical to small heat-shock proteins
β (beta)- γ (gamma)	all vertebrates so far studied		related to *Myxococcus xanthus* (a bacterium) protein S (spore coat protein); *Physarum polycephalum* (a slime mould) spherulin 3a (a cytoplasmic structural protein)
Taxon-specific enzyme crystallins			
δ (delta)	reptiles most birds	Reptilia Aves	identical to arginosuccinate lyase (amino acid catabolism)
ε (epsilon)	crocodiles some birds	Reptilia Aves	identical to lactate dehydrogenase (anaerobic glycolysis)
ζ (zeta)	guinea pig, camel, llama	Mammalia, Eutheria	identical to NADP.2H:quinone oxidoreductase (respiratory chain)
η (eta)	elephant, shrews	Mammalia, Eutheria	identical to aldehyde dehydrogenase I (oxidation of acetaldehyde to acetate)
λ (lambda)	rabbits, hares	Mammalia, Lagomorpha	related to hydroxyacyl CoA dehydrogenase (fatty acid catabolism)
μ (mu)	kangaroos, quoll	Mammalia, Metatheria	related to ornithine cyclodeaminase (amino acid catabolism)
ρ (rho)	frogs	Amphibia	related to NADP.2H-dependent reductases
τ (tau)	lamprey, turtle	Vertebrata	identical to α-enolase (glycolytic pathway)
S	squid	Cephalopoda	related to glutathione S-transferase (destroys ROS)
Ω (omega)	octopus	Cephalopoda	related to aldehyde dehydrogenase

There appears to be little constraint on what gene product may be recruited as a crystallin — a protein that needs, on the face of it, a specific set of properties. Furthermore, enzymes appear to be predisposed to being recruited as crystallins. It may be that the robust regulation of enzyme synthesis lends itself to production of large quantities of protein in a restricted location. The utilization of enzymes as crystallins by both cephalopods and vertebrates suggests that there is some common basis. The evolutionary implications of such a strategy are interesting, since selective pressure favouring enzymatic activity might be deleterious to the protein's function as a crystallin, or vice versa.

5.4.3 VISUAL PIGMENTS AND PHOTORECEPTION

Key molecules in all eyes are the visual pigments (photopigments), the molecules that absorb the light. The retina of vertebrate eyes contains two types of photoreceptors, rods and cones, in proportions appropriate to the lifestyle and typical environment of the animal (Section 4.6.2).

○ How do rods and cones differ with respect to the visual pigments they contain?

● Rods contain a single photopigment, rhodopsin, while cones contain two or three pigments, depending on the species.

All visual pigments consist of a lipid chromophore bound to a protein called opsin (Section 4.6.2). Comparative sequence analysis shows that the opsin gene family appears to have evolved by a series of gene duplications.

Opsin evolution in vertebrates has been studied in detail, following the molecular cloning of the genes that encode the polypeptide, and found to result from several instances of gene duplication followed by sequence divergence. Figure 5.6 shows how the sequences of fish (a teleost, the Mexican cave fish *Astanyx fasciatus*) and human red- and green-sensitive opsins are related. In this diagram, we are comparing two types of photopigment from two species. The fish from which these data were obtained has two green-sensitive photopigments.

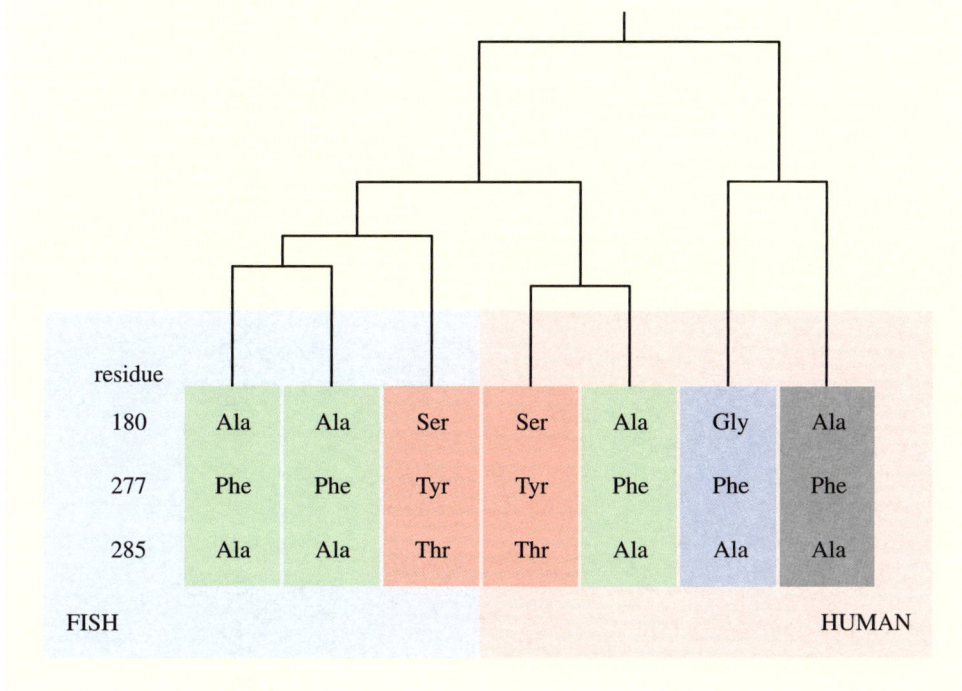

Figure 5.6 Sequence relationship between opsins of a teleost fish (the Mexican cave fish *Astanyx fasciatus*) and humans. In this diagram, key amino acid residues at positions 180, 277 and 285 in the opsin polypeptides are shown. These amino acid residues are known to be significant for determining the sensitivity of opsins to specific wavelengths of light (shown as correspondingly-coloured boxes) when combined with the chromophore. The grey box represents rhodopsin.
Ala = alanine; Ser = serine; Gly = glycine; Phe = phenylalanine; Tyr = tyrosine; Thr = threonine.

residue							
180	Ala	Ala	Ser	Ser	Ala	Gly	Ala
277	Phe	Phe	Tyr	Tyr	Phe	Phe	Phe
285	Ala	Ala	Thr	Thr	Ala	Ala	Ala

FISH   HUMAN

In Figure 5.6, the human red- and green-sensitive opsins are grouped together, separately from the fish red- and green-sensitive opsins, indicating that the divergence between the genes for the red- and green-sensitive opsins occurred after the divergences of the two lineages leading to humans and fish. The gene duplication that produced the two opsins must have happened independently within the two lineages after their divergence. Note that a second duplication event has resulted in the evolution of two green-sensitive opsins in fish. While the red and green-sensitive opsins originated from a gene duplication event after the last common ancestor of fish and humans, we can also see from Figure 5.6 that the origin of the human blue-sensitive opsin lies in a much earlier duplication. Interestingly, when the amino acid residues known to determine the absorbance specificity of these opsins were examined, they proved to be the same. Three key amino acid residues, at positions 180, 277 and 285, are shown: all the green-absorbing opsins have the same residues at these positions, while all the red-absorbing opsins have another set of residues at the same positions.

○ How can we reconcile the similarity of these particular amino acid residues with the evolutionary relationships between the opsins, in view of the *differences* in the rest of the protein sequences?

● The correspondences of these amino acids probably reflect the convergent evolution of short but functionally important sequences, rather than common ancestry of the genes.

Similar analyses have shown that the divergence between the red- and green-sensitive opsins of primates occurred around 40 Ma ago.

The case of the evolution of visual pigments shows how new genes can arise by gene duplication followed by divergence and convergence of sequence and function.

We can see some of the molecular mechanisms that have given rise to the range of opsins in vertebrates in the way in which colour-vision deficiency * in humans can arise. In humans, the gene that encodes the blue-sensitive opsin lies on chromosome 7, while those encoding the red-sensitive and green-sensitive opsins are located on the X chromosome, and have arisen following a relatively recent gene duplication event. In fact, humans with normal colour vision have from one to three adjacent copies of the green-sensitive opsin gene on the X chromosome.

During meiotic recombination, the two homologues may pair incorrectly, as shown in Figure 5.7. Recombination in the mispaired region can result in a number of alternative gene combinations, as shown in the figure. Such illegitimate recombination can give rise to the normal and defective phenotypes as shown, and to variable numbers of green-sensitive opsin genes.

○ What are the implications of the genes for red- and green-sensitive opsins being located on the X chromosome for the mode of inheritance of red–green colour-vision deficiency?

* Colour-vision deficiency is also popularly known as 'colour-blindness'.

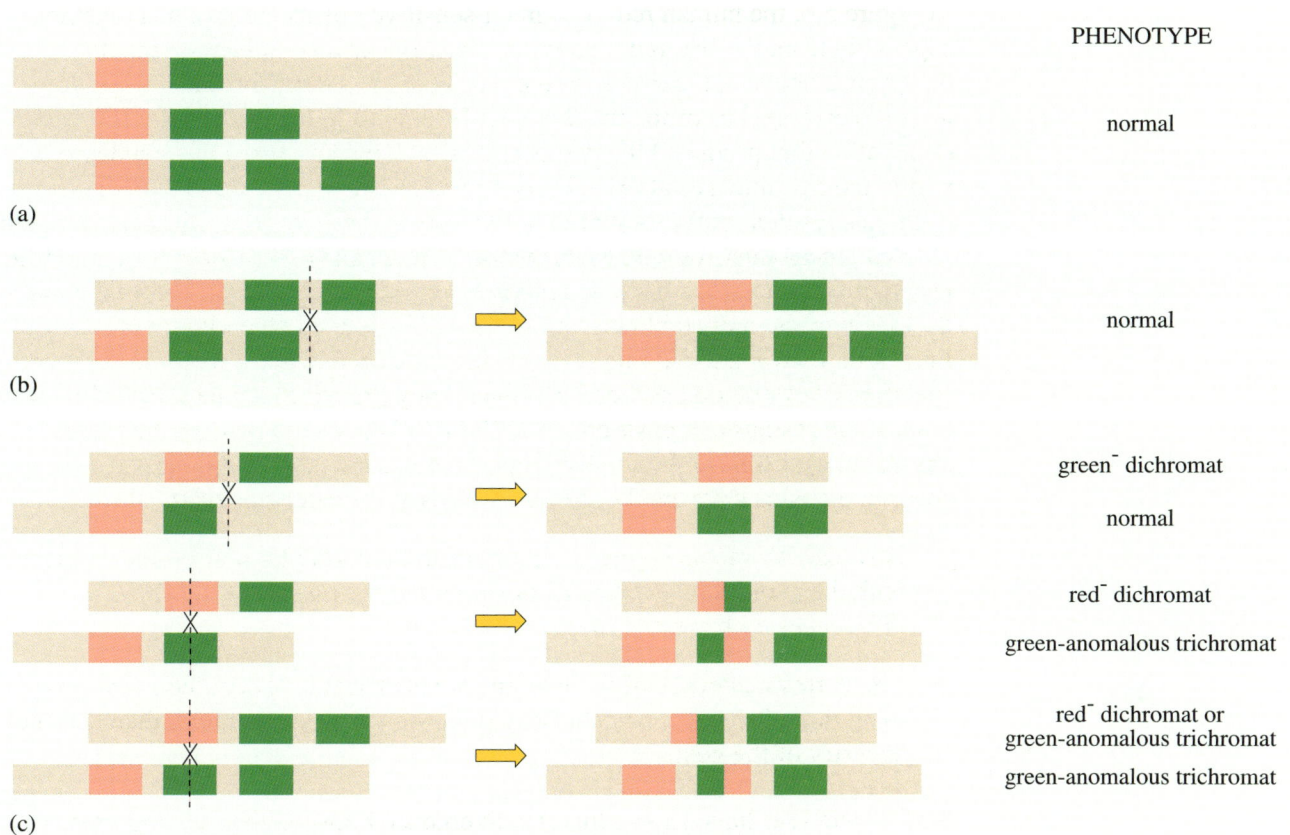

PHENOTYPE

Figure 5.7   Rearrangements of the genes encoding the human red- and green-sensitive opsins (shown as red and green bars, respectively). (a) Three arrays of human red- and green-sensitive opsin genes found in people with normal colour vision. Note that different chromosomes have a variable number of green-sensitive opsin genes. (b) The consequence of abnormal meiotic recombination between two chromosomes carrying two copies of the green-sensitive opsin gene, yielding one gamete in which the array has been shortened to one copy, and the other in which the array has been lengthened to three copies of the green-sensitive opsin gene. Variable numbers of opsin gene copies can arise by this mechanism. (c) Similar recombination events yielding gametes with various opsin gene arrays (some of which occur in people who have functionally normal colour vision, and some of which cause colour-vision deficiency). In the second and third examples, recombination occurs *between* a red- and a green-sensitive opsin gene — crossing over at this point is possible because of the high sequence similarity between the two genes, and their physical proximity in the genome. The terms denoting the various phenotypes are explained in Section 4.6.2. 'Trichromat' means there are *three* functional photopigments (blue-, red- and green-sensitive), conferring normal colour vision; 'dichromat' means there are *two* — in this case, blue- and red-sensitive, or blue- and green-sensitive. (The blue-sensitive photopigment gene is on another chromosome.) Thus 'green⁻ dichromat' has no green-sensitive photopigment gene but still has red and blue; 'red⁻ dichromat' has no red-sensitive photopigment gene but still has green and blue; 'green-anomalous' means that recombination between a red- and a green-sensitive photopigment gene forms hybrid genes which produce phenotypes with unpredictable properties, depending on where the recombination event occurred.

● Because these genes are located on the X chromosome, the pattern of inheritance is **sex-linked**. Defective phenotypes are therefore far more common among males.

5.4.4 THE EVOLUTION OF LACTATION

The evolution of mammary glands and the significance of lactation for mammalian reproductive biology, growth processes and ecology were discussed in Section 4.4.1. Darwin realized the importance of lactation for the success of mammals and their great diversity of habits and habitats. He envisaged four stages in the evolution of lactation:

1   increased proximity between mother and egg;
2   secretion from maternal glands helps to support the young;
3   offspring utilize maternal secretion as food source;
4   a gradual improvement in nutritional composition of the secretion and the means of its delivery to the offspring.

Modern molecular biology enables us to refine our understanding of exactly what genetic changes were necessary for the evolution of the biochemical pathways that synthesize milk.

○   What component of milk would require new enzymes for its synthesis?

●   Lactose, which is synthesized only in the mammary glands.

Lactose is not produced in significant quantities in any other animal (or plant) tissue. As a disaccharide, lactose is much more soluble than polysaccharides such as glycogen, but being a larger molecule, it exerts a smaller osmotic pressure in solution than a similar concentration of a monosaccharide such as glucose. The constituents of lactose, glucose and galactose, are widespread, but a specific enzyme, **lactose synthase**, is required to join them together to form the disaccharide. This new enzyme presumably evolved along with the evolution of the many other metabolic and structural features associated with lactation (described in Section 4.5.1).

○   What biochemical processes involve attaching simple sugars such as galactose to larger molecules such as proteins? Where do such processes occur?

●   Glycosylation covalently attaches chains of sugar residues to proteins. This process occurs in the Golgi apparatus of cells.

Glycosylation is very widespread in eukaryotes and is catalysed by various enzymes, including **galactosyltransferase** which transfers galactose units from UDP (uridine diphosphate)-galactose to *N*-acetylglucosamine linked to proteins, as shown in Figure 5.8a. This enzyme can also attach galactose to glucose *in vitro* but does so only very slowly unless the concentration of glucose is about 500 times higher than ever normally occurs in mammalian tissues.

○   Is it physiologically possible to have such a high concentration of glucose around metabolically active animal cells?

●   No. Too much glucose upsets the delicate balance of energy metabolism and ATP production. Also the osmotic effect of such a high concentration of solute in the tissues would be lethal!

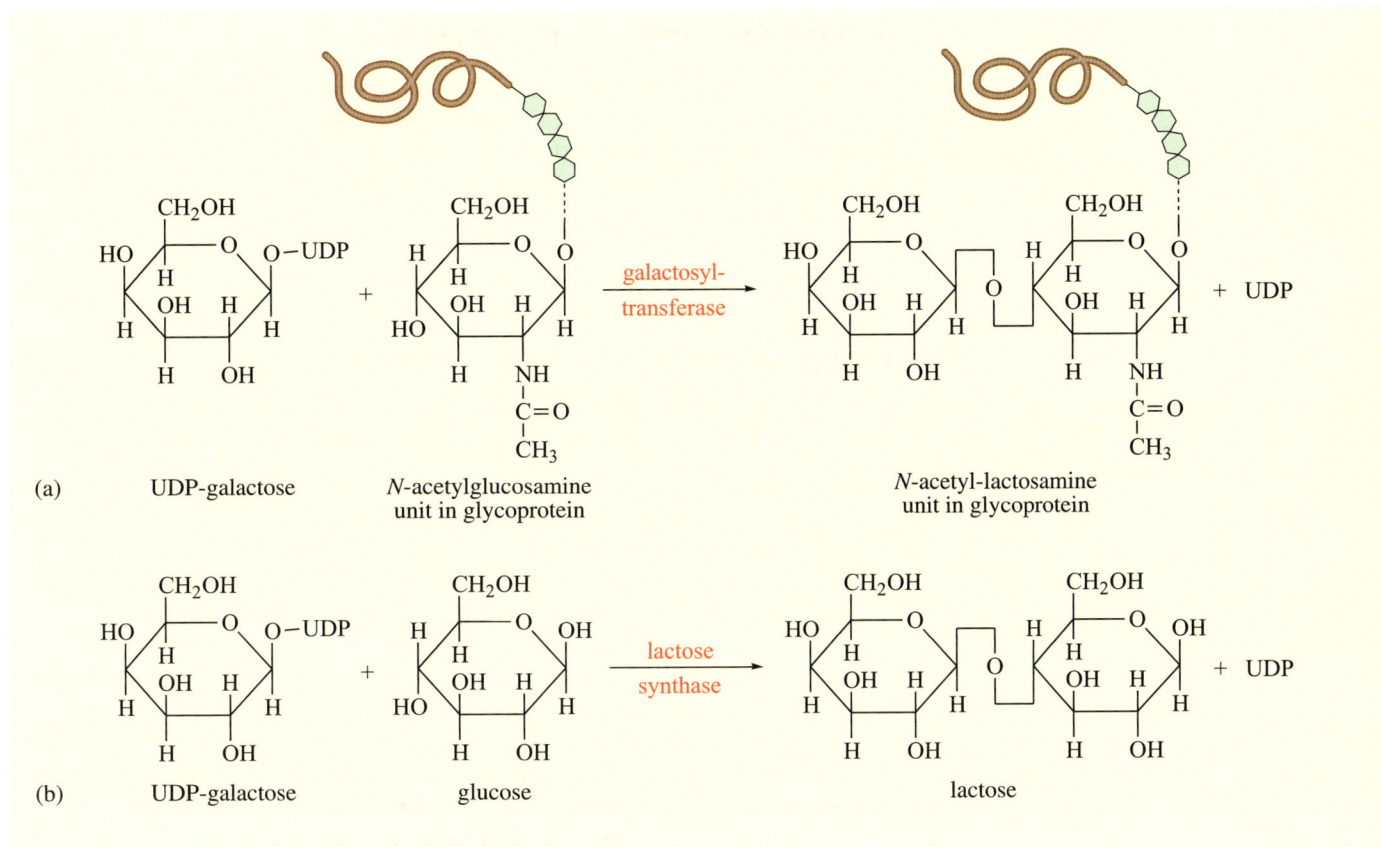

(a)   UDP-galactose   N-acetylglucosamine   N-acetyl-lactosamine
                        unit in glycoprotein   unit in glycoprotein

(b)   UDP-galactose   glucose   lactose

Under the influence of insulin, the concentration of glucose in the blood is tightly regulated and never changes by more than 10-fold (usually less in mammals).

Detailed analysis of lactose synthase extracted from lactating mammary glands show that the enzyme is composed of two subunits: one is identical to galactosyltransferase and the other is a protein with no independent enzymatic activity, called **α-lactalbumin**. When complexed with galactosyltransferase, α-lactalbumin alters the enzyme's substrate binding specificity, so that its affinity for glucose is increased over 1000-fold and its ability to add galactose units to glycosylated proteins is inhibited. Thus coupling these two components together creates a mechanism for a new biochemical reaction, the formation of lactose (shown in Figure 5.8b), which was essential for the evolution of lactation.

○ What other changes affecting the genes for lactose synthase are necessary for them to contribute efficiently to milk synthesis?

● The genes must be expressed in mammary glands, but only in response to the hormones and other cellular signals produced at the end of pregnancy and during lactation.

One such hormone is prolactin, which was so named because it was first identified in the blood of lactating mammals. However, prolactin is also secreted by the pituitary gland of the brain of most vertebrates, indicating that it is in fact much more ancient than the origin of lactation. Prolactin controls various aspects

Figure 5.8 Lactose synthesis in mammary glands is similar to the widespread and very ancient process of protein glycosylation. (a) The reaction catalysed by galactosyltransferase, in which a molecule of galactose is transferred from UDP-galactose to protein-linked N-acetylglucosamine to form an N-acetyl-lactosamine residue on a glycoprotein. (b) The same enzyme in conjunction with the protein α-lactalbumin has a related, but distinct activity: galactose is covalently linked to free glucose to form the disaccharide lactose, which is secreted in milk. The protein complex that catalyses this reaction is therefore called lactose synthase.

of growth and reproductive biology, including parental care in higher teleost fishes, and egg-brooding behaviour in birds. Presumably it was coopted into regulating milk-producing enzymes and their genes in mammals when the lactation habit evolved.

Cloning and sequencing the genes encoding the α-lactalbumin of various species of mammals revealed a significant similarity of these genes to those for certain kinds of lysozymes — ubiquitous antibacterial proteins which are also secreted in milk. Lysozymes may confer some protection against bacterial pathogens in the mouth and gut of the otherwise vulnerable neonate, and/or may be digested and their amino acids used for synthesizing other proteins, as in the case of any other proteinaceous food.

○ Why is it not surprising that lysozyme has an affinity for oligosaccharides and catalyses their hydrolysis to monosaccharides?

● Because the enzyme binds to peptidoglycans in the bacterial cell wall and splits glycosidic bonds between *N*-acetylglucosamine units (which causes fatal punctures).

Sequence comparisons of lysozymes from a great variety of organisms reveal that the genes that produce them fall into several groups, of which the most important in this context are the calcium-binding *c*-lysozymes (the '*c*' stands for chicken — this class of lysozymes was first isolated from bird eggs). These antibacterial proteins are found in all vertebrates, especially in the skin and its secretions. About 40% of the amino acid sequence of α-lactalbumin is identical (and the rest very similar) to that of Ca^{2+}-binding *c*-lysozymes.

Sequence analysis studies suggest that genes for these different forms of lysozyme arose by gene duplication. At least some such events are very ancient: current estimates indicate that the duplication that gave rise to the Ca^{2+}-binding and non-Ca^{2+}-binding *c*-lysozymes happened before the divergence of the lineages leading to present-day fish and mammals.

Incorporation of the sequence of α-lactalbumin into this tree suggests it is more closely related to the Ca^{2+}-binding *c*-lysozymes than to other kinds of *c*-lysozyme (see Figure 5.9). The evolutionary appearance of α-lactalbumin must have arisen from a subsequent gene duplication event which eventually allowed for the evolution of specialized *c*-lysozymes such as Ca^{2+}-binding *c*-lysozymes.

Because α-lactalbumin is specific to milk production, and is present in nearly all mammals tested, the gene duplication that provided the raw material for the evolution of the gene encoding this protein must pre-date the evolution of mammals. Statistical comparison of the DNA sequences of Ca^{2+}-binding *c*-lysozymes from several different kinds of vertebrates and α-lactalbumins indicates that this gene duplication event must pre-date the last common ancestor of mammals and birds, about 300 Ma ago.

○ How would you investigate any intermediate stages in the evolution of α-lactalbumin from lysozymes?

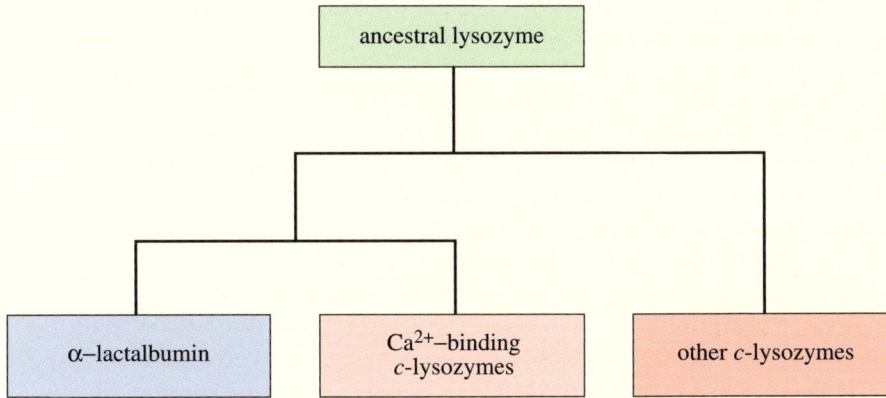

Figure 5.9 The relationship between c-lysozymes and α-lactalbumin. Sequence analysis of α-lactalbumin indicates an evolutionary relationship between it and c-lysozymes, particularly Ca^{2+}-binding c-lysozymes. By calculating the divergence from several lysozyme sequences, a tree of descent can be constructed, as shown here.

● Since it is impossible to study genes in organisms that have long been extinct, the next best strategy is to examine the genes for lysozymes and closely related proteins in the most primitive living mammals.

The most primitive living mammals are the monotremes (subclass Prototheria), egg-laying mammals found only in Australia and neighbouring New Guinea; there are only three species, the duck-billed platypus and two species of echidna (also called spiny anteaters). The gross structure of their mammary glands is somewhat different from that of metatherian (marsupial) or eutherian (placental) mammals, but lactation is copious and the hatchlings grow fast on the rich milk. Echidna milk is low in lactose and the mammary glands do not contain α-lactalbumin but they secrete a lysozyme that has both antibacterial properties (i.e. breaks glycosidic bonds) and weak activity in the lactose synthase system (i.e. forms glycosidic bonds). These observations suggest that the milk secreted by the earliest mammary glands included a lysozyme, and that α-lactalbumin evolved subsequently, making lactose synthesis more efficient.

So it seems that the synthesis of lactose in large quantities became possible by combining galactosyltransferase with a modified form of another very ancient protein, lysozyme. This evolutionary acquisition of a new synthetic pathway is of interest because it involves the recruitment of an antibacterial protein to act within a complex with a pre-existing enzyme to change its catalytic specificity.

The main milk protein, **casein**, is highly soluble, with a loosely packed, flocculent structure that resembles some of the proteins involved in blood clotting. Casein-coding genes may have arisen from those that produce this large family of proteins by processes similar to those described for lysozyme. As well as being easily secreted from the mother's glands, proteins with loose tertiary structure are more easily attacked by digestive enzymes than insoluble, highly ordered molecules. Mammalian neonates can digest the proteins and absorb the resulting amino acids quickly and thoroughly.

SUMMARY OF SECTION 5.4

1   Vertebrates have many different collagens, with a wide range of structural roles. The genes that encode the collagens arose by duplication and subsequent diversification, especially in the vertebrate lineage.

2   Crystallins are proteins that form transparent crystalline arrays. Most of the many different crystallins are derived from enzymes that have a variety of different catalytic roles.

3   The power of gene recombination to produce proteins with novel characteristics is illustrated by the visual pigments. Abnormal recombination of opsin genes produces several different syndromes of colour-vision deficiency. Gene duplication followed by divergence can produce photopigments that are sensitive to novel ranges of wavelengths (i.e. colours).

4   The ability to synthesize lactose is fundamental to lactation in mammals and depends upon the evolution of α-lactalbumin, which alters the substrate specificity of the enzyme galactosyltransferase; α-lactalbumin appears to have arisen from lysozyme by gene duplication and functional divergence in mammalian ancestors.

5.5 CONTROL GENES

During the development of a multicellular animal from the single-celled zygote, a great many events unfold in strict temporal sequence. Examples include the establishment of the major axes of the organism, such as the anterior–posterior and dorsal–ventral axes, the subdivision of the body into segments, and the direction of each of the segments to specific developmental fates. These processes are determined by **control genes**. Rather than encoding proteins that are utilized directly in the body structures or in metabolism (i.e. enzymes), control genes encode proteins that regulate the expression of other genes. One such protein, Pax-6, is discussed below.

○   What biological property would you expect a protein encoded by a control gene to have?

●   Proteins encoded by control genes act as transcription factors, which bind to the promoter regions of other genes and determine whether or not that gene is expressed in a given cell type.

Clearly, animal development is complex, and as a consequence, control genes act as part of a hierarchy of interacting transcription factors. Furthermore, a control gene may regulate more than one developmental process. A mutation of a control gene can fundamentally alter body structure, as we shall see in Section 5.6.

What information about the evolution of diversity can we gain from the study of control genes? The analysis of natural and artificially induced mutations of control genes is the main route by which their normal function in development is determined. As we shall see, the defects observed in laboratory animals where normal gene function is experimentally impaired can often mimic the normal development of other species. Finally, techniques of genetic manipulation of model organisms are sufficiently advanced that the experimenter can transfer genes from one species to another, and create specific sequence changes. In other words 'experimental' evolutionary biology is now possible, in contrast to the traditional approach of 'observational' evolutionary biology.

5.5.1 *PAX-6* AND THE EVOLUTION AND DEVELOPMENT OF EYES

A basic eye consists of little more than a patch of photosensitive tissue, with the appropriate connections to the animal's nervous system. Many animals have various modifications to this 'minimal' eye, including morphological improvements to enhance light capture and image formation. These features include lenses tailored for specific environments and photopigments adapted to be responsive to particular wavelengths (colours) of light (as discussed in Sections 4.6.2 and 5.4.3).

In the animal kingdom, there is a wide variety of kinds of eye, from simple photoreceptors to the compound eyes of arthropods such as insects to the elaborate eyes of vertebrates. Representative examples of eye structure, drawn from several contrasting phyla, are shown in Figure 5.10. The structural diversity of eyes, together with the distribution of eye patterns amongst the phyla, had given rise to the idea that eyes had arisen independently many times during evolution. In other words, eyes are polyphyletic, with some authors considering eyes to have evolved some 40–60 times since the appearance of animal life. The alternative viewpoint, that the eye had a unique origin (i.e. monophyletic) was held to be unlikely.

Eyes are often remarkably elaborate structures. The types of eye shown in Figure 5.10 include a number of structural features that collect and transduce light into electrical signals for interpretation by the brain, as described in Section 4.6.2. Eyes apparently require such an intricate set of morphological features that it is hard to see how they can evolve by natural selection. After all, how much use is a partially formed eye to the organism? How do such organs evolve? In the 19th century, explaining the evolution of the eye by means of Darwin's theory of evolution by natural selection presented a challenge to scientists. Indeed, it is the very example that many people, to this day, beginning with the 19th century theological writer William Paley, use to support the belief in divine design in preference to evolution by natural selection. As we shall see in this section, recent molecular genetic studies have shed some light on both the developmental processes that form the eye, and the evolution of eyes.

Can we relate the developmental processes that give rise to each type of eye? Here we discuss the insights into the evolution of eyes offered by comparative molecular biology.

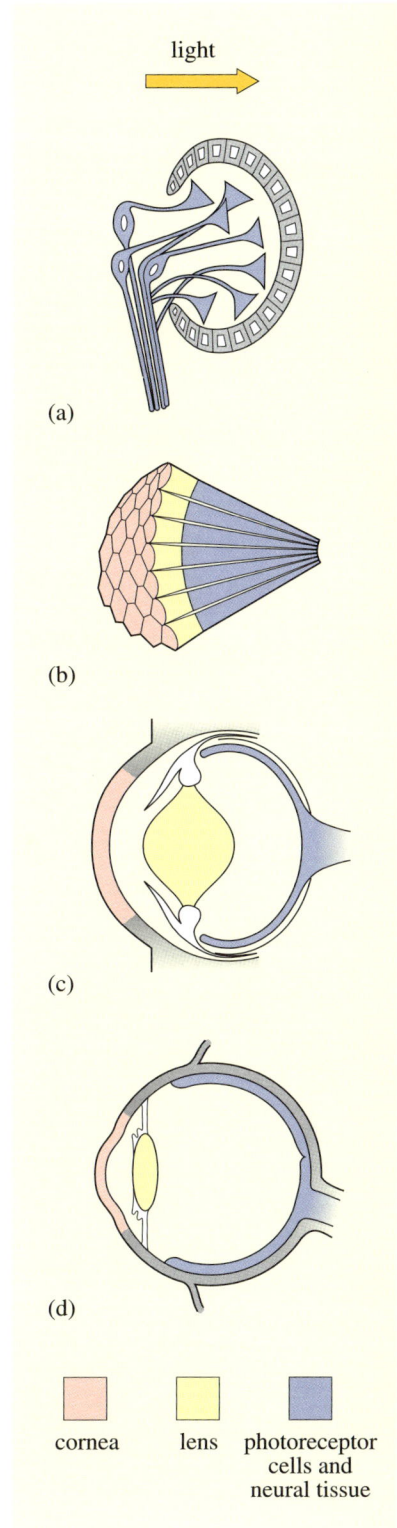

Figure 5.10 The different forms of eyes: (a) the primitive eye of planarians; (b) the insect compound eye; the simple eyes of (c) a cephalopod mollusc (squid) and (d) vertebrates. Note the structural similarity between squid and vertebrate eyes.

Pax-6 is a transcription factor which was first isolated from the mouse on the basis of sequence similarity with sections of the *paired* gene of the fruit-fly *Drosophila melanogaster*. (The *paired* gene has a role in the specification of segment identity during development.) The principle of cloning genes from one species on the basis of sequence similarity with another is described in Box 5.1.

Mouse *Pax-6* was identified by hybridization with the *Drosophila paired* gene sequence. *Pax-6* contains a *paired* domain and a so-called homeobox domain (the latter is explained in Section 5.6) The various groups of *Pax* genes are named for this feature: '*Pax*' is simply a contraction of '*Paired box*'. It is known that both the paired and homeobox domains of the Pax-6 protein bind to DNA, an appropriate property for a transcription factor.

Box 5.1   Techniques for cloning homologous genes

Hybridization with sequences from related species

Species regarded as closely related on morphological criteria usually have very similar gene sequences. We can take advantage of this sequence similarity to isolate homologous genes, using the method of nucleic acid hybridization. The mouse homologue of a human gene can generally be identified by hybridizing the human DNA sequence to mouse DNA, which is possible because the two genes are likely to be very similar in sequence. This approach becomes more difficult if the two species are more distantly related, because the sequences have diverged more and so hybridize less readily.

Amplification by PCR

DNA amplification by the polymerase chain reaction (PCR) requires two oligonucleotide primers, complementary to the DNA fragment to be amplified. PCR can be used to amplify a gene from a species from which we have no relevant sequence information.

Sequence conservation between two related species enables suitable primers to be designed using sequence information from another species. An example would be amplification of a gene from the rat where only the sequence of the mouse homologue is known (see Figure 5.11). Appropriate primers can be designed using the mouse sequence. Of course, this approach suffers similar problems to the hybridization approach above. Efficient primers are complementary to regions with high sequence conservation, such as those encoding protein domains important in function. A domain is a region of a protein with a defined biochemical role (e.g. DNA-binding, enzyme active site) or the DNA sequence that encodes it. A particular protein domain can often be recognized on the basis of amino acid sequence, and therefore of DNA sequence (here called a DNA domain). DNA domains are often conserved between many genes of related function both within a species and between related species.

○ What difficulty do you envisage in designing primer sequences from the amino acid sequence of the protein encoded by the gene?

● The redundancy of the genetic code means that the nucleotide sequence cannot be unambiguously predicted from the amino acid sequence.

Another point is that there is no way of locating intron/exon boundaries (splice junctions) from the amino acid sequence because introns are not translated. If the primer sequence spans a splice junction, it is unlikely to function in a PCR reaction.

Whole-genome sequence analysis

The genome sequences of an increasing number of species have been determined. Some examples are given in Table 5.1. These sequences are maintained in computer databases, which are publicly accessible through the internet. There are many computer programs that enable these databases to be searched for DNA sequences similar to a gene of interest. The availability of complete genome sequences for an increasing number of species, together with these powerful software tools, means that homologous sequences may be found in databases for a wide range of species.

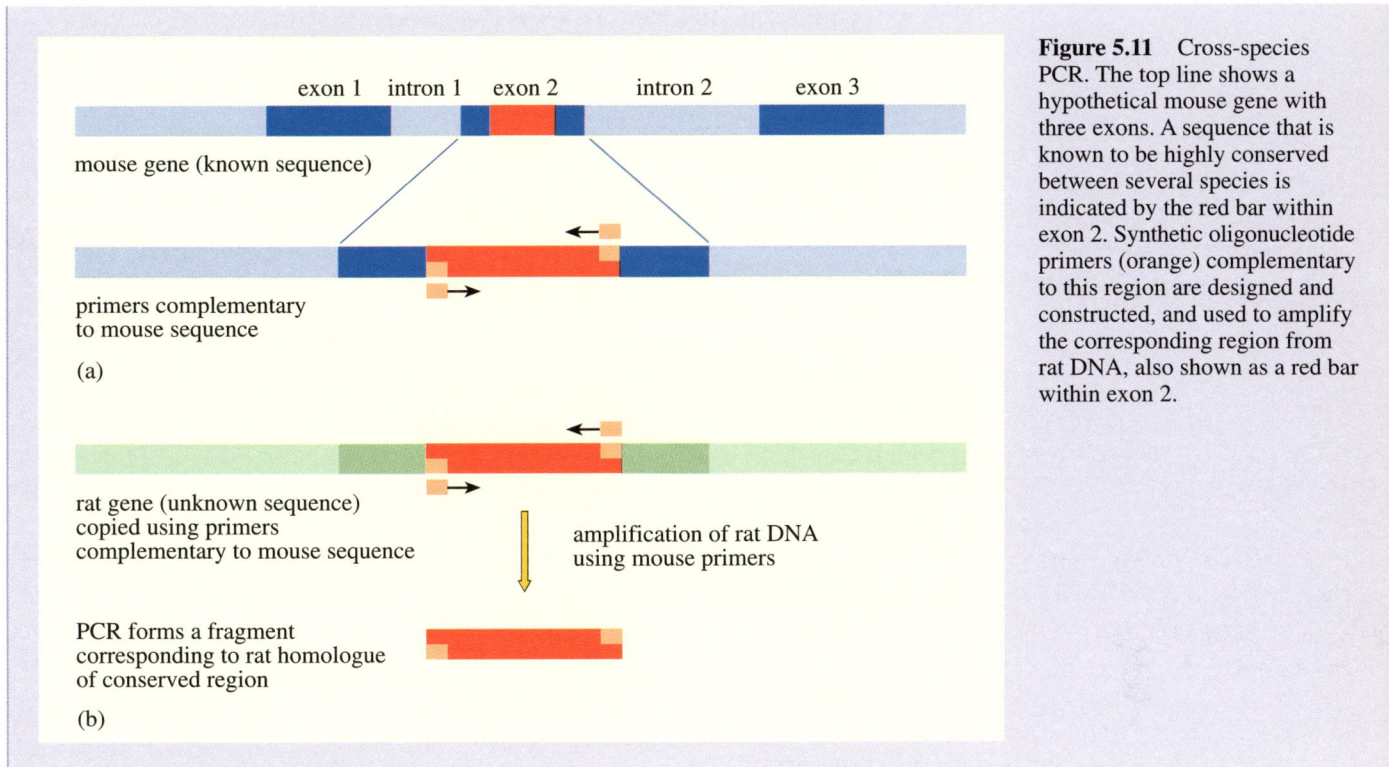

Figure 5.11 Cross-species PCR. The top line shows a hypothetical mouse gene with three exons. A sequence that is known to be highly conserved between several species is indicated by the red bar within exon 2. Synthetic oligonucleotide primers (orange) complementary to this region are designed and constructed, and used to amplify the corresponding region from rat DNA, also shown as a red bar within exon 2.

○ How might a role in DNA binding for the paired and homeobox domains of Pax-6 be demonstrated?

● By *in vitro* studies using a cell-free system, in which polypeptides expressed from cloned genes are used in a DNA-binding assay that measures binding to target DNA sequences corresponding to genes known or suspected to be regulated by Pax-6.

Further homologues of *Pax-6* were isolated from several other species, including human and zebrafish. Subsequently, mouse *Pax-6* was found to be the gene altered in the *small eye* mutant phenotype, in which eye development is abnormal. In an interesting correspondence, mutation of the human *Pax-6* gene is the basis for the genetic disorder aniridia, in which abnormal development forms eyes without the iris (this structure is shown in Figure 4.19).

The conclusions from these observations are that not only is the gene sequence conserved between these species but also their functions are similar: both genes encode transcription factors involved in the normal development of eyes. However, the picture is not quite as straightforward as just described, since these genes appear to be involved in other biological processes in addition to the initiation of eye development.

5.5.2 EXPERIMENTAL INVESTIGATION OF AN EVOLUTIONARY PROCESS

Subsequent to the discovery of *Pax-6* in vertebrates, the *Drosophila Pax-6* homologue was also cloned, and identified as corresponding to a gene known from mutational analysis as *eyeless* (abbreviated as *ey*: the *ey* mutant was first identified in 1915!). The obvious conclusion from these observations is that not only do development of the mouse and human eye require *Pax-6* activity, but that the same is true in insects. Furthermore, arthropod and vertebrate eyes may share an evolutionary origin and *Pax-6* is perhaps a universal master control gene for eye development. By using transgenic techniques, *Drosophila* bearing the *ey* gene controlled by a promoter that directs expression in a variety of different tissues other than those in which it is normally expressed, can be obtained. (Procedures by which transgenic *Drosophila* are produced are outlined in Box 5.2.) Such flies develop ectopic (i.e. out-of-place) eyes in sites corresponding to expression of the *ey* transgene. Figure 5.12 shows the head of such a transgenic *Drosophila* in which expression of *ey* has occurred in a group of cells that normally give rise to parts of the adult antenna, resulting in the development of a section of compound eye upon the antenna. The eye tissue clearly resembles the normal compound eye, complete with distinct ommatidia. Surprisingly, electrophysiological recordings indicate that these ectopic eyes are functional. Similar results were obtained when the mouse *Pax-6* gene was expressed in *Drosophila* in the same manner.

Box 5.2 MAKING TRANSGENIC *DROSOPHILA*

One technique widely used to investigate physiological functions of genes and for identifying homologous genes in different lineages of organisms is that of generating transgenic organisms, in which a gene of interest has been stably integrated into the genome. The integrated gene, or transgene, need not be derived from the same species; indeed, it may be derived from a completely unrelated species of plant or animal. As well as being used for purely laboratory-based investigations into gene function, transgenic plants and animals expressing transgenes of commercial interest are being developed by biotechnology companies. The analysis of *Pax-6/eyeless* function in *Drosophila* utilizes a well developed system for the introduction of DNA into essentially random locations in the genome, based on the *P* element, a naturally occurring transposable DNA element of *Drosophila* (Saffrey and Metcalfe, 2001).

The *P* element is used as a vector to insert segments of DNA, of any origin, into the *Drosophila* genome. The DNA of interest is cloned in a *P* element engineered for the purpose: the *P* element typically contains promoter sequences to direct expression of the transgene when incorporated into the fly genome, together with a reporter gene to allow recognition of flies bearing the transgene (the reporter gene may confer antibiotic resistance or modify eye colour) and the terminal repeats which are essential for integration. The gene encoding the enzyme that catalyses the mobilization of the wild-type *P* element is not contained within this element, but rather is supplied by a second DNA molecule, the helper element. The helper element expresses transposase, but cannot itself integrate into the genome as it lacks

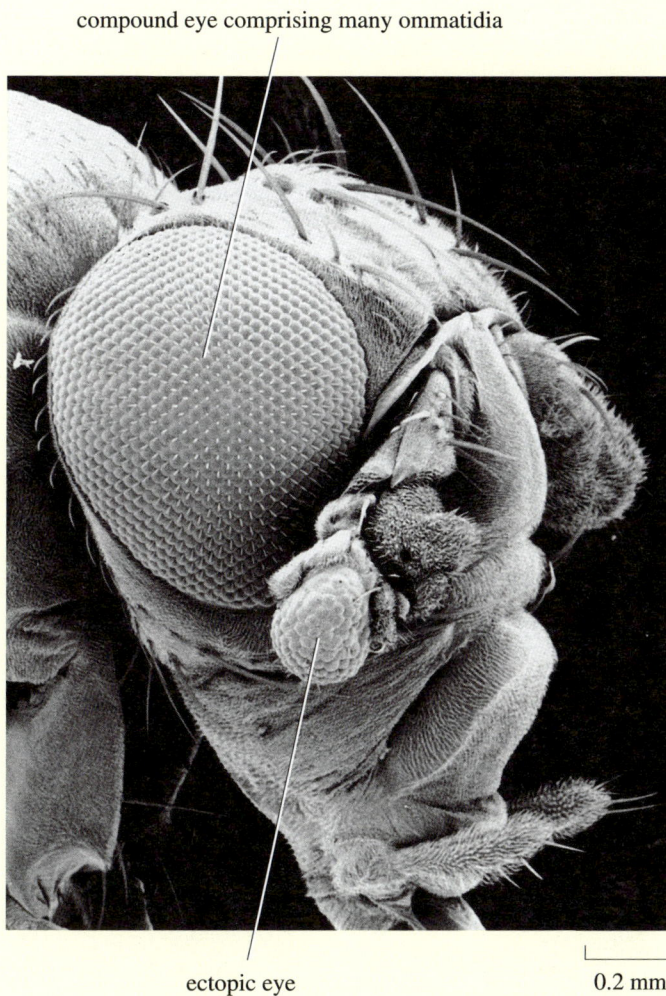

one of the terminal repeats. A mixture of the two DNAs is injected into the posterior pole of the developing embryo. A variety of *P* element vectors are available to the researcher, so the DNA fragment or gene under investigation may be placed under the control of any of a wide variety of promoters, to direct expression that is specific and tightly controlled in both time and place.

The procedures for creating transgenic *Drosophila* are simple and highly efficient. Their eggs are surprisingly large, about 0.5 mm long, are easy to obtain in large numbers and don't require complex equipment to manipulate. Furthermore, injected eggs develop easily in culture. Transgenic mammals are rather more difficult to make, partly because of the difficulty in obtaining and manipulating their eggs and because they must be implanted *in utero* to develop.

compound eye comprising many ommatidia

ectopic eye

0.2 mm

Figure 5.12 Scanning electron micrograph of the head of a *Drosophila* in which a transgenic copy of the *eyeless* (*ey*) gene has been ectopically expressed during development of the antenna. Note the growth of compound eye tissues on the antenna. In such cases, the ommatidia are found to be electrophysiologically functional, and to contain normal eye pigmentation.

The evolutionary relationships between *Pax-6* homologues from a number of species have been examined by aligning their sequences. The *Pax* gene family is large, and defined by many sequences, obtained from many different species of organisms. The family is divided into six classes on the basis of the presence or absence of specific domains. *Pax-6* is a member of *Pax* Class VI, and, as described above, contains both *paired* and homeobox domains, which are found to be particularly well conserved when alignments of *Pax-6* sequences are examined. For a number of these sequences, the position of the splice junctions are known, and they provide information about the ancestry of *Pax-6*.

○ How might splice junctions be mapped?

● By comparison of the sequence of genomic and cDNA sequences. cDNA clones are derived by reverse transcription of mRNA, and therefore lack introns. By alignment of cDNA and genomic sequences, the splice junctions can be accurately determined.

Aligning the four splice junctions of *Pax-6* that lie within the *paired* and homeobox domains shows that they are nearly all coincident.

The genetic analysis of eye development in *Drosophila* has identified, amongst other genes, a further homologue of *Pax-6*, known as *twin of eyeless* (abbreviated to *toy*), which is involved in regulation of *ey* activity. A network of interacting genes required for eye development in *Drosophila* has been identified, and it remains to be seen whether the features and interactions of this network are conserved in vertebrates.

5.5.3 HOW MANY TIMES DID EYES EVOLVE?

The extensive conservation of a transcription factor, Pax-6, and of one of its developmental roles in a wide range of species from contrasting phyla suggest that the evolution of eyes may have been a unique occurrence, and that the eyes of all extant species, no matter how morphologically diverse, share an evolutionary history. How then can we integrate the many kinds of eye structure within such a scheme?

One proposition is that eye evolution has occurred by an *intercalary* system: the incorporation of additional steps in the primitive pathway initiated by *Pax-6*, as illustrated in Figure 5.13. A prerequisite for this hypothesis is the presence of the genes that encode the protein component of the visual pigment rhodopsin (Section 5.4.3) and the transcription factor *Pax-6*. Once a prototype eye, perhaps a basic light-sensitive pit, has evolved using the genes encoding rhodopsin and Pax-6, selection can act to add further complexity to the structure. In this hypothetical process, rhodopsin lies at the end and *Pax-6* at the beginning of a genetic cascade, within which additional components are added during evolution. Additional gene functions are recruited between these end-points, perhaps following duplication of genes of appropriate function and subsequent divergence of regulation and function. According to this hypothesis, expression of *Pax-6* remains a requirement for the initiation of eye development, while the intervening components evolve separately in different lineages — for example, towards compound eyes in arthropods, or simple eyes in vertebrates and cephalopod molluscs.

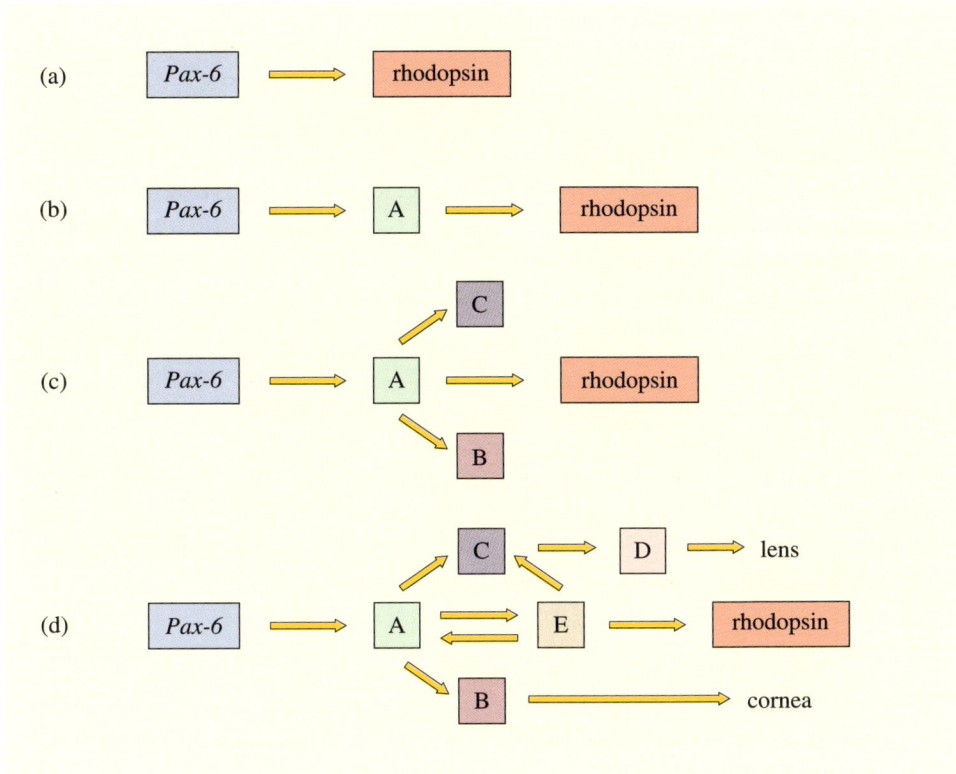

Figure 5.13 Diagrammatic representation of the intercalary hypothesis for the evolution of eyes: (a) a primitive eye results from a pathway consisting of the *Pax-6* gene product directing the formation of a simple eye which has cells expressing the photopigment rhodopsin; (b)–(d) various more complex eyes, can be derived by intercalating novel gene functions (A–E) which form auxiliary structures such as the lens and cornea. Note that this hypothesis can explain the evolution of very different types of eyes from one primitive form.

Is the conclusion that eye development is monophyletic justified? The conventional view, based upon the existence of varied kinds of eye within the animal kingdom, is that eyes have arisen many times during evolution. This hypothesis would suggest that the role of *Pax-6* in eye development in different phyla is related to Pax-6 function rather than evidence of a single evolutionary origin of eyes. Pax-6 is a transcription factor, with a number of different roles in normal development. It is possible that the requirement for a specific transcription factor activity was most easily filled by Pax-6 each time that eyes evolved.

There is a crucial difference between the evolution of crystallins (Section 5.4.2) and the evolution of Pax-6. In the case of crystallins, very *different* proteins have been recruited to serve the *same* optical function (transparency and capacity to refract light). In the case of Pax-6, the *same* protein is used for the *same* task in the control of development in widely disparate groups of animals.

5.5.4 SIGNALLING PATHWAYS AND THE DEVELOPMENT OF AIRWAYS

Most organs develop from simple structures, or primordia, composed of undifferentiated populations of cells which, through a process of proliferation and differentiation, eventually form the mature organ. A fundamental aspect of the development of animals is signalling between cells. Without the ability to send and receive signals, a cell can pick up no information about its location relative to other cells and tissues, information crucial to setting in train the correct programme of genetic activities required to form an organ. In this section, we will consider the role of signalling pathways important in the development of airways in both vertebrates and invertebrates, as exemplified by vertebrate alveoli and insect tracheae.

○ How do air-breathing vertebrates and terrestrial insects differ in their respiratory systems?

● Air-breathing vertebrates carry out gas exchange using internal lungs. Oxygen is carried from the lungs to the organs via the blood, inside the red blood cells which contains the oxygen-binding protein, haemoglobin (see Sections 4.3.2 and 5.2.1). In contrast, terrestrial insects deliver oxygen to cells and tissues through the tracheae (Section 2.1.2), which lead to the tracheoles, a system of fine branching tubes from the ends of which gases diffuse into and out of the cells.

Insects and vertebrates carry out gas exchange in very different ways. The vertebrate lung consists of a branching network of bronchioles, which eventually terminate in the alveoli, small sacs where gases are exchanged between the lung and the blood. Insects utilize a network of branching tracheae to pass oxygen to the organs. Can there be any similarity between the processes by which these two types of organs develop? A clue may be gleaned from Figure 5.14, which compares the development of tracheae and alveoli. In both cases, a progressive series of branching events take place as the structure develops.

We will again take *Drosophila* as an example of insect development. The tracheal system develops from small groups of ectodermal cells known as *placodes* on both sides of each segment of the embryo. The first step in the development of tracheae is the simple invagination of each placode to form a sac, which then undergoes a series of branching events (Figure 5.14a). Primary branches form at specific sites, and correspond to lead cells and stalk cells. Lead cells form the tips of the branches, and may later produce secondary branches.

Interestingly, tracheal development carries on without any further cell division after the establishment of the placode — all the branching and elongation of the developing tubes is solely a result of changes in cell shape and cell migration. In normal *Drosophila*, the pattern of branching and elongation is invariant, and evidently under strict control. The rate of branching in particular must be tightly regulated: too little branching and the system provides too low a surface area for sufficient gas exchange, but ubiquitous branching would result in a large bag-like structure that does not bring the gases in close enough contact with the cells for gas exchange to be efficient.

A number of genes that, when mutated, alter the development of the tracheal system have been identified in *Drosophila*. The detailed molecular analysis of these genes has revealed clues as to how the development of airways is regulated. In particular, *branchless* and *breathless* are mutations that cause the embryo to develop with unbranched sacs rather than normal tracheae.*

* Note that, in general, *Drosophila* genes (at least those identified through normal mutagenesis screens) are named for their mutant phenotype rather than the normal function of their product. Thus *white* is named for its mutant alleles' effect on eye pigmentation (white eyes instead of the normal brick-red colour), and the product encoded by the wild-type *white* gene is required for normal uptake of pigment into the eye.

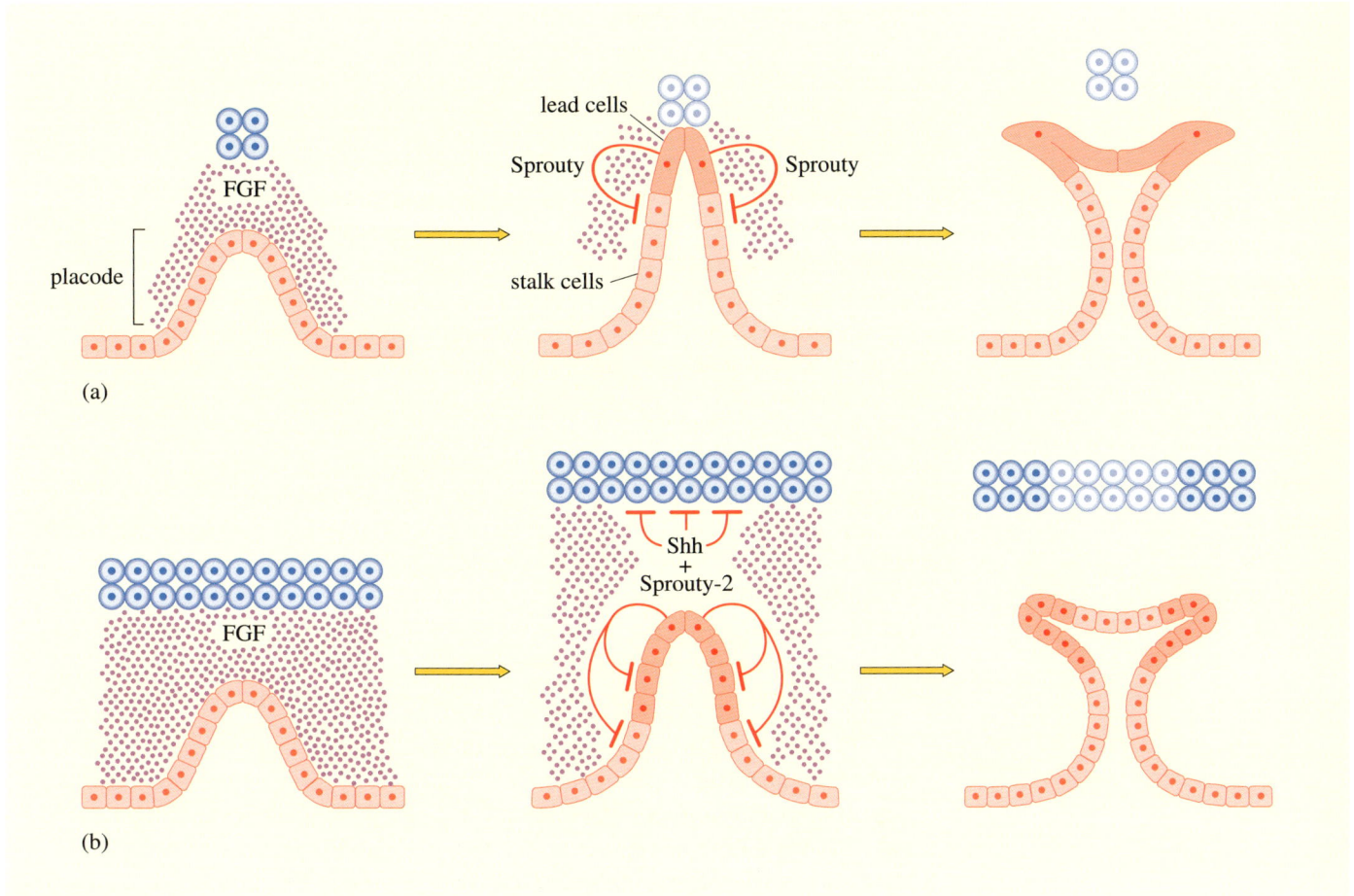

Figure 5.14 The branching patterns of insect tracheae and vertebrate alveoli develop in a similar manner and involve homologous proteins. (a) The formation of insect tracheae. The fibroblast growth factor (FGF) encoded by the gene *branchless* guides the migration of cells to form a primary branch. The lead cells secrete an inhibitor of FGF signalling, Sprouty, a protein encoded by the gene *sprouty,* which restricts further branching to those cells closest to the source of FGF, forming the simple pattern shown on the right. (b) The molecular controls of the development of alveoli in the mouse lung are similar to those shown in (a). The developing lung epithelium grows towards the source of FGF, and is induced to secrete both Shh (the product of the *sonic hedgehog* gene) and Sprouty. Shh curtails further secretion of FGF, thus promoting the formation of a secondary branch. The contrasts with (a) are that in the mouse, Shh locally inhibits secretion of FGF while, as in *Drosophila*, the product of the *sprouty-2* gene (Sprouty-2) restricts further branching. Adapted from Placzek and Skaer (1999).

○ What is the likely function of a gene, such as *branchless*, that when mutated, produces a phenotype that has abnormally reduced tracheal branching?

● It is likely that the normal function of the gene is to promote the branching of tracheae.

How is branching controlled in the development of vertebrate alveoli? An apparently surprising discovery was that in the mouse, this process is directed by gene products that are homologous to those that control tracheal development in *Drosophila*. This rather striking result bears careful consideration. As stated above, in the case of Pax-6 and normal eye development, opinion is divided as to whether the central role of the *Pax-6* reflects a common evolutionary origin of eyes, independent of morphological criteria. In the case of the development of lung alveoli and tracheae, can we conclude that both organs, with similar physiological functions, have a common origin?

The most likely explanation lies in the signalling pathways concerned. We have described above how important signalling is in development. What is striking, however, is the fact that components of one signalling cascade are repeatedly reused in other signalling pathways that control the formation of quite different tissues or biochemical pathways. Often the distinction between these pathways lies downstream of the signal, within the cell to which the signal is transmitted and causes a specific effect. It is most likely that the apparent similarity in the development of these two structures (alveoli and tracheae) lies in the recruitment of a pathway that promotes invagination and branching of tissues. In fact, all the factors described above are not expressed solely in the developing airways, and are used in many tissues and at many developmental stages.

SUMMARY OF SECTION 5.5

1   The discovery that homologues of *Pax-6* initiate eye development in widely different taxa has led to a revision of theories about the origin and evolution of eyes. The functional equivalence of these genes can be elegantly demonstrated by transgenic experiments. However, the ubiquity of *Pax-6* involvement in initiation of eye development is not necessarily an indication of monophyly.

2   Evolutionary biology can now be studied experimentally, as well as by the analysis of observations. The unravelling of the role of *Pax-6* in eye development by a combination of mutation and transgenic technology suggests solutions to important evolutionary problems.

3   The development of airways in insects and vertebrates is another case in which two structures that are of distinct developmental origin (but of related function) have similar organization and are formed as a result of the activities of homologous gene products.

5.6   HOX GENES AND THE EVOLUTION OF BODY PLANS

Section 5.2 showed how additional genes form and existing ones acquire new functions. We have also seen how genes that control the formation of certain groups of tissues are conserved in evolution. This section describes briefly what is known of the molecular basis of the changes in body plan that are seen to have occurred in evolution.

How does the development of *Drosophila* compare to that of other organisms, such as vertebrates? One can think of a segmental animal as consisting of a series of repeated units. In some animals, such as annelids (Section 1.2.2), many of the segments are functionally and morphologically very similar or identical. In the case of flies, there are significant morphological differences between segments (leg-bearing segments of the thorax are clearly different from abdominal segments, which bear no appendages). Figure 2.14 shows such segmental differences in the crayfish (a crustacean).

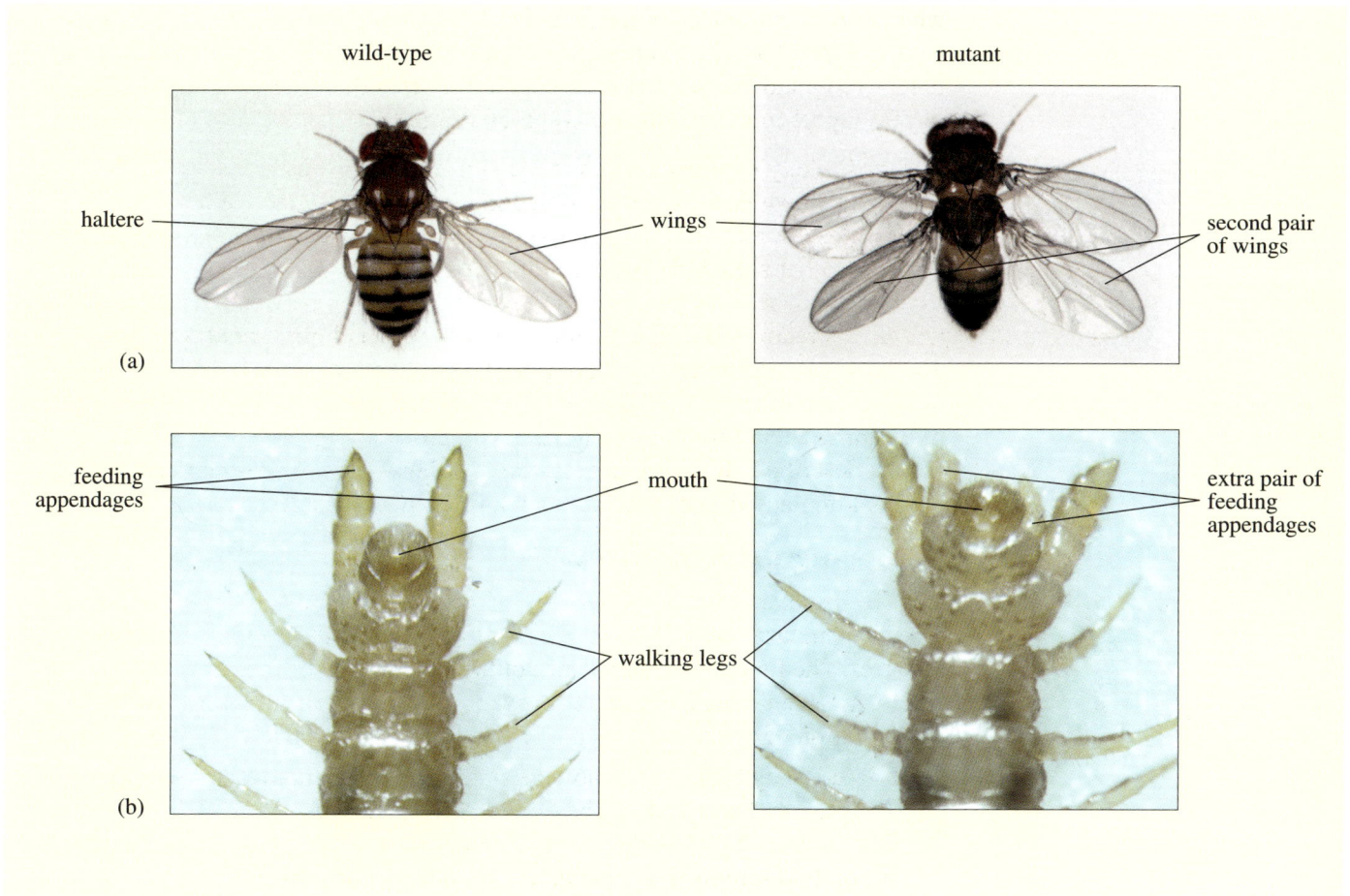

wild-type   mutant

haltere   wings   second pair
                                                   of wings

(a)

feeding   mouth   extra pair of
appendages   feeding
                                                   appendages

                        walking legs

(b)

As explained in Section 4.2, vertebrate dorsal tissues are also arranged segmentally. The segmentation in vertebrates is most apparent in fish, tailed amphibians (and tadpoles) and reptiles, and in the embryonic stages of birds and mammals. Certain genetic mutations alter the development of a particular segment in such a way that it acquires the morphological characteristics normally found in another segment. This change in segmental identity is known as **homeotic transformation**, and in fact the term was coined to describe segmental transformations seen in human vertebrae.

You can see two quite extreme examples of homeotic transformation in Figure 5.15. Figure 5.15a (right) shows a fruit-fly in which a combination of several mutations has caused the third segment of the thorax to develop structures characteristic of the second segment of the thorax. Normally, the third thoracic segment of dipterans does not bear wings, but rather two small balancing organs known as halteres, which are homologous to wings. In the fly pictured here, these organs have been transformed into fully developed wings. This kind of effect is not confined to flies: Figure 5.15b shows a centipede in which a homeotic transformation has caused an additional pair of feeding appendages to form.

Figure 5.15 Homeotic transformations in arthropods. (a) A normal fruit-fly (left) and a mutant, four-winged *bithorax* fly (right). The mutation changes the segmental identity of the third thoracic segment, which normally bears one pair of legs and one pair of halteres (balancing organs), to one corresponding to the second thoracic segment, which normally bears one pair of legs and one pair of wings. (b) Wild-type centipede (left) and mutant centipede with segmental transformation (right).

Extensive research on the development of *Drosophila* (Box 5.3) has identified the cascade of genetic and molecular interactions that together determine the body plans of animals. Consequently, much of this discussion of the molecular genetics of development is centred on this species. However, many genes identified in *Drosophila* have homologues in a wide variety of animals, including vertebrates. Indeed, the identification of such genes in *Drosophila* has driven the discovery of developmental genes in vertebrates.

BOX 5.3   OVERVIEW OF *DROSOPHILA* DEVELOPMENT

In order to discuss the molecular genetics of development, we must first briefly review the process of development in *Drosophila*, which are true flies, or Diptera — exopterygote insects with a grub-like larva.

Following fertilization, the *Drosophila* zygote undergoes about 10 very rapid cycles of nuclear division (about 10 minutes per cycle), without any accompanying division of the cytoplasm and membranes, to form a large cell containing many identical nuclei, i.e. a syncytium.

○ Would there be gene transcription during these rapid cycles of nuclear division?

● No. Gene transcription is not possible because each nuclear cycle requires accurate copying of the entire nuclear genome, condensation of chromosomes, assembly of the mitotic spindle, and segregation of the chromosomes. A 10-minute cycle leaves insufficient time for transcription to occur, and restricts access of the transcriptional apparatus to the DNA.

Gene products required for biological processes in early embryogenesis are therefore derived from the mother. For example, during the early stages of development, the anterior–posterior and dorsal–ventral axes of the embryo are specified by the activity of maternally provided proteins present in anterior–posterior and dorsal–ventral concentration gradients as appropriate. These gradients are established maternally, as described in Saffrey (2001).

During subsequent development, the embryo is subdivided into increasingly small sections, ultimately resulting in the segmental body plan of the larva. Each of these segments is subsequently assigned a specific identity. The first thoracic segment, for example, is morphologically distinct from the other thoracic segments and the abdominal segments, and in fact each larval segment can be recognized by a variety of cuticular features and sense organs. The determination of segmental identity is achieved through the action of several homeotic genes.

How does the development of the larva relate to the development of adult structures? By the time the larva hatches from the egg, it already carries within it groups of cells that proliferate during metamorphosis to yield adult structures. These groups of cells are known as the **imaginal discs** (Figure 5.16), and have their segmental identity determined at the same time as the larval tissues within which they are embedded, and by the same genetic interactions. During the pupal stage, metamorphosis occurs and the adult structures are assembled. Only a few larval organs are retained, the nervous system being an example (though it is subject to considerable remodelling). Each imaginal disc is in fact a flattened sac with a narrow mouth. During metamorphosis, the imaginal disc cells proliferate and evert through the opening of the disc. The proliferating cells together form the tissues of the adult fly.

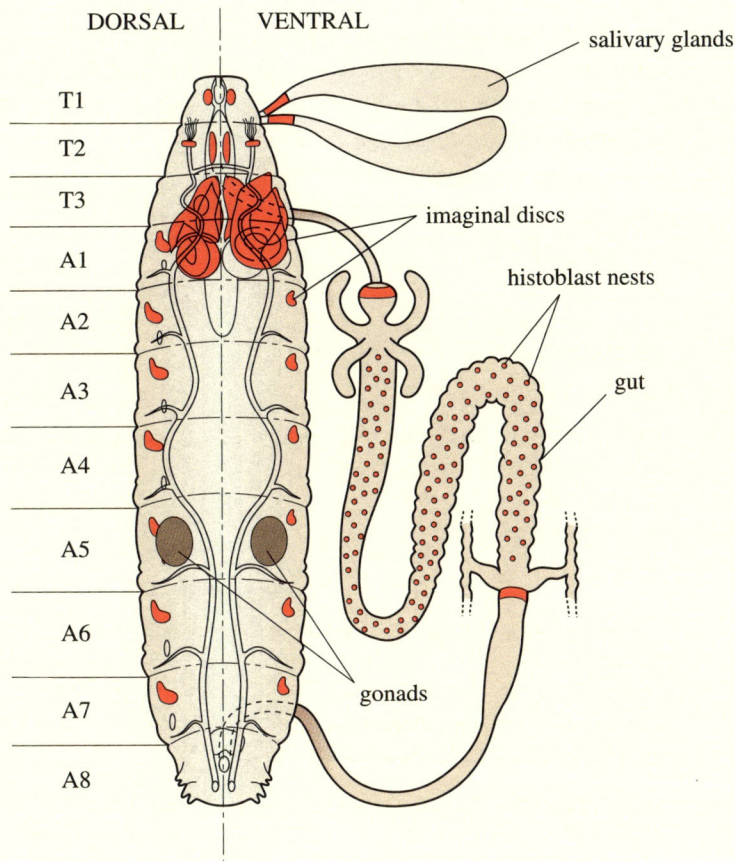

Figure 5.16 The imaginal discs within the *Drosophila* larva (shown in red). Dorsal features are shown on the left and ventral structures on the right — the salivary glands and gut have been displaced for clarity. During metamorphosis, each disc develops into a specific set of adult structures. The abdominal segments carry looser aggregates of cells, the histoblast nests (also red), which carry out the same function. The precise developmental fates of the cells of each disc are determined by the pattern of expression of members of the Hox gene clusters. T1, T2, T3: thoracic segments; A1 to A8: abdominal segments.

The first complex of homeotic genes to be subjected to detailed mutational analysis was the *bithorax* complex (abbreviated as *BX-C*). (A **gene complex** is a group of closely linked genes with related functions and coordinated activity.) The *bithorax* analysis represented many years work by Ed Lewis and colleagues (for which Lewis, together with two other major researchers in the field, Christiane Nüsslein-Volhard and Eric Weischaus, later received a Nobel Prize). The *BX-C* appears to determine segment identity from the second thoracic segment to the tip of the abdomen. The evidence for this conclusion comes from the study of mutants. For example, if the entire *BX-C* gene complex is deleted, the morphology of each segment of the abdomen and the third thoracic segment resembles that of the second thoracic segment. Of course, such gross disturbance to development is lethal. How then can such analyses be carried out? The answer lies in the developmental cycle of *Drosophila*.

As mentioned in Box 5.3, maternal provision of various important gene products enables development to proceed in the absence of early gene transcription. Embryonic development therefore carries on more or less independently of zygotic genotype. In the case of *BX-C* deletion, homozygous mutants are derived from eggs laid by heterozygous (and therefore phenotypically wild-type mothers).

These embryos develop into abnormal larvae, which fail to hatch from the eggshell. These larvae can be manually removed from the eggs and their cuticles examined. Because each segment has its own unique combination of cuticular structures (such as sense organs and denticle belts) which enables unambiguous identification of its identity, the effects of total deletion of the *BX-C* can be assessed. Figure 5.17 shows some of the cuticular features used to study segments in *Drosophila* larvae. The denticle belts are groups of tooth-like projections used to grip the surface during locomotion. Many other features are sense organs.

Figure 5.17 Each segment of the *Drosophila* larva can be unambiguously identified by a number of cuticular structures, such as denticle belts and sense organs. (a) Wild-type larva with the normal arrangement of cuticular features in each segment. T1, T2, T3: thoracic segments; A1 to A8: abdominal segments. (b) Larva that is homozygous for a deletion of the entire *bithorax* complex (*BX-C*), and therefore lacks all the gene functions of the complex. All the segments posterior to the second thoracic segment resemble the second thoracic segment.

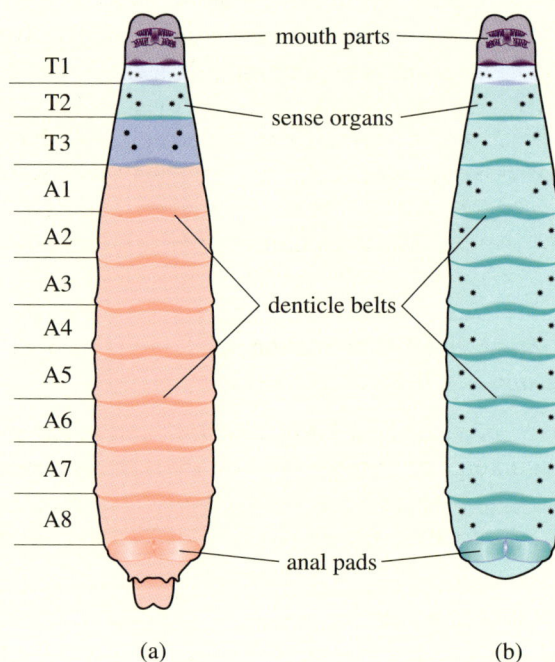

○ How can we maintain and study animals carrying lethal mutations?

● These mutations are recessive, so organisms heterozygous for a lethal allele are phenotypically wild-type. *Drosophila*, in common with most animals, is diploid. Recessive lethal mutations are maintained in the laboratory as heterozygotes.

By the early 1980s, the analysis of mutants suggested that the *BX-C* is an array of many tightly linked genes, each determining one segment's identity. Subsequently, the complex was cloned by dint of much effort, and found to consist of three genes, *Ultrabithorax*, *Abdominal-A* and *Abdominal-B*. It turned out that these three genes represented very complex units, each encoding several different transcripts, by virtue of differential mRNA splicing (Silva-Fletcher, 2001). The transcripts encode transcription factors which in turn control the expression of a cascade of other genes and thereby specify the developmental fate of the segment. Many of the *BX-C* mutations in fact turned out to be mutations in sequences that direct the expression of these three genes in their correct time and place, and in their correct splice variant. A complex indeed!

The molecular analysis of other *Drosophila* homeotic gene complexes followed, notably the *Antennapedia* complex (*ANT-C*). Whereas the *BX-C* is known to affect segments posterior to the thorax, the *ANT-C* affects more anterior segments. *Antennapedia* is named for the transformation of antennae into leg structures, in a similar way as the haltere to wing transformation shown in Figure 5.15a.

○ What does the homeotic transformation of antenna to leg suggest?

● It suggests that antennae and legs are homologous structures.

Detailed molecular studies soon revealed that certain sections of these developmentally important genes are well conserved between animals of widely disparate phyla. It turned out that these conserved sections, or **homeoboxes**, which are characteristic of many homeotic genes, each encode a DNA-binding protein domain. (When aligning base sequences of DNA domains from different sources, those that match are highlighted by drawing boxes around them, which is why such sequences are referred to as 'boxes'.) A gene containing a homeobox sequence is called a **Hox gene** (see Figure 5.16). We can identify families of Hox genes within a species, and correlate them with homologous families in other, very different, species. Homeoboxes are not confined to homeotic genes. For example, the *Pax-6* gene, which we considered in some detail in Section 5.5.1, contains both a homeobox and another DNA binding domain, the *paired* domain.

Several clusters of Hox genes are now known in *Drosophila*, and molecular studies have identified their counterparts in other species, from many phyla. The complete sequence of the *C. elegans* genome has revealed the full repertoire of Hox genes. Similar analysis will soon be available for the *Drosophila* and human genomes, both recently sequenced.

Hox genes play a vital role in the development of all kinds of segmental animals. Can we draw any conclusions about when Hox genes first appeared, and their evolutionary history since that time? The 'pre-Cambrian explosion' refers to the sudden appearance of a widely disparate set of body plans during the Cambrian period (approximately 500–540 Ma ago). At that time, multicellular organisms become apparent in the fossil record, largely because of the presence of hard parts, such as the arthropod cuticle. All extant phyla are represented in the Cambrian fauna, as exemplified by the famous Burgess Shale fossils: chordates, annelids and arthropods are all found in this assemblage. (Section 2.1.3 describes the appearance of some of the arthropods in the Burgess Shale fauna.) Since all these groups contain Hox gene clusters apparently sharing a common ancestry, the most parsimonious explanation is that the ancestral Hox gene evolved before the appearance of the extant phyla.

5.6.1 HOX GENES IN THE EVOLUTION OF LIMBLESS VERTEBRATES

Changes that have arisen in Hox gene clusters during evolution are illustrated in Figure 5.18. Clearly, the genes have evolved by the processes of duplication and divergence. How do such changes in the functions of specific Hox genes, or in patterns of Hox gene expression, lead to radical departures in body form?

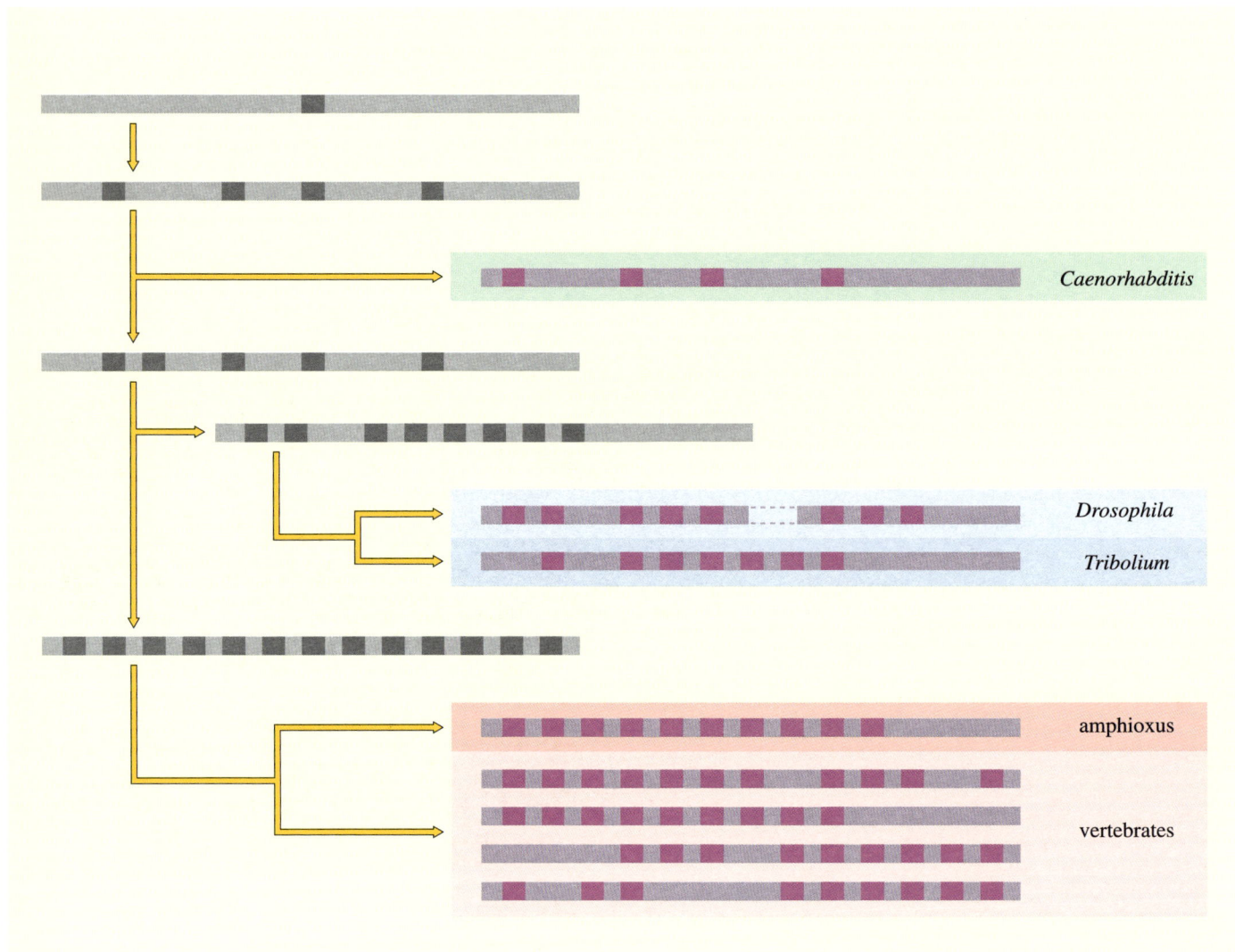

Figure 5.18 The organization of Hox gene clusters in organisms of different phyla. Different taxa are shown in separate boxes: Chordata, including Hox gene arrangements typical of vertebrates and the acraniate chordate, amphioxus; Arthropoda, including a beetle, *Tribolium* (Coleoptera) and *Drosophila* (Diptera); *Caenorhabditis* (Nematoda). Actual Hox genes of living species are shown as dark purple boxes, while hypothetical intermediates are shown as arrays of dark grey boxes. Repeated gene duplication events gave rise to Hox gene arrays of increasing size, while cluster duplication has resulted in the appearance of four clusters in vertebrates. Note that the *Drosophila* Hox gene complement is divided into two clusters, albeit on the same chromosome, presumably as a result of chromosome rearrangement.

Snakes and lizards are closely related groups of reptiles. The most notable difference between them is the limblessness of snakes (though there are several lineages of limbless lizards). This feature, probably an adaptation to subterranean life, has evidently arisen more than once in vertebrate evolution — for example, limbless lizards such as slow-worms, and amphibians such as the caecilians (order Gymnophiona). Cetaceans (whales and dolphins) and sirenians (sea-cows) have lost the hindlimbs, while the forelimbs are much reduced in ostriches, cassowaries and rheas, and lost entirely in kiwis and the extinct moa.

Some primitive snakes, notably the large constrictors (pythons, boas and anacondas), have vestigial hindlimbs. Rudimentary pelvic bones embedded in the musculature (Figure 5.19a) and modified scales in the skin form a tiny spur-like appendage on both sides of the body, near the anus (Figure 5.19b). As 19th century comparative anatomists first realized, these structures and the proportions of the major body regions (head, neck, thorax, abdomen and tail) offer clues about the genetic and evolutionary processes that produce limblessness and the elongated (*serpentiform*) morphology. This line of enquiry has recently been extended to molecular biology.

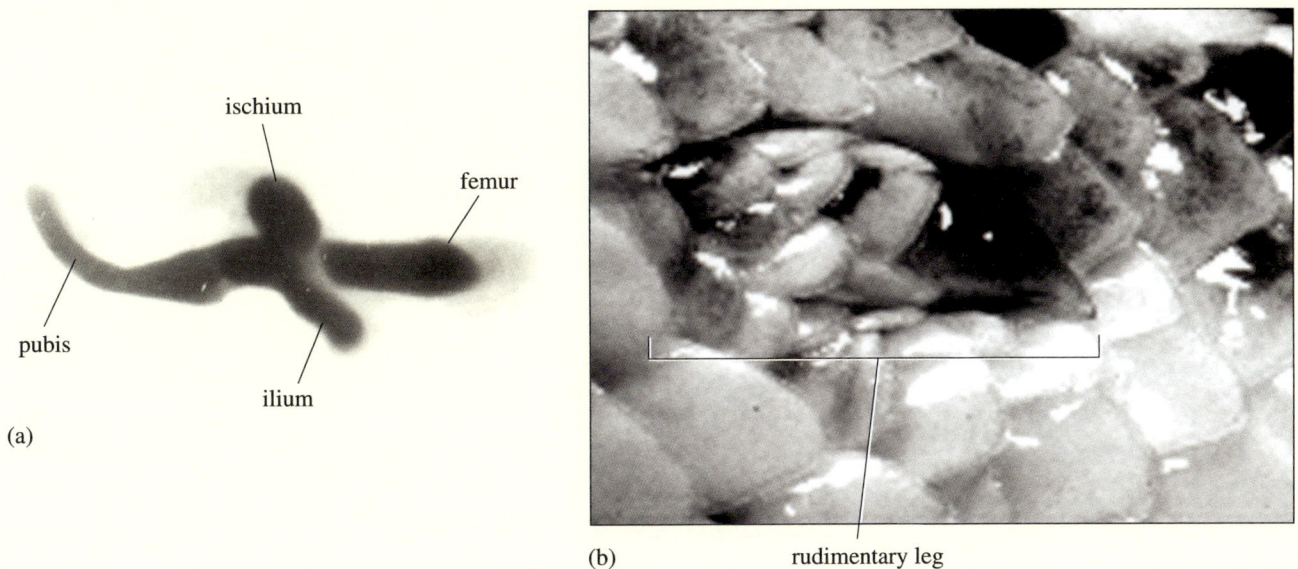

Figure 5.19 (a) The rudiments of the three pelvic bones (pubis, ilium and ischium) and the femur (thigh bone) in a python embryo. (b) In adult snakes, the tiny rudimentary legs are recognized by their covering of distinctive scales.

Embryos of limbed and limbless vertebrates are examined for regions of expression of particular Hox genes. Hox gene products are identified by the binding of specific antibodies which recognize the protein against which they were raised. In this procedure, the embryo is treated so that proteins are fixed but antibodies are allowed to penetrate the tissues. The specific binding of antibodies reveals precisely in which cells those proteins are expressed. Figure 5.20 shows an interpretation of just such an experiment, in which the patterns of expression of three Hox genes are examined in embryos of chicks and pythons. The entire body trunk of a typical snake resembles an elongated thorax, with typical thoracic features such as ribs and muscle blocks, but without girdles or limbs.

○ Where would you expect a thorax-specific Hox gene to be expressed in the python embryo compared to that of the chick?

● The gene would be expressed over the full length of the body segments of the elongated thorax in the python, but over a much shorter region in the chick.

Figure 5.20 shows that expression of Hox genes normally expressed in the thorax extends further posterior in the python embryo than in the chick embryo. In the chick embryo, expression of the Hox genes *Hox-C8* and *Hox-C6* is limited to a short region that ultimately forms the thorax. *Hox-B5* is expressed in a much longer region, extending further anterior and posterior than either *Hox-C6* or *Hox-C8*. As in the insect, it is the relative expression of different Hox genes in each segment that determines the developmental fate of that segment. In contrast, these two genes in python embryos are expressed in a larger region that extends much further towards the posterior of the embryo. The extended thorax is greatly extended at the expense of limbs and girdles, which are greatly reduced or absent.

Figure 5.20 Schematic illustration of the results of an investigation that reveals the distribution of Hox-B5, Hox-C8 and Hox-C6 proteins in the embryos of (a) a chick and (b) a python. The regions of expression of the thoracic Hox genes *Hox-C8* and *Hox-C6* are much extended in the python embryo, resulting in the extension of the thorax relative to that of the chick embryo.

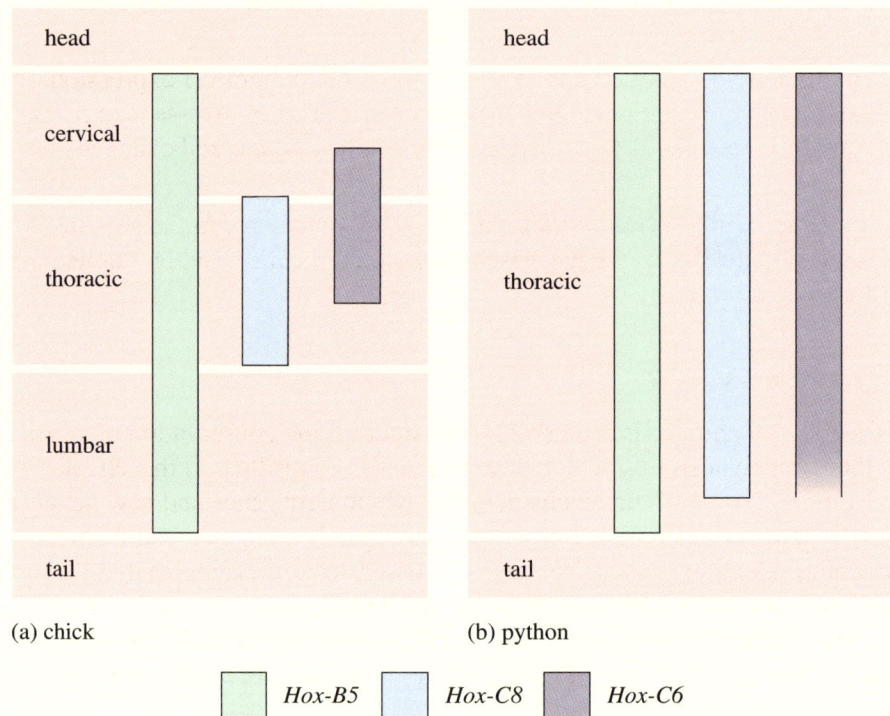

(a) chick   (b) python

Hox-B5   Hox-C8   Hox-C6

It is possible that a series of simple changes in the pattern of expression of Hox genes, such as the extension of expression along the anterior–posterior axis, could have led to the development of the serpentiform body plan in legless lizards and amphibians. These hypothetical events would have been accompanied by a progressive increase in the numbers of thoracic segments, at the expense of the cervical and lumbar regions, and reduction and eventual loss of the limbs and girdles.

Similar evolutionary changes might be expected to account for the various arthropod body plans. Section 2.1 provides an overview of the diversity of arthropods. For all this diversity, arthropods share many morphological features. All have an exoskeleton, and are divided along the anterior–posterior axis into a series of segments. In all arthropods, you should be able to recognize that each segment has the potential to form a set of appendages. For example, a segment of a fly may develop a pair of legs, or of antennae, or no appendages at all, depending on its position along the body. Another example is the biramous crustacean limb (Section 2.1.3), which has the potential to develop into several different structures, again depending on which segment it is in. A further example is the segmental arrangement of the centipede compared with that of spiders or insects. Most of the centipede segments possess pairs of legs, while many segments in spiders and insects are limbless.

SUMMARY OF SECTION 5.6

1   The proteins encoded by homeobox-containing (Hox) genes function as transcription factors, activating cascades of gene functions, and determining segmental identity.

2   Both mutational analysis and examination of the patterns of expression of Hox genes in normal development of morphologically diverse taxa reveal that changes to patterns of Hox gene activity can lead to marked changes in morphology.

3   Hox genes with related functions are generally clustered in the genome. Hox gene families may have arisen by gene duplication and by duplication of entire gene clusters.

5.7 OVERVIEW

The diversity of animals is remarkable, considering the conservation of genes and the genetic mechanisms by which development is controlled. In this chapter we have examined some of the mechanisms by which new genes and new networks of genetic interactions produce novel morphological or biochemical features, upon which selection can act. We have discussed several examples that illustrate the different routes by which novel or altered genetic functions can arise, and which have been elucidated using the power of modern molecular genetics. Indeed, it is clear that modern biology allows us to explore evolutionary mechanisms by direct experimentation.

REFERENCES

Placzek, M. and Skaer, H. (1999) Airway patterning: a paradigm for restricted signalling, *Current Biology*, **9**, pp. R506–R510.

Pond, C. (2001) Biological investigation, in *Introduction to Diversity*, I. Ridge and C. M. Pond (eds), The Open University, Milton Keynes, pp. 93–130.

Ridge, I. (2001) Ordering diversity, in *Introduction to Diversity*, I. Ridge and C. M. Pond (eds), The Open University, Milton Keynes, pp. 1–53.

Saffrey, J. (2001) Life and death of cells, in *The Core of Life*, *Vol. II*, J. Saffrey (ed.), The Open University, Milton Keynes, pp. 438–486.

Saffrey, J. and Metcalfe, J. (2001) DNA and genome evolution, in *The Core of Life*, *Vol. II*, J. Saffrey (ed.), The Open University, Milton Keynes, pp. 345–388.

Silva-Fletcher, A. (2001) Control of gene expression, in *The Core of Life*, *Vol. II*, J. Saffrey (ed.), The Open University, Milton Keynes, pp. 389–438.

Stevens, V. (2001) Genetic diversity, in *Generating Diversity*, M. Gillman (ed.), The Open University, Milton Keynes, pp. 89–110.

Walker, C. and Swithenby, M. (2001) Proteins: structure and catalytic function, in *The Core of Life*, *Vol. I*, J. Saffrey (ed.), The Open University, Milton Keynes, pp. 53–106.

Wistow, G. (1993) Lens crystallins: gene recruitment and evolutionary dynamics, *Trends in Biochemical Sciences*, **18**, pp. 301–306.

FURTHER READING

Gerhart, J. and Kirschner, M. (1997) *Cells, Embryos and Evolution*, Blackwell Science, Oxford. [Best for molecular and developmental perspectives on invertebrates and their evolution. Quite a difficult read!]

Lawrence, P. A. (1992) *The Making of a Fly*, Blackwell Science, Oxford. [Although a little out of date, this book provides a good overview of the genetics and embryology of *Drosophila*.]

ACKNOWLEDGEMENTS

Grateful acknowledgement is made to the following sources for permission to reproduce material in this book:

CHAPTER 1

FIGURES

Figures 1.1a, b, 1.4a–d, 1.5a, b, 1.6a, 1.9a, 1.11c and d, 1.12a, 1.13a, 1.14b, 1.17a, 1.18a, c, 1.19b, c, 1.20 a, c, 1.21a, b, 1.24a, b, 1.25a: Courtesy of Pat Morris; *Figure 1.2a–d*: Courtesy of Dr J. Vacelet; *Figures 1.3a, b, 1.8c, 1.9b, 1.12b, 1.13b, 1.14d, 1.23, 1.26d*: Courtesy of Verina Waights; *Figure 1.6b*: Mark Deeble and Victoria Stone/Oxford Scientific Films; *Figures 1.6c and 1.27b*: Karen Gowlett-Holmes/Oxford Scientific Films; *Figures 1.7, 1.30a, b*: Peter Parks/ Oxford Scientific Films; *Figures 1.8d, 1.11a, 1.17b*: Courtesy of James Mendelssohn; *Figure 1.10b*: Quillin, K. J. (1998) 'Ontogenetic scaling of hydrostatic skeletons: Geometric, static stress and dynamic stress scaling of the earthworm *Lumbricus terrestris*', *Journal of Experimental Biology*, **201**, pp. 1871–1883, Company of Biologists Ltd; *Figures 1.11b, 1.14c, 1.15b, 1.16, 1.18b, 1.19a, 1.21c, 1.26a, 1.28a, c–e, 1.31, 1.32a–d*: Courtesy of Caroline Pond; *Figure 1.14a*: Excerpt from Edward Ruppert and Robert D. Barnes (1994) *Invertebrate Zoology,* 6th edn, copyright © 1994 by Saunders College Publishing, reproduced by permission of the publisher; *Figure 1.15a*: Courtesy of Professor J. D. Currey; *Figure 1.20b*: F. S. Westmorland/ Science Photo Library; *Figures 1.22a, b*: Seibel, B. A., Thuesen, E. V. and Childress, J. J. (2000) 'Light-limitation on predator-prey interactions: Consequences for metabolism and locomotion of deep-sea cephalopods', *Biological Bulletin*, **198**, pp. 284–298; *Figure 1.25b, c*: Ellers, O., Johnson, A. S. and Moberg, P. E. (1998) Structural strengthening of urchin skeletons by collagenous sutural ligament, *Biological Bulletin,* **195**, pp. 136–144; *Figure 1.27a*: Heather Angel; *Figure 1.28b*: Zig Leszczynski/Oxford Scientific Films; *Figure 1.29*: Jonathan Bird/www.oceanicresearch.org.

TABLES

Table 1.1: Mokady, O., Lazar, B. and Loya, Y. (1996) 'Echinoid bioerosion as a major structuring force of Red Sea coral reefs', *Biological Bulletin*, **190**, pp. 367–372.

CHAPTER 2

FIGURES

Figure 2.1: Winchester, N. N. (1997) Arthropods of coastal old-growth *Sitka* spruce forests, in: Watt, A. D., Stork, N. E. and Hunter, M. D. (eds) *Forests and Insects,* © 1997 The Royal Entomological Society; *Figure 2.4*: Reproduced by permission of the Royal Society of Edinburgh and H. B. Whittington from *Transactions of the Royal Society of Edinburgh: Earth Sciences* (1985), **76**, pp. 149–160; *Figures 2.5a and 2.6 a–d, 2.14a and 2.39*: Kershaw, D. R. (1983) *Animal Diversity,*

University Tutorial Press Limited; *Figures 2.7 and 2.16*: Hickman, C. P. Jr, Roberts, L. S. and Larson, A. (1996) *Integrated Principles of Zoology,* 11th edn, The McGraw-Hill Companies Inc; *Figure 2.8a*: Michael Leach/OSF; *Figures 2.8b and 2.11a, b*: Figure from *Invertebrate Zoology,* 2nd edn, by Robert D. Barnes, copyright © 1968 by Saunders College Publishing, reproduced by permission of the publisher; *Figure 2.9*: from *An Atlas of Invertebrate Structure* by Freeman & Bracegirdle, reprinted by permission of Heinemann Educational Publishers; *Figures 2.10, 2.12b, 2.17, 2.18, 2.26b, c, 2.29, 2.30, 2.36*: Gullan, P. J. and Cranston, P. S. (1994) *The Insects*: *An outline of entomology,* Chapman & Hall by permission of Penelope J. Gullan; *Figures 2.12a and 2.14b*: Barrington, E. J. W. (1979) *Invertebrate Structure and Function,* Nelson Thornes Ltd; *Figure 2.13*: Fincham, A. A. (1977) 'Larval development of British prawns and shrimps', *Bulletin British Museum Natural History,* **32** (1), The Natural History Museum; *Figures 2.19, 2.28 and 2.32*: Romoser, W. and Stoffolano J. (1998) *The Science of Entomology,* reproduced by permission of The McGraw-Hill Companies; *Figure 2.20b*: Oxford Scientific Films; *Figure 2.21*: Roberts, M. J. (1995) *Spiders of Britain and Northern Europe,* HarperCollins Publishers Ltd; *Figure 2.22*: Hickman, C. P., Roberts, L. S. and Larson, A. (1996) *Integrated Principles of Zoology,* reproduced by permission of the McGraw-Hill Company; *Figure 2.24*: Poteser, M. and Kral, K. (1995) 'Visual distance discrimination between stationary targets in praying mantis: An index of the use of motion parallax', *Journal of Experimental Biology,* **198**, pp. 2127–2137, Company of Biologists Ltd; *Figures 2.25 and 2.31*: Chapman, R. F. (1969) *The Insects*: *Structure and Function*, The English Universities Press Ltd; *Figure 2.26a*: Ragge, D. R. (1965) *Grasshoppers, Crickets and Cockroaches of the British Isles,* Frederick Warne & Co. Ltd; *Figure 2.27*: L. M. Crowhurst/Oxford Scientific Films; *Figure 2.33a, b*: Young, D. (1989) *Nerve Cells and Animal Behaviour*, Cambridge University Press; *Figure 2.33c*: Wilson, D. M. and Weis-Fough. T. (1962) *Journal of Experimental Biology,* **40**, pp. 643–667, Company of Biologists Ltd; *Figure 2.34*: Burrows, M. (1975) in Usherwood, P. N. R. and Newth, D. R. (eds) *'Simple' Nervous Systems,* Edward Arnold Publishers; *Figure 2.37*: Figure from *Biological Evolution* by Peter W. Price, copyright © 1996 by Saunders College Publishing, reproduced by permission of the publisher; *Figure 2.38*: © 1977 O. W. Richards and R. G. Davies, from *Imms' General Textbook of Entomology,* Chapman and Hall.

CHAPTER 3

FIGURES

Figure 3.6: Peter Jordan and Gerald Webbe (1982) *Schistosomiasis Epidemiology, Treatment and Control*, reprinted by permission of Butterworth Heinemann; *Figure 3.7*: NIBSC/Science Photo Library; *Figure 3.8*: Supplied by Dr H. D. Blankespoor, Museum of Zoology, University of Michigan, USA; *Figures 3.10 and 3.19*: Matthews, B. E. (1998) *An Introduction to Parasitology,* Cambridge University Press; *Figure 3.16*: Piekarski, G. (1987) *Medical Parasitology,*

Springer-Verlag; *Figure 3.20*: Noble, E. R. and Noble G. A. (1964) *Parasitology*: *The Biology of Animal Parasites*, courtesy of National Naval Medical Center, Bethesda, Maryland; *Figure 3.22*: Andy Crump/TDR/WHO/Science Photo Library; *Figure 3.24b*: Riley, J., Banaja, A. A. and James, J. L. (1978) *International Journal for Parasitology*, **8**, p. 246, Pergamon Press; *Figure 3.27a–c*: © J. R. Busvine (1980) *Insects and Hygiene,* Chapman and Hall; *Figure 3.30*: Godfray, H. C. J. (1994) *Parasitoids*: *Behavioral and Evolutionary Ecology*, Oxford University Press; *Figure 3.31*: Edward Ruppert and Robert D. Barnes, *Invertebrate Zoology*, 6th edn, copyright © 1994 by Saunders College Publishing, reproduced by permission of the publisher; *Figure 3.34*: Dynes, R. A. (1993) 'Factors causing feed intake depression in lambs infected by gastrointestinal parasites', Ph.D. thesis, Lincoln University, New Zealand; *Figure 3.35*: Mercer, J. G. and Chappell, L. H. (2000) 'Appetite and parasite', *Biologist*, **47** (1), Institute of Biology; *Figure 3.37*: Kennedy, C. R. (1975) *Ecological Animal Parasitology,* Blackwell Scientific Publications; *Figure 3.38*: Bowers, E. A., Bartoli, P., Russell-Pinto, F. and James, B. L. (1996) 'The metacercariae of sibling species of *Meiogymnophallus*, including *M. rebecqui* comb. nov. (Digenea: Gymnophallidae), and their effects on closely related *Cerastoderma* host species (Mollusca: Bivalvia)' *Parasitology Research*, **82**, pp. 505–510, Springer-Verlag.

CHAPTER 4

FIGURES

Figures 4.1b, 4.8a, 4.10a, 4.12a: Harvey Pough, F. *et al.* (1999) *Vertebrate Life*, Pearson Education; *Figure 4.1c*: Courtesy of Pat Morris; *Figure 4.6*: Hill, R. W. (1976) *Comparative Physiology of Animals*: *An Environmental Approach*, Copyright © 1976, Addison Wesley Longman Inc.; *Figures 4.9a, 4.16, 4.19, 4.20, 4.25, 4.26a, 4.28, 4.34, 4.35*: Kardong, K. V. (1998) *Vertebrates*: *Comparative Anatomy, Function, Evolution*, 2nd edn, The McGraw-Hill Companies, Inc.; *Figure 4.13a*: Shadwick, R. E. *et al.* (1992) *Philosophical Transactions*: *Biological Sciences*, Figures 7 & 11, p. 423, The Royal Society, London; *Figure 4.15*: Stoddart, D. M. (1980) *The Ecology of Vertebrate Olfaction*, Kluwer Academic Publishers, B.V.; *Figure 4.17*: Bradbury, J. W. and Vehrencamp, S. L. (1998) *Principles of Animal Communication*, Copyright © Sinauer Associates Inc; *Figure 4.21*: © Eade Creative Services Inc; *Figure 4.22*: J. A. L. Cooke/OSF; *Figure 4.24*: Kalmijn, J. A. (1971) 'The electric sense of sharks and rays', *Journal of Experimental Biology*, **55**, pp. 371–383, in Schmidt-Nielsen, K. (1997) *Animal Physiology*: *Adaptation and Environment*, 5th edn, Cambridge University Press, © 1971 Company of Biologists Ltd; *Figures 4.31, 4.32, 4.33*: Rosen, S. and Howell, P. (1997) *Signals and Systems for Speech and Hearing* © Academic Press; *Figure 4.35*: Romer, A. S. (1960) *Man and the Vertebrates*, University of Chicago Press; *Figure 4.39b*: Prentis, T. (1981) 'The hearing of the barn owl', ed. E. Knudsen, *Scientific American*, **245**(6), pp. 113, © Tom Prentis.

CHAPTER 5

FIGURES

INDEX

Note: Entries in **bold** are key terms. Page numbers referring to information that is given only in a figure or caption are printed in *italics*.